T0136385

Transdermal
Drug Delivery

DRUGS AND THE PHARMACEUTICAL SCIENCES

Executive Editor

James Swarbrick
PharmaceuTech, Inc.
Pinehurst, North Carolina

Advisory Board

DRUGS AND THE PHARMACEUTICAL SCIENCES

A Series of Textbooks and Monographs

ADDITIONAL VOLUMES IN PREPARATION

Transdermal Drug Delivery

Second Edition, Revised and Expanded

edited by

Richard H. Guy
*Universities of Geneva and Lyon, Archamps, France
and University of Geneva, Geneva, Switzerland*

Jonathan Hadgraft
NRI, University of Greenwich, Chatham, England

CRC Press
Taylor & Francis Group
Boca Raton London New York

CRC Press is an imprint of the
Taylor & Francis Group, an **informa** business

CRC Press
Taylor & Francis Group
6000 Broken Sound Parkway NW, Suite 300
Boca Raton, FL 33487-2742

First issued in paperback 2019

© 2003 by Taylor Francis Group, LLC
CRC Press is an imprint of Taylor & Francis Group, an Informa business

No claim to original U.S. Government works

ISBN-13: 978-0-8247-0861-0 (hbk)
ISBN-13: 978-0-367-39570-4 (pbk)

Library of Congress Cataloging-in-Publication Data
A catalog record for this book is available from the Library of Congress.

**Visit the Taylor & Francis Web site at
http://www.taylorandfrancis.com**

**and the CRC Press Web site at
http://www.crcpress.com**

Preface

The previous edition of our book was compiled in the late '80s at the height of academic and industrial activity in transdermal research. While this route of administration continues to be limited by the number of suitable drug candidates available, it still attracts considerable worldwide interest, and, importantly, the pharmaceutical industry is now prepared to consider new chemical entities for transdermal delivery. This will reinvigorate the field, which, until now, has depended on the development of existing compounds that typically do not possess optimal physicochemical properties for dermal delivery.

The second edition reflects our increased knowledge of the mechanisms of absorption and how these can be used to advantage in the development of medicinal agents and formulations for both dermal and transdermal delivery. The barrier properties of the skin, thanks to the use of sophisticated biophysical techniques, are much better understood and their modulation by both chemical and physical techniques has achieved impressive results. Consequently, the manner in which the basic physicochemical properties of a drug determine the amount that can be transported across the stratum corneum can now be explained in detail. The revised text shows the importance of these properties and how predictive models can be established to examine the feasibility of delivering molecules into and through the skin. Over the last 15 years, considerable advances have been made in the use of physical approaches to promote absorption. These techniques, which include electrical, ultrasound, and other minimally invasive strategies, are reviewed here in some detail. Chemical enhancers, on the other hand, have been fully examined in multiple other texts and are therefore not covered in this edition.

The fact that the skin is a metabolically active organ means that its barrier properties can be modulated by interference with lipid synthesis. This novel approach is reviewed in this book, as is the application of supersaturation as a mechanism for enhanced delivery. Finally, as no transdermal system can be successful until it has passed through stringent regulatory control, the final chapter considers the steps required for the registration of dermal delivery systems.

In summary, this text attempts to achieve two broad objectives. The first is to provide a "snapshot" of the field and an evaluation of some creative ideas under examination. The second objective is to serve as a reference work that summarizes the state of the art and can be used to guide the interested reader into the fascinating world (and associated challenges) of transdermal drug delivery.

Richard H. Guy
Jonathan Hadgraft

Contents

Contributors

Annette L. Bunge   Chemical Engineering Department, Colorado School of Mines, Golden, Colorado, U.S.A.

Adrian Davis   GlaxoSmithKline Consumer Healthcare, Weybridge, England

M. Begoña Delgado-Charro   Centre Interuniversitaire de Recherche et d'Enseignement, Universities of Geneva and Lyon, Archamps, France, and School of Pharmacy, University of Geneva, Geneva, Switzerland

James A. Down   BD Technologies, Research Triangle Park, North Carolina, U.S.A.

Peter M. Elias   Departments of Dermatology and Medicine, University of California, San Francisco, San Francisco, California, U.S.A.

Kenneth R. Feingold   Departments of Dermatology and Medicine, University of California, San Francisco, San Francisco, California, U.S.A.

Richard H. Guy   Centre Interuniversitaire de Recherche et d'Enseignement, Universities of Geneva and Lyon, Archamps, France, and School of Pharmacy, University of Geneva, Geneva, Switzerland

Jonathan Hadgraft   Skin and Membrane Transfer Research Center, NRI, University of Greenwich, Chatham, England

Noel G. Harvey   BD Technologies, Research Triangle Park, North Carolina, U.S.A.

Victor Meidan   New Jersey Center for Biomaterials, Newark, New Jersey, U.S.A.

Gopinathan Menon   Avon Products, Inc., Suffern, New York, U.S.A.

Mark Pellett   Wyeth Consumer Healthcare, Havant, England

Véronique Préat   Unité de Pharmacie Galénique, Université Catholique de Louvain, Brussels, Belgium

S. Lakshmi Raghavan   Skin and Membrane Transfer Research Center, NRI, University of Greenwich, Chatham, England

Vinod P. Shah   Center for Drug Evaluation and Research, Food and Drug Administration, Rockville, Maryland, U.S.A.

Carl Thornfeldt   Cellegy Pharmaceuticals, Inc., Foster City, California, U.S.A.

Janice Tsai   Department of Dermatology, University of California, San Francisco, San Francisco, California, U.S.A.; Department of Clinical Pharmacy, National Cheng Kung University, Taiwan; and Cellegy Pharmaceuticals, Inc., Foster City, California, U.S.A.

Rita Vanbever   Unité de Pharmacie Galénique, Université Catholique de Louvain, Brussels, Belgium

Brent E. Vecchia*   Chemical Engineering Department, Colorado School of Mines, Golden, Colorado, U.S.A.

---

* *Current affiliation*: Blakely Sokoloff Taylor & Zafman LLP, Denver, Colorado, U.S.A.

1

Feasibility Assessment in Topical and Transdermal Delivery: Mathematical Models and In Vitro Studies

Jonathan Hadgraft
NRI, University of Greenwich, Chatham, England

Richard H. Guy
Centre Interuniversitaire de Recherche et d'Enseignement, Universities of Geneva and Lyon, Archamps, France, and University of Geneva, Geneva, Switzerland

I. INTRODUCTION

Fortunately for us, the skin has evolved into an extremely efficient barrier, which prevents both excessive water loss from the body and the ingress of xenobiotics. It enables us to withstand a considerable range of environmental challenges. The reasons for this are manifold and may be summarized simply for the purposes of this chapter.

The outer layer of the skin, the stratum corneum, forms the rate-controlling barrier for diffusion for almost all compounds (Fig. 1).

It is composed of dead, flattened, keratin-rich cells, the corneocytes. These dense cells are surrounded by a complex mixture of intercellular lipids. They comprise ceramides, free fatty acids, cholesterol, and cholesterol sulphate. Their most important feature is that they are structured into ordered bilayer arrays. The predominant diffusional path for a molecule crossing the stratum corneum ap-

1

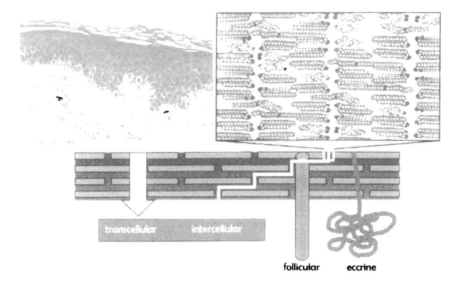

Figure 1 A schematic representation of the skin. Top left is a photomicrograph of the skin, clearly showing the stratum corneum, viable epidermis, and underlying dermis. Top right is a representation of the structured lipids found in the intercellular channels. The bottom illustration shows the different routes that a diffusant can take in its passage through the stratum corneum, although the predominant one is usually intercellular.

pears to be intercellular (1). The diffusant therefore follows a tortuous route and has to cross, sequentially, and repeatedly, a number of hydrophilic and lipophilic domains. It is not surprising, therefore, that the lipid–water partitioning characteristics of the permeant is a dominant determinant of its penetration or that mathematical models developed to predict percutaneous absorption include a term to describe partitioning. Fick's laws of diffusion describe the diffusional step, and although the skin is a heterogeneous membrane, simple solutions can often be applied. This will be appreciated later when these laws and their limitations are described.

There are a number of reasons why it is important to predict the absorption rate. For dermal drug delivery, it is desirable to choose a drug structure that has the best possible chance of arriving at the site of action. It should be remembered that compounds such as the corticosteroids have a bioavailability of only a few percent (2). If drugs were chosen with more favorable partition and diffusion characteristics, this bioavailability could be improved considerably. An accurate and descriptive mathematical model would enable this rational choice and facilitate the design of novel topical agents. For transdermal delivery, sufficient drug must permeate the various strata of the skin so as to build up a plasma concentration, which elicits a systemic effect. This route has a number of attractions, and an accurate and predictive model would be invaluable in the selection and evolution of appropriate transdermal drug candidates. Equally, there are also chemicals, the absorption of which in significant amounts is clearly undesirable. Compounds such as pesticides are obvious examples, but there are other materials, present perhaps as formulation excipients, that could also be detrimental. An appropriate mathematical model would allow a reliable risk assessment to be made before in vivo evaluations are conducted.

There are different considerations to be taken into account depending on whether the drug is to be delivered for local action or for systemic action. Since this book concerns primarily transdermal delivery, the major emphasis will be how to ensure the transport of drug through the skin into the underlying dermal vasculature and hence the systemic circulation. For a drug to be administered transdermally, it has to be very potent, as it is unlikely that more than a few tens of milligrams per day can be delivered. To a first approximation, feasibility can be assessed from the daily dose. But, as will be seen, even for a compound like nitroglycerine, which has ideal physicochemical properties for transdermal delivery from a reasonable patch area, no more than 40 to 50 mg per day can be delivered.

In some ways, it is more difficult to assess the feasibility of topical drug delivery, as the levels required in the skin for therapeutic effect are usually unknown. For transdermal delivery, there is a well-documented and determinable end point, the plasma level required for efficacious therapy. Advances in noninvasive monitoring and microdialysis can be helpful in determining the target skin concentration for topical therapy, but data are limited, and the reliability of the methodologies involved is still in question, as the techniques remain in very much a developmental stage.

Validated mathematical models represent an economically advantageous approach for the assessment of skin permeation, and their use is recommended before full-blown in vitro and in vivo experiments are conducted. The purpose of this chapter is to examine the limitations of mathematical modeling and to consider appropriate in vitro models prior to full clinical testing.

II. FICK'S LAWS OF DIFFUSION

Considering that the skin is such a heterogeneous membrane, it is surprising that simple diffusion laws can be used to describe the percutaneous absorption process (3). Since transdermal delivery involves the application of a device over a long period of time, it is generally assumed that steady-state conditions have been reached and that the most relevant law of diffusion is therefore Fick's first law. The second law describes non–steady state diffusion and can be used to analyze the rates of release from matrix type transdermal patches, to evaluate the lag phase prior to the establishment of steady-state conditions, and to describe concentration profiles across the skin as they evolve towards linearity.

The most quoted form of Fick's first law of diffusion describes steady-state diffusion through a membrane:

$$J = \frac{KD}{h}(c_o - c_i) \tag{1}$$

where J is the flux per unit area, K is the stratum corneum-formulation partition coefficient of the drug, and D is its diffusion coefficient in the stratum corneum of path length h; c_o is the concentration of drug applied to the skin surface, and c_i is the concentration inside the skin. In most practical situations, $c_o \gg c_i$, and Eq. (1) simplifies to

$$J = k_p c_i \tag{2}$$

where k_p ($= DK/h$) is the permeability coefficient, which has units of velocity (often quoted as cm h^{-1}), i.e., it is a heterogeneous rate constant and encodes both partition and diffusional characteristics. The input rate of the drug into the systemic circulation, from a patch of area A, is therefore given by the product

$$\text{Input rate} = A \times k_p \times c_o \tag{3}$$

The output or elimination rate from the systemic circulation equals the clearance (Cl) multiplied by the plasma concentration at steady state ($c_{p,ss}$)

$$\text{Output rate} = Cl \times c_{p,ss} \tag{4}$$

Hence Eqs. (3) and (4) may be combined to predict the drug's plasma concentration following transdermal delivery:

$$c_{p,ss} = \frac{Ak_p c_o}{Cl} \tag{5}$$

The plasma concentration achieved therefore depends directly on the area of the device, the skin permeability, and the applied concentration and is inversely related to the drug's clearance (4).

For a given drug, the clearance and the target plasma level are likely to be known, so to examine the feasibility of delivery, one needs the drug's skin permeability and its solubility, as this will give an indication of the maximum concentration that can be applied. These parameters can be estimated from basic physicochemical properties, which are typically measured during preformulation.

III. ESTIMATION OF PERMEABILITY COEFFICIENT FROM SIMPLE PHYSICOCHEMICAL PARAMETERS

Over the years, a database of skin permeability values has been consolidated (e.g., Ref. 5). To avoid confusion and to simplify interpretation, permeabilities from aqueous solutions alone are considered. The use of skin permeabilities from other vehicles is complicated in that most other solvents are capable of altering the barrier function of skin. Analysis of the existing database enabled Potts and Guy to formulate an empirical relationship between k_p and two simple characteristics of the permeant: the octanol–water partition coefficient (K_{oct}) and the molecular weight (MW) (6).

$$\log k_p \text{ (cm/h)} = -2.72 + 0.71 \log K_{oct} - 0.0061 \text{ MW} \tag{6}$$

If $\log K_{oct}$ has not been measured, or is not available, it can easily be predicted by commercially available algorithms. The latter are increasingly more reliable and accurate.

Equation 6 shows that, as $\log K_{oct}$ increases, the permeability also increases, whereas the greater the molecular weight the smaller is k_p. Permeants that are best absorbed through the skin are therefore small and lipophilic. This analysis often gives rise to confusion, as it must be realized that it is not the permeability coefficient alone that determines the efficiency of topical and transdermal delivery. Rather, it is the flux across the skin, which is the product of the permeability coefficient and the drug concentration in the vehicle. As the dataset under discussion concerns percutaneous transport from aqueous solution, the maximum achievable flux is therefore k_p multiplied by the aqueous solubility (S_w). Often, this value is also known from preformulation studies; if not, it can also be estimated from equations such as (7):

$$\log S_w = -\log K_{oct} - \frac{1.11 \, \Delta S_f(mp - T)}{4.577(273 + T)} - 0.54 \tag{7}$$

where ΔS_f is the entropy of fusion of the permeant (which can also be estimated), mp is its melting point, and T is the temperature at which the solubility is required. It is immediately apparent that (a) as K_{oct} increases, the aqueous solubility decreases, and (b) the lower the melting point, the higher the solubility (i.e., a reflection of the role of intermolecular forces on solubility). Solubilities can also be estimated using commercially available software; typically, the algorithms employed require knowledge of the melting point and the octanol–water partition coefficient. Since the maximum flux is the product of k_p and S_w, simply increasing the drug's K_{oct} results in a "battle" between two competitive factors, as shown in Fig. 2.

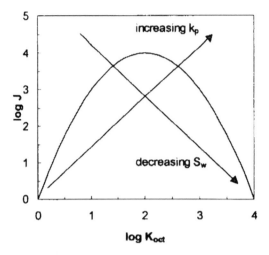

Figure 2 Schematic diagram showing the competitive effects of permeability coefficient and solubility on the maximum flux as a function of the octanol–water partition coefficient.

Such a parabolic relationship between the maximum flux, or the amount of drug permeated in a certain time, and log K_{oct} is not unusual (see, e.g., Refs. 8 and 9). In general, furthermore, it is observed that compounds, which diffuse through the skin most readily, are those having log K_{oct} around 2 and a low melting point (e.g., nicotine and nitroglycerine).

To illustrate the points made here, let us consider nitroglycerine. The drug has a molecular weight of 227 Da, a log K_{oct} of 2.05, and an aqueous solubility of 5 mg/mL. The Potts–Guy equation predicts $k_p = 2.3 \times 10^{-3}$ cm/h, from which (with S_w) one can estimate a maximum transdermal flux of 11.5 µg/cm²/h. It follows that a dose approaching 3 mg would be deliverable from a 10 cm² patch in 24 hours. In fact, the systems on the market typically perform a little better, achieving rates of 0.2 mg/h (~5 mg per day) across a 10 cm² surface. Nevertheless, the calculation serves to validate the power of this approach to the assessment of TDD feasibility.

An obvious limitation of this method is that the calculation is based on an aqueous formulation. The permeability coefficient will, of course, change with the nature of the formulation placed on the skin, as will the drug's solubility. From a theoretical standpoint, a saturated solution in any solvent should result in the same steady-state flux across the skin, because the chemical potential of the drug in any saturated solution is maximal. However, in reality, such equivalence is not always observed, because many solvents are themselves able to change the barrier properties of the skin and lead to flux rates that do not conform to simple thermodynamic arguments. Attempts have been made to predict the way in which a solvent will interact with the skin, and the use of solubility parameters has been proposed (10). To date, however, there has not evolved any consistent relation between prediction and experiment, and the field remains without a means with which to estimate how a solvent will influence the permeability coefficient of the skin.

IV. DESIGN OF TRANSDERMAL DRUGS

The majority of the drugs that have been considered for transdermal delivery have been designed with another route of administration in mind. Their properties are therefore not optimum for passage across the skin. That is, the ideal characteristics for gastrointestinal absorption are not necessarily the same as those that promote skin penetration. There are, however, some guidelines that can be used to select transdermal candidates with optimum properties. As discussed above, the molecules should be small, have a low melting point (good solubility properties), and have a log K_{oct} of ~2. Another factor that needs to be taken into consideration is the nature of the functional groups on the molecule. A further analysis of the skin permeability database has separated partition and diffusion effects (11,12). The number of hydrogen bonding groups attached to the molecule has a significant effect on the diffusion of the permeant through the skin. Addition of a single hydrogen bonding group (equivalent, for example, to changing benzene into phenol) reduces D by an order of magnitude; a further hydrogen bonding group (e.g., phenol to resorcinol) decreases D by a further order of magnitude. A third one reduces D further but not by as much; thereafter, there appears to be a plateau. Presumably this is a reflection of the ability of the permeant to interact with the polar head groups of the structured lipids in the skin. Transdermal drugs will not, therefore, contain a large number of pendant hydrogen bonding groups. It should also be noted that molecules with a significant number of such groups are hydrophilic in nature and hence fail as transdermal candidates because their log K_{oct} values are much smaller than the optimum.

A further consideration is the ionization potential, and hence pK_a, of the drug. Many drugs are weak acids or bases. Dogma states that the free acid or the free base should be chosen, as unionized species partition better into the lipophilic membrane. However, a number of factors need to be taken into account. The skin has a surface pH of around 4 to 5 and a good buffer capacity, probably owing to the free fatty acids that make up an important component of the stratum corneum lipids. When a drug is applied to the skin surface, its ionization state could change because of the acidic environment. Alternatively, the permeant may be able to form an ion pair and hence partition into the skin in its ionized form (13–16). In any case, the maximum flux of the drug will be given by Eq. (2), modified to take into account the relative amounts of ionized and unionized drug:

$$J_{max} = k_{p(u)}S_{(u)} + k_{p(i)}S_{(i)} \qquad (8)$$

where $k_{p(u)} = K_{(u)}D_{(u)}/h$ and $k_{p(i)} = K_{(i)}D_{(i)}/h$, and $S_{(u)}$ and $S_{(i)}$ are the solubilities of the unionized and ionized species, respectively. The dominant term will depend on the relative magnitude of the solubilities and partition coefficients of the ionized and unionized species. Little systematic work, however, has been conducted on the role of pH and partitioning into skin. Those studies that have been conducted suggest that solubility increases with ionization more than partitioning into the skin decreases. This means that it may be more effective to choose the salt of a transdermal drug rather than its unionized form. Analysis of the limited data set available also suggests that it is possible to estimate values of the permeability coefficient for the ionized species using relationships similar to the Potts and Guy equation, in which K_{oct} is replaced by a "distribution coefficient" (17).

V. OTHER MODELING APPROACHES TO PREDICT FEASIBILITY OF DERMAL DELIVERY

Roberts et al. have presented a comprehensive review of mathematical models in percutaneous absorption (18). Non-steady-state solutions of Fick's laws of diffusion have been used to give insight into the mechanisms of skin penetration and its modulation (e.g., Refs. 19–22). They have also been used for interpreting concentration profiles of the permeant across the stratum corneum (23). This is particularly useful in the data analysis of tape stripping experiments (24,25). The solutions to Fick's second law of diffusion are complex. It should be remembered that data from skin permeation studies are subject to considerable biological variation, and sometimes it is difficult to have confidence in fitting the data to such complex equations. A simplified approach is to treat the delivery system, the skin, and the blood as separate compartments that are linked using first-order rate constants. This is an extension of pharmacokinetic modeling and was developed by Guy and Hadgraft from the earlier work of Riegelman (26) and Wallace and Barnett (27). These kinetic models (28,29) (see, e.g., Fig. 3) use simple first-order rate constants to model diffusion ($k = D/l^2$) and a ratio of first-order rate constants to describe partitioning between the different barriers encountered, such as drug partitioning between the stratum corneum and the viable tissue.

The models can be adapted to allow for subsidiary processes, such as loss

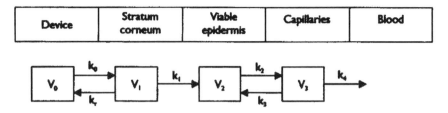

Figure 3 A representative "kinetic" model of the skin (30).

of a volatile permeant (31) and metabolism of the permeant as it diffuses through the skin (32). The models have been used successfully to examine the transdermal delivery of a number of drugs, e.g., nitroglycerine (33), estradiol, rolipram (34), and theophylline in preterm infants (35).

A further extension of the pharmacokinetic approach is the physiological modeling proposed by McDougal and coworkers (36). This has largely been employed in the prediction of toxicological effects of dermally absorbed substances (37).

Further examination of the mathematical modeling approach is given in later chapters.

VI. ENHANCEMENT STRATEGIES

Mathematical modeling, whether used simply to estimate the permeability coefficient or to develop a more complex compartmental model, should indicate the feasibility of delivering enough drug through the skin. If the estimates are less than a factor of 10 to 100 of the desired value, enhancement strategies will need to be considered. There are a number of approaches that can be used, but it is very difficult to predict the exact degree of enhancement that can be achieved.

There are chemicals known to improve diffusion through the skin. In general, these compounds have similar structures in that they possess a polar head group and a long alkyl chain. This indicates that they tend to have surface-active properties. The polarity of the head groups, however, tends to be quite low (they are not charged) and they are therefore less irritant than cationic and anionic surfactants. Table 1 lists some excipients used in transdermal devices.

Compounds such as oleic acid are known to intercalate into the structured lipids of the skin (38,39). Here they reduce the diffusional resistance to permeation and aid drug transport. Other excipients in Table 1 that are expected to act similarly are methyl laurate, glyceryl mono-oleate, etc. It is clear that there are

Table 1 Examples of Transdermal Patches
and Excipients Used Therein

Fentanyl	Duragesic®	Ethanol
GTN	Nitrodisc®	Isopropyl palmitate
	Nitrol®	Isopropyl palmitate
		Lauryl alcohol
	NTS®	Isopropyl palmitate
Estradiol	Estraderm®	Ethanol
	Climara	Isopropyl myristate
		Glyceryl monolaurate
	Trial® Sat	Oleic acid
		Propylene glycol
	Vivelle®	Oleic acid
	Menorest®	1,3 butylene glycol
		Lecithin
		Dipropylene glycol
		Propylene glycol
Testosterone	Androderm®	Glyceryl mono-oleate
		Methyl laurate
		Ethanol

excipients present that do not possess a polar head group and a long alkyl chain. For example, ethanol and propylene glycol are small "solvent" molecules that will, in their own right, permeate into the skin. Their presence in the skin will change the partitioning of the drug between the delivery device and the skin. If partitioning is increased, there will be an enhanced flux. It is also possible for compounds like ethanol to extract lipids from the skin (40). This will affect the barrier properties of the stratum corneum. It should also be noted that the skin responds quite rapidly to lipid extraction, barrier reformation occurring in a matter of hours.

The two fundamental approaches to permeation enhancement as a result of added excipients are therefore to improve D or K. If an enhancer (or combination) is used that can alter both D and K, it is possible to obtain synergistic effects (41). Simple inspection of Eq. (1) shows that the combined effect will be multiplicative rather than additive.

Another mechanism for enhancing the concentration in the outer layers of the skin and hence improving flux is to use supersaturation (Fig. 4).

Although thermodynamically unstable, supersaturated states can be produced that can be transiently stabilized using agents such as antinucleant poly-

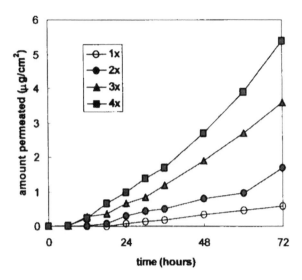

Figure 4 The permeation of piroxicam through human epidermis in vitro, showing the effect of the degree of supersaturation. (Data adapted from Ref. 42.)

mers (43). The high chemical potential of these systems drives the permeant into the skin, and the obtained flux is highly correlated with the degree of supersaturation. This concept will be dealt with more fully in a later chapter. It is also possible to obtain synergistic effects with penetration enhancers that improve diffusion through the skin (44).

Very often, transdermal patches are designed with as small an area as possible. In order to have a high drug loading, the active is often in the patch close to its solubility limit. Patches are often made using solvent evaporation techniques, and this can lead to devices in which the drug is supersaturated. During storage, crystallization can occur, which may compromise the efficacy of drug delivery. Some changes on long storage have been identified, for example, in estradiol patches (45). Crystallization (and supersaturation) can be induced as a result of the evaporation of a volatile component from the patch. For example, diethyl *m*-toluamide (DEET) has a relatively high vapor pressure and will be lost from an open patch. During use, DEET can be absorbed into the outer layers of the skin. If its concentration drops significantly in the patch to less than that required to sustain the solubility of the active, supersaturation or crystallization may occur. Another mechanism for creating supersaturated states in vivo involves the uptake of water from the skin. Water loss through the skin is well documented. Under an occlusive patch, water can be taken up into hydrophilic adhesives. With this increase in hydration, the nature of the adhesive will change so that its ability to hold the active in solution is affected significantly; that is, a higher chemical potential of the active can be induced, which increases drug flux.

In the discussion above, the significance of pK$_a$ was addressed. In general, charged molecules permeate the skin poorly. However, the delivery of charged entities can be significantly improved using a small electric current as the driving force (46). Typical iontophoretic currents (<0.5 mA/cm^2) are achieved with a modest potential difference of, at most, a few volts. The current is carried across the skin by a combination of (a) the charged "active," (b) other ions that are present in the electrode formulations, and (c) counterions electrotransported out of the skin in the opposite direction. The transport number of the active is therefore the key determinant of delivery and is often much smaller than that of small ions such as sodium or chloride, which are typically present. These factors and the underlying theory will be dealt with in a later chapter. It is pertinent to note that the structure of the drug can also be important. If it contains structural elements that allow binding to channels in the skin through which the charge is passing, strong adsorption can occur and change the electrical properties of the membrane (47). This can lead to a reduced overall flux through the skin. Some charged molecules (e.g., certain peptides and small proteins, such as insulin) are zwitterionic, and the net charge will be dependent on the surrounding pH. The surface pH of the skin is around 4 to 5, while the viable tissue is at 7.4. If the pI of the permeant lies between these two values, it will be cationic at the skin surface and anionic within the viable tissue. Such a permeant that starts at the surface being repelled by the anode will, at some point in its passage across the skin barrier, become anionic and will therefore be attracted back to the anode. It is possible, then, to obtain a band of the active in the skin at which the attraction and repulsion counteract one another (48). Clearly this is an unsuitable situation in which the drug is required to permeate all the way through the skin to the systemic circulation.

Other physical techniques that can be used to enhance permeation include electroporation (49,50) and sonophoresis (51), and these approaches are considered in later chapters. It is also possible to breach the stratum corneum using high-velocity particles (52) or microneedles (53) or to ablate the skin thermally using different technologies (54). A final chapter addresses the potential of some of these so-called minimally invasive techniques.

VII.   IN VITRO ASSESSMENTS

Mathematical predictions provide a useful indication of drug uptake and permeation across the skin, but refinement is necessary before there can be total confidence in the models. Prior to any clinical assessment, it is essential to have appropriate in vitro models available. Numerous studies have examined animal models, and none can be said to be perfectly predictive of human skin. Most vehicles have an effect on the barrier properties of the skin, and the effect varies with the animal species. Of all the species examined, the pig appears to be most representative, but whenever possible human skin should be the membrane of choice. In the literature, a number of protocols have been proposed for measuring drug flux across skin, and standard methodology has been suggested (55,56). Following such guidelines, reasonable agreement between in vitro and in vivo permeation is normally found. Problems can arise if there are metabolic processes in the skin in vivo that are absent in vitro, but it is possible to conduct in vitro experiments on freshly harvested tissue, the viability of which can be maintained for a limited time (57). Generally speaking, however, an in vitro experiment, correctly performed, provides a good indication of a drug's ability to cross the skin barrier.

The availability of human skin tissue can be a problem, and considerable effort has been made to culture skin from human cell lines (e.g., Ref. 58). Culturing the cells does not appear to be a problem. Epidermal sheets with almost normal histology have been produced, but to date the permeability barrier is not as robust as that of genuine human tissue.

The various guidelines [e.g., that endorsed by the U.S. Food and Drug Administration (55)] concerning in vitro permeation testing suggest that the concentration in the receptor medium must not exceed 10% of the solubility of the permeant. This is to ensure that transport is not limited by solubility constraints. It does, however, pose some experimental problems, particularly for testing lipophilic drugs. Sometimes the solubility constraint can be addressed by modifying the receptor phase. Typically, phosphate buffered saline or a similar physiological buffer is used (pH 7.4), and it is possible to add solvents such as ethanol, or nonionic surfactants, to boost the drug's solubility. A word of caution, though: it is important that any additive does not alter the barrier properties of the skin, as it may diffuse down its concentration gradient from the receptor phase, across the skin, and back into the donor compartment. The problem of low solubility can also be circumvented by the use of flow-through cells, provided that a sensitive and specific analytical assay for the drug is available.

In vitro permeation tests are also valid for monitoring physical methods of permeation enhancement with reproducible results reported for both iontophoresis and electroporation. Less extensive investigations have also been made using skin that has been subject to physical damage such as laser ablation (54).

Ex vivo experiments have also been used to monitor the role of uptake into the blood supply and metabolism. The most successful of these is the isolated perfused porcine skin flap model that has given insight into absorption mechanisms and has been used for both passive delivery and iontophoresis (59,60).

VIII. IN VIVO ASSESSMENTS

Clearly, the gold standard for measuring in vivo dermal or transdermal absorption
is the use of living human tissue. In the case of transdermal delivery, the goal
is to obtain an appropriate blood level of the active. This is usually monitored
using a straightforward analysis of blood samples over the time course of the
application. It is interesting to continue blood sampling after the device has been
removed. Data analysis of the concentration decay from the blood can give information about any depot effect in the skin, since some transdermally applied drugs
do appear to bind to skin components. Since the skin is metabolically active, it
is also important to determine the levels of metabolites in the blood and compare
these to those found using conventional delivery. One of the perceived advantages
of transdermal delivery is avoidance of first-pass metabolism. The balance of the
metabolites, particularly if they are active, can make an important contribution
to the overall bioactivity of the transdermal device.

Another way of monitoring transdermal uptake involves the measurement
of the amount of residual active in the patch. This approach has been used to
examine the delivery of nitroglycerine, and the results correlate well with in vitro
data (61); see Fig. 5.

Variability between individuals can be used to determine the relative degree
to which drug delivery is controlled by either the device or the skin. Intersubject
variability in skin permeability is high, and if the skin controls drug input into

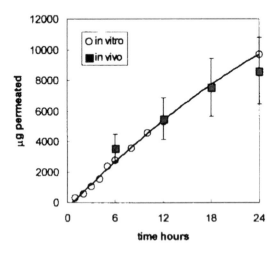

Figure 5 A comparison of the permeation of nitroglycerine from a transdermal patch
as measured both in vivo and in vitro. (Data adapted from Ref. 62.)

the body, considerable scatter in the amount of drug delivered (as determined from the residual amounts in the patch) may be anticipated. If the patch controls input, on the other hand, the residual amount in the patches is much more uniform (63); see Fig. 6.

Assessing the in vivo delivery of agents that act within the skin is much more complex. Relative estimates can be made for compounds that elicit a visible physiological response, e.g., nicotinic acid esters that cause vasodilatation, or the corticosteroids, which vasoconstrict. Much of the pioneering work on formulation effects resulted from experiments in which the degree of pallor induced by the steroid was monitored as a function of the vehicle composition (65,66). Developments in techniques to measure blood flow, such as laser Doppler velocimetry and with the chromameter, have allowed more objective evaluations to be made (67). However, as most topical agents do not elicit a simple visible response, how can one measure drug levels in the skin? A biopsy is perhaps the most direct, but also the most invasive, approach. Microdialysis (7,8) is more sophisticated, albeit hardly noninvasive either, and can be quite useful for permeants of low lipophilicity. A schematic representation of cutaneous microdialysis is shown in Fig. 7.

Reproducible positioning of the microdialysis probe can be achieved with ultrasound imaging, and the temporal profile of drug concentration in the skin can be obtained, as well as information on the release of biochemical markers,

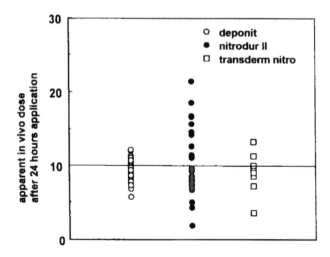

Figure 6   The variability in apparent dose from three transdermal patches. The smallest spread in the data is seen for the patch that has the highest control of GTN input into the systemic circulation. (Data adapted from Ref. 64.)

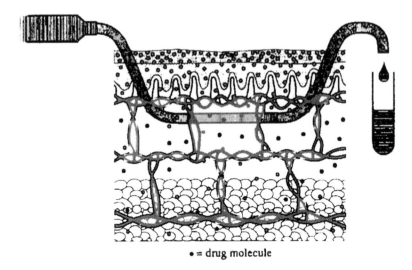

• = drug molecule

Figure 7 A representation of cutaneous microdialysis. (Redrawn from Ref. 68.)

such as histamine (6). However, the concentration in the dialyzate is only related, but not necessarily equal, to the concentration in the skin. The procedure is good, therefore, for examining relative effects and for indicating the clearance of the active from its site of action.

A far less invasive procedure is tape-stripping (69). The amount of active in the stratum corneum is determined at set time intervals after application and is assumed to be directly related to the level in the underlying viable tissue. Recent research has demonstrated that accurate concentration versus distance profiles within the stratum corneum can be obtained, and that the data can be deconvoluted to give measures of the diffusion coefficient of the drug within the stratum corneum and its partition coefficient between the vehicle and the stratum corneum (70), i.e., the fundamental parameters that control transdermal delivery.

A number of rapidly developing biophysical techniques [e.g., OTTER (optothermal transient emission radiometry) (71) and photoacoustic spectroscopy (72)] may also be used to quantify drug concentrations in the skin. If the drug has a suitable chromophore, confocal microscopy may also prove useful (73).

IX. TOXICITY

Major problems in developing a transdermal system are caused by irritation and sensitization. Testing for such has to be examined, therefore, at a very early stage of a development programme. To date there are no algorithms available that can predict skin toxicity on the basis of chemical structure. Cultured cells can be used to provide some guidance (74). Specific biochemical indicators of toxicity such as IL-1α can be measured; however, the approach needs considerable validation before it can be used routinely as a predictive screen. Animal models do not appear to be very reliable, partly for the reasons discussed above. Their skin structure does not provide the same barrier as human skin, and their biochemical responses to toxins can be totally different.

It would be useful to have answers to very simple questions such as: Is the toxicity of a potent irritant, but poor permeant, equivalent to that of a poor irritant but rapid permeant? The underlying mechanisms involved are being explored, and it may be possible to develop means of suppressing the unwanted effect during the application of the drug. If a safe mechanism for this could be identified, it would open a large range of drugs to topical and/or transdermal administration.

X. CONCLUSIONS

There have been significant developments in the ability to predict skin penetration, and the various algorithms available allow direct estimation of the feasibility of delivering a drug transdermally. However, more confidence in the modeling approach must be established to take into account the effects of formulation, etc. Mathematical modeling must be supplemented with in vitro experiments, the design of which must be carefully considered so that artifactual results can be avoided. A combined mathematical prediction plus confirming in vitro data should provide the information needed to determine the feasibility of delivering a drug transdermally. This will not provide any indication of potential skin toxicity, however; the latter requires careful and early attention in the development program.

REFERENCES

1. J Hadgraft. Skin, the final frontier. International Journal of Pharmaceutics 2001; 224(1–2):1–18.
2. RJ Feldmann, HI Maibach. Regional variation in percutaneous penetration of 14C cortisol in man. Journal of Investigative Dermatology 1967; 48(2): 181–183.
3. BW Barry. Reflections on transdermal drug delivery. Pharmaceutical Science and Technology Today 1999; 2(2):41–43.
4. VM Knepp, J Hadgraft, RH Guy. Transdermal drug delivery—problems

and possibilities. CRC Critical Reviews in Therapeutic Drug Carrier Systems 1987; 4(1):13–37.

5. GL Flynn. Physicochemical determinants of skin absorption. In: TR Gerrity, CJ Henry, eds. Principles of Route-to-Route Extrapolation for Risk Assessment. New York: Elsevier, 1990, pp 93–127.

6. RO Potts, RH Guy. Predicting skin permeability. Pharmaceutical Research 1992; 9(5):663–669.

7. SC Valvani, SH Yalkowsky, TJ Roseman. Solubility and partitioning. 4. Aqueous solubility and octanol–water partition-coefficients of liquid non-electrolytes. Journal of Pharmaceutical Sciences 1981; 70(5):502–507.

8. T Yano, A Nakagawa, M Tsuji, K Noda. Skin permeability of various non-steroidal anti-inflammatory drugs in man. Life Sciences 1986; 39:1043–1050.

9. T Degim, J Hadgraft, WJ Pugh. Modelling the transdermal delivery of non-steroidal anti-inflammatory agents. Proceedings of the International Symposium on Controlled Release of Bioactive Materials 1995; 22:652–653.

10. RF-JB Groning. Threedimensional solubility parameters and their use in characterising the permeation of drugs through the skin. Pharmazie 1996; 51:337–341.

11. MS Roberts, WJ Pugh, J Hadgraft. Epidermal permeability: penetrant structure relationships. 2. The effect of H-bonding groups in penetrants on their diffusion through the stratum corneum. International Journal of Pharmaceutics 1996; 132(1–2):23–32.

12. WJ Pugh, MS Roberts, J Hadgraft. Epidermal permeability—penetrant structure relationships. 3. The effect of hydrogen bonding interactions and molecular size on diffusion across the stratum corneum. International Journal of Pharmaceutics 1996; 138(2):149–165.

13. PG Green, J Hadgraft. Facilitated transfer of cationic drugs across a lipoidal membrane by oleic acid and lauric acid. International Journal of Pharmaceutics (Amsterdam) 1987; 37(3):251–256.

14. M Kadono, K Kubo, H Miyazaki, N Tojyo, S Nakagawa, K Miyashita et al. Enhanced in vitro percutaneous penetration of salicylate by ion pair formation with alkylamines. Biological and Pharmaceutical Bulletin 1998; 21(6):599–603.

15. RA Nash, DB Mehta, JR Matias, N Orentreich. The possibility of lidocaine ion pair absorption through excised hairless mouse skin. Skin Pharmacology 1992; 5(3):160–170.

16. C Valenta, U Siman, M Kratzel, J Hadgraft. The dermal delivery of lignocaine: influence of ion pairing. International Journal of Pharmaceutics 2000; 197(1–2):77–85.

17. J Hadgraft, C Valenta. PH, pK(a) and dermal delivery. International Journal of Pharmaceutics 2000; 200(2):243–247.

18. MS Roberts, YG Anissimov, RA Gonsalvez. Mathematical models in percutaneous absorption. In: RL Bronaugh, HI Maibach, eds. Percutaneous Absorption. New York: Marcel Dekker, 1999, pp 3–55.
19. WJ Albery, J Hadgraft. Percutaneous absorption: theoretical description. J Pharm Pharmacol 1979; 31:129–139.
20. WJ Albery, J Hadgraft. Percutaneous absorption: in vivo experiments. Journal of Pharmacy and Pharmacology 1979; 31(3):140–147.
21. WJ Albery, RH Guy, J Hadgraft. Percutaneous absorption: transport in the dermis. International Journal of Pharmaceutics 1983; 15:125–148.
22. F Pirot, YN Kalia, AL Stinchcomb, G Keating, A Bunge, RH Guy. Characterization of the permeability barrier of human skin in vivo. Proceedings of the National Academy of Sciences of the United States of America 1997; 94(4):1562–1567.
23. AC Watkinson, AL Bunge, J Hadgraft, A Naik. Computer simulation of penetrant concentration-depth profiles in the stratum corneum. International Journal of Pharmaceutics 1992; 87:175–182.
24. YN Kalia, I Alberti, A Naik, RH Guy. Assessment of topical bioavailability in vivo: the importance of stratum corneum thickness. Skin Pharmacology and Applied Skin Physiology 2001; 14:82–86.
25. I Alberti, YN Kalia, A Naik, JD Bonny, RH Guy. In vivo assessment of enhanced topical delivery of terbinafine to human stratum corneum. Journal of Controlled Release 2001; 71(3):319–327.
26. S Riegelman. Pharmacokinetics: pharmacokinetic factors affecting epidermal penetration and percutaneous absorption. Clinical Pharmacology and Therapeutics 1974; 18(5, part 2):873–883.
27. SM Wallace, G Barnett. Pharmacokinetic analysis of percutaneous absorption: evidence of parallel penetration pathways for methotrexate. Journal of Pharmacokinetics and Biopharmaceutics 1978; 6(4):315–325.
28. RH Guy, J Hadgraft, HI Maibach. A pharmacokinetic model for percutaneous absorption. International Journal of Pharmaceutics 1982; 11:119–129.
29. RH Guy, J Hadgraft, HI Maibach. Percutaneous absorption in man: a kinetic approach. Toxicology and Applied Pharmacology 1985; 78(1):123–129.
30. RH Guy, J Hadgraft. The prediction of plasma levels of drugs following transdermal application. Journal of Controlled Release 1985; 1:177–182.
31. RH Guy, J Hadgraft. A theoretical description of the effects of volatility and substantivity on percutaneous absorption. International Journal of Pharmaceutics 1984; 18:139–147.
32. RH Guy, J Hadgraft. Pharmacokinetics of percutaneous absorption with concurrent metabolism. International Journal of Pharmaceutics (Amsterdam) 1984; 20(1–2):43–52.
33. RH Guy, J Hadgraft. Kinetic analysis of transdermal nitroglycerin delivery. Pharmaceutical Research 1985; 2:206–211.

34. J Hadgraft, S Hill, M Humpel, LR Johnston, LR Lever, R Marks, et al. Investigations on the percutaneous absorption of the antidepressant rolipram in vitro and in vivo. Pharmaceutical Research 1990; 7(12):1307–1312.

35. NJ Evans, N Rutter, J Hadgraft, G Parr. Percutaneous administration of theophylline in the preterm infant. Journal of Pediatrics 1985; 107(2):307–311.

36. JN McDougal, HJ Clewell, ME Andersen, GW Jepson. Physiologically based pharmacokinetic modelling of skin penetration. In: RC Scott, RH Guy, J Hadgraft, eds. Prediction of Percutaneous Penetration Methods, Measurements, Modelling, 1990. London: IBC Technical Services, 1990, pp 263–272.

37. GW Jepson, JN McDougal. Physiologically based modeling of nonsteady state dermal absorption of halogenated methanes from an aqueous solution. Toxicology and Applied Pharmacology 1997; 144(2):315–324.

38. B Ongpipattanakul, RR Burnette, RO Potts, ML Francoeur. Evidence that oleic acid exists in a separate phase within stratum corneum lipids. Pharmaceutical Research 1991; 8(3):350–354.

39. A Naik, L Pechtold, RO Potts, RH Guy. Mechanism of oleic acid–induced skin penetration enhancement in vivo in humans. Journal of Controlled Release 1995; 37(3):299–306.

40. D Bommannan, RO Potts, RH Guy. Examination of the effect of ethanol on human stratum corneum in vivo using infrared spectroscopy. Journal of Controlled Release 1991; 16(3):299–304.

41. PK Wotton, B Mollgaard, J Hadgraft, A Hoelgaard. Vehicle effect on topical drug delivery. 3. Effect of azone on the cutaneous permeation of metronidazole and propylene glycol. International Journal of Pharmaceutics 1985; 24(1):19–26.

42. MA Pellett, AF Davis, J Hadgraft. Effect of supersaturation on membrane transport. 2. Piroxicam. Int J Pharm 1994; 111:1–6.

43. AF Davis, J Hadgraft. Effect of supersaturation on membrane transport. I. Hydrocortisone acetate. International Journal of Pharmaceutics 1991; 76(1–2):1–8.

44. MA Pellett, AC Watkinson, KR Brain, J Hadgraft. Diffusion of flurbiprofen across human stratum corneum using synergistic methods of enhancement. In: KR Brain, VJ James, KA Walters, eds. Perspectives in Percutaneous Penetration. Cardiff: STS Publishing, 1997, p 86.

45. KR Brain, J Hadgraft, VJ James, VP Shah. In vitro assessment of skin permeation from a transdermal system for the delivery of oestradiol. International Journal of Pharmaceutics 1993; 89(2):R13–R16.

46. PG Green, M Flanagan, B Shroot, RH Guy. Iontophoretic drug delivery.

In: KA Walters, J Hadgraft, eds. Pharmaceutical Skin Penetration Enhancement. New York: Marcel Dekker, 1993, pp 311–333.

47. AM Rodriguez Bayon, RH Guy. Iontophoresis of nafarelin across human skin in vitro. Pharmaceutical Research 1996; 13(5):798–800.

48. BH Sage, RA Hoke, inventors; Becton Dickinson and Company, assignee. Molecules for iontophoretic delivery. US patent 5843015, Dec 1 1998.

49. R Vanbever, E LeBoulenge, V Preat. Transdermal delivery of fentanyl by electroporation. I. Influence of electrical factors. Pharmaceutical Research 1996; 13(4):559–565.

50. R Vanbever, N De Morre, V Preat. Transdermal delivery of fentanyl by electroporation. II. Mechanisms involved in drug transport. Pharm Res 1996; 13:9:1360–1366.

51. S Mitragotri, DA Edwards, D Blankschtein, R Langer. Mechanistic study of ultrasonically-enhanced transdermal drug delivery. Journal of Pharmaceutical Sciences 1995; 84(6):697–706.

52. DF Sarphie, BJ Bellhouse, Y Alexander, KR Brain, J Hadgraft, VJ James, et al. Gas propulsion of microprojectiles for the transdermal delivery of powder form drugs. In: KR Brain, VJ James, KA Walters, eds. Prediction of Percutaneous Penetration, 1993. Cardiff: STS Publishing, 1993, pp 360–368.

53. S Henry, DV McAllister, MG Allen, MR Prausnitz. Microfabricated microneedles: a novel approach to transdermal drug delivery. Journal of Pharmaceutical Sciences 1998; 87(8):922–925.

54. JS Nelson, JL McCullough, TC Glenn, WH Wright, LH Liaw, SL Jacques. Mid-infrared laser ablation of stratum corneum enhances in vitro percutaneous transport of drugs. Journal of Investigative Dermatology 1991; 97(5):874–879.

55. JP Skelly, VP Shah, HI Maibach, RH Guy, RC Wester, G Flynn, et al. FDA and AAPS report of the workshop on principles and practices of in vitro percutaneous penetration studies—relevance to bioavailability and bioequivalence. Pharmaceutical Research 1987; 4(3):265–267.

56. D Howes, R Guy, J Hadgraft, J Heylings, U Hoeck, F Kemper, et al. Methods for assessing percutaneous absorption—the report and recommendations of ECVAM Workshop 13. ATLA-Alternatives to Laboratory Animals 1996; 24(1):81–106.

57. SW Collier, NM Sheikh, A Sakr, JL Lichtin, RF Stewart, RL Bronaugh. Maintenance of skin viability during in vitro percutaneous absorption/metabolism studies. Toxicology and Applied Pharmacology 1989; 99(3):522–533.

58. O Doucet, N Garcia, L Zastrow. Skin culture model: a possible alternative

to the use of excised human skin for assessing in vitro percutaneous absorption. Toxicology in Vitro 1998; 12(4):423–430.

59. MP Carver, PL Williams, JE Riviere. The isolated perfused porcine skin flap. III. Percutaneous absorption pharmacokinetics of organophosphates, steroids, benzoic acid, and caffeine. Toxicology and Applied Pharmacology 1989; 97(2):324–337.

60. JE Riviere, BH Sage, NA Monteiro-Riviere. Transdermal lidocaine iontophoresis in isolated perfused porcine skin. Journal of Toxicology—Cutaneous and Ocular Toxicology 1989; 8(4):493–504.

61. J Hadgraft, D Lewis, D Beutner, HM Wolff. In vitro assessments of transdermal devices containing nitroglycerin. International Journal of Pharmaceutics 1991; 73(2):125–130.

62. J Hadgraft, D Beutner, HM Wolff. In vivo–in vitro comparisons in the transdermal delivery of nitroglycerin. International Journal of Pharmaceutics 1993; 89(1):R1–R4.

63. RH Guy, J Hadgraft. Rate control in transdermal drug delivery? International Journal of Pharmaceutics 1992; 82:R1–R6.

64. J Hadgraft. Pharmaceutical aspects of transdermal nitroglycerin. International Journal of Pharmaceutics 1996; 135(1–2):1–11.

65. I Sarkany, JW Hadgraft, GA Caron, CW Barrett. The role of vehicles in the percutaneous absorption of corticosteroids. An experimental and clinical study. British Journal of Dermatology 1965; 77:569–575.

66. CW Barrett, JW Hadgraft, GA Caron, I Sarkany. The effect of particle size and vehicle on the percutaneous absorption of fluocinolone acetonide. British Journal of Dermatology 1965; 77:576–578.

67. RH Guy, E Tur, S Bjerke, HI Maibach. Are there age and racial differences to methyl nicotinate–induced vasodilatation in human skin? Journal of the American Academy of Dermatology 1985; 12(6):1001–1006.

68. E Benfeldt. In vivo microdialysis for the investigation of drug levels in the dermis and the effect of barrier perturbation on cutaneous drug penetration—studies in hairless rats and human subjects—Preface. Acta Dermato-Venereologica 1999:3–59.

69. A Rougier, D Dupuis, C Lotte, R Roguet. The measurement of the stratum corneum reservoir. A predictive method for in vivo percutaneous absorption studies: influence of application time. Journal of Investigative Dermatology 1985; 84(1):66–68.

70. I Alberti, YN Kalia, A Naik, JD Bonny, RH Guy. Effect of ethanol and isopropyl myristate on the availability of topical terbinafine in human stratum corneum, in vivo. International Journal of Pharmaceutics 2001; 219(1–2):11–19.

71. P Xiao, RE Imhof. Optothermal measurement of water distribution within the stratum corneum. Skin Bioengineering 1998; 26:48–60.

72. BD Hanh, RHH Neubert, S Wartewig, A Christ, C Hentzsch. Drug penetration as studied by noninvasive methods: Fourier transform infrared-attenuated total reflection, Fourier transform infrared, and ultraviolet photoacoustic spectroscopy. Journal of Pharmaceutical Sciences 2000; 89(9):1106–1113.

73. PJ Caspers, GW Lucassen, EA Carter, HA Bruining, GJ Puppels. In vivo confocal Raman microspectroscopy of the skin: noninvasive determination of molecular concentration profiles. Journal of Investigative Dermatology 2001; 116(3):434–442.

74. E Boelsma, H Tanojo, HE Bodde, M Ponec. An in vivo in vitro study of the use of a human skin equivalent for irritancy screening of fatty acids. Toxicology in Vitro 1997; 11(4):365–376.

2
Evaluating the Transdermal Permeability of Chemicals

Brent E. Vecchia* and Annette L. Bunge
Colorado School of Mines, Golden, Colorado, U.S.A.

I. INTRODUCTION

Contaminated terrestrial waters and chemically disinfected waters frequently contact human skin, and waterborne contaminants (environmental pollutants and disinfection byproducts, respectively) can be dermally absorbed. Reasonable estimates of the amount of chemical absorption by the skin is important for determining safe levels of contamination. Similarly, dermal absorption estimates are helpful in evaluating the potential for transdermal drug delivery.

The stratum corneum (SC) permeability coefficient is one of the key parameters for estimating dermal absorption. Fortunately, permeability coefficients for many compounds have been measured from aqueous solutions, and informed decisions about dermal uptake can now be made. When decisive experimental measurements are not available, predictive equations based on measurements for other compounds can provide a reasonable substitute. Many equations have been developed, but not all provide adequate estimates for all chemicals of concern. In this chapter, 22 different equations are compared to the large number of permeability coefficient values collected by Flynn (1).

II. BACKGROUND

Most attempts to develop predictive equations for permeability coefficients have recognized contributions from molecular size and from the solubility within the

---

* *Current affiliation*: Blakely Sokoloff Taylor & Zafman LLP, Denver, Colorado, U.S.A.

SC lipids. The availability, appealing physical interpretation, and relative success in describing a wide range of biological processes make molecular weight (MW) and logarithmically transformed octanol–water partition coefficients (log K_{ow}) by far the most widely used parameters for predicting skin penetration. Equations based on log K_{ow} alone have assumed either that the effect of MW is minor (which may be correct if the MW range of the regressed data set was narrow) or that log K_{ow} and MW are so strongly correlated that the effect of MW is represented by log K_{ow} alone. All previously published equations presented in this chapter predict permeability coefficients with one or more of these parameters.

One of the dominant questions in modeling dermal absorption is whether hydrophilic compounds and ionized species penetrate the SC by a different pathway from that of uncharged lipophilic chemicals. It is widely held that lipophilic compounds partition into and penetrate the lipid domains of the SC. A number of investigators have examined permeability coefficient and log K_{ow} data with a more complex model in which the transport mechanism changes for penetrants of differing polarity. In particular, the apparent independence of permeability coefficients upon lipophilicity (i.e., log K_{ow}) for polar compounds has led to the hypothesis that these compounds penetrate the SC by a polar pathway (1,2).

III. RESULTS AND DISCUSSION

A. The Correlations

Table 1 lists 22 published equations for estimating permeability coefficients for chemicals penetrating human SC from aqueous vehicles, P_{cw} (cm h^{-1}). Five of these (A through E) are in the form of previously published equations (4, 10, 11, 15, and 17) but were regressed on a different set of data (20). In some cases, the equations have been rearranged or are condensed (but equivalent) forms of the original. The chemical classes and range of MW and log K_{ow} upon which each equation was developed are also listed in Table 1.

Since most of the permeability coefficients used in developing these equations are determined with intact epidermis consisting of the SC and the viable epidermis (VE), the equations actually predict the permeability coefficient of the entire epidermis, P_w, which is related to the permeability coefficients through the SC (P_{cw}) and the VE (P_{ew}) as:

$$\frac{1}{P_w} = \frac{1}{P_{cw}} + \frac{1}{P_{ew}} \tag{1}$$

Except for very lipophilic penetrants (approximately log $K_{ow} > 4$), the permeability coefficient through the entire epidermis (i.e., P_w) is not different from that through the SC alone (i.e., P_{cw}), so that these equations are often considered to predict SC permeability coefficients. This issue was addressed more quantitatively by Vecchia (4), who collected several published experimental values of the ratio P_{cw}/P_{ew}, which he used to derive correlations based on log K_{ow}. One of the equations in Table 1 (Model 5), explicitly incorporates the resistance due to the VE tissues (10) by incorporating the ratio of SC to VE permeability coefficients, called B (i.e., $B = P_{cw}/P_{ew}$). Other equations in Table 1 (e.g., Models 10 and E) have included separate estimates for P_{cw} and P_{ew}. Occasionally, permeability coefficients are measured with the dermis present, which presents an additional barrier to the penetration of lipophilic compounds. This barrier may affect in vitro penetration of the more lipophilic chemicals but probably does not contribute during in vivo absorption in which the dermis layer is normally well-perfused by capillary blood flow.

All of the equations examined in this work consist of parameters developed from regressions of experimental measurements. Some permeability coefficient equations are appropriately termed semitheoretical because the chosen functional form of their parameters conforms to a theoretical model. For example, the Potts and Guy and the Kasting et al. correlations (Models 15 and 10) were developed assuming a theoretically expected exponential dependency upon molecular size as represented by MW. By contrast, in using Model 8, it is assumed that permeability coefficients vary linearly with molecular size as represented by MW (13). Consequently, calculations of permeability coefficients using Models 8 and 15 or 10 over a wide range of molecular size give very different results. Unlike the other equations in Table 1, Model 7 assumes that permeability coefficients vary with log K_{ow} squared.

Figures 1 through 3 compare the model equations of Table 1 (evaluated at MW $\approx$ 100, 300, and 500, respectively) with experimental values from the Flynn database (1) at comparable MW ranges: 50 > MW > 150, 250 > MW > 350, and 450 > MW > 550, respectively. The log K_{ow} values presented by Flynn (1) were used as one of the regression parameters. The five revised equations (Models A through E) are plotted as dashed curves. According to Wilschut et al. (20), of these five equations, Model E (i.e., the revised Model 10) produced the best regression of the permeation data they examined. Model E is emphasized in Figs. 1 through 3 by the heavy dashed curve. Like Model E, Model 5 separately represents the resistance of the VE, and it is designated for comparison to Model E by the heavy solid curve. The chief difference between Models 5 and E are at low values of log K_{ow} (< about 0), because Model E includes a polar pathway while Model 5 does not.

Figure 1 clearly illustrates, with some notable exceptions, that most of the 22 equations can adequately predict the available skin permeability coefficient data for chemicals of low MW. Though the permeability coefficient equations displayed in Fig. 1 are in relative agreement, the data uncertainty is nearly three orders of magnitude. It will be shown later that a portion of this variability can

Table 1 Permeability Coefficient Correlations Based on K_{ow} and MW*

Model No.	Equation source	Chemical class[a]	Permeability correlation (P_w in cm/h)	Data source[b]	Data range
1	Abraham et al. (3)	Misc ($n = 43$)	$\log P_w = -2.184 + 0.851 \log K_{ow} - 0.012\,\mathrm{MW}$	NS[b]	NS[b]
2	Vecchia (4)	Misc ($n = 170$)	$\log P_w = -2.44 + 0.514 \log K_{ow} - 0.0050\,\mathrm{MW}$	4	$18 < \mathrm{MW} < 585$ $-3.1 < \log K_{ow} < 4.6$
3[c]	Bronaugh and Barton (5)	Flynn Database ($n \approx 90$)	$\log P_w = -2.61 + 0.67 \log K_{ow} - 0.0061\,\mathrm{MW}$	1	$18 < \mathrm{MW} < 765$ $-3 < \log K_{ow} < 5.5$
4	Brown and Rossi (6)	Alkanols, phenols, drugs ($n = 39$)[d]	$\log P_w = -1.0 + 0.75 \log K_{ow} - \log(120 + K_{ow}^{0.75})$	7–9	$32 < \mathrm{MW} < 765$ $-0.8 < \log K_{ow} < 4.6$
5	Cleek and Bunge (10)[e]	Flynn Database ($n \approx 90$)	$\log P_{cw} = -2.8 + 0.74 \log K_{ow} - 0.006\,\mathrm{MW}$ $$P_w = \frac{P_{cw}}{1 + P_{cw}\sqrt{\mathrm{MW}/2.6}}$$	1	$18 < \mathrm{MW} < 765$ $-3 < \log K_{ow} < 6$
6	El Tayer et al. (11)	Steroids ($n = 11$)	$\log P_w = -5.32 + 0.8 \log K_{ow}$	12	$403 < \mathrm{MW} < 503$ $1.4 < \log K_{ow} < 5.5$
7	El Tayer et al. (11)	Phenolics ($n = 18$)	$\log P_w = -5.15 + 2.39 \log K_{ow} - 0.37\,(\log K_{ow})^2$	8	$94 < \mathrm{MW} < 197$ $0.8 < \log K_{ow} < 3.7$
8	Flynn and Amidon (in (13))	Flynn Database ($n \approx 90$)	$\log P_w = -1.44 + 0.79 \log K_{ow} - 1.45 \log \mathrm{MW}$	1	$18 < \mathrm{MW} < 765$ $-3 < \log K_{ow} < 6$

9	Flynn (1)	Flynn Database $(n \approx 90)$	For MW $\leq$ 150 $\log P_w = -3.0$ for $\log K_{ow} < 0.5$ $\log P_w = -3.5 + \log K_{ow}$ for $0.5 \leq \log K_{ow} \leq 3.0$ $\log P_w = -0.5$ for $\log K_{ow} > 3.0$ For MW $>$ 150 $\log P_w = -5.0$ for $\log K_{ow} < 0.5$ $\log P_w = -5.5 + \log K_{ow}$ for $0.5 \leq \log K_{ow} \leq 3.5$ $\log P_w = -1.5$ for $\log K_{ow} > 3.5$	1	$18 < \text{MW} < 765$ $-3 < \log K_{ow} < 6$
10	Kasting et al. (2)	Misc.[f]	$P_w = \left[\dfrac{1}{P_{lip} + P_{pol}} + \dfrac{1}{P_{aq}}\right]^{-1}$ $\log P_{lip} = -2.87 + \log K_{ow} - 0.0078\text{MW}$ $P_{pol} = 1 \times 10^{-5}\ \sqrt{300/\text{MW}}$ $P_{aq} = 0.15\ \sqrt{300/\text{MW}}$	2	$18 < \text{MW} < 518$ $-1.4 < \log K_{ow} < 6.3$
11	McKone and Howd (14)	Misc. $(n = 51)$[g]	$\log P_w = -2.40 - 0.6 \log \text{MW}$ $+ \log(0.24 + 3\,K_{ow}^{0.8})$ $- \log(1 + 0.0040\,K_{ow}^{0.8})$	14	$18 < \text{MW} < 227$ $-1.4 < \log K_{ow} < 4.6$
12	Michaels et al. (9)[h]	Drugs $(n = 10)$	$\log P_w = -3.6 + \log K_{mw}$ $+ \log(1.16 + 3 \times 10^{-6}\,K_{mw})$ $- \log(0.16 + 2 \times 10^{-3}\,K_{mw})$ $K_{mw} = $ mineral oil – water partition coefficient (plotted results assumed $K_{mw} \approx K_{ow}$)	9	$165 < \text{MW} < 765$ $1 < \log K_{ow} < 4$
13	Morimoto et al. (15)	Drugs $(n = 16)$	$\log P_w = -4.0 + \log(4.21\,K_{ow}^{0.75} + 0.983)$	15	$130 < \text{MW} < 358$ $-4.7 < \log K_{ow} < 4.0$

Table 1 Continued

Model No.	Equation source	Chemical class[a]	Permeability correlation (P_w in cm/h)	Data source	Data range
14	Potts and Guy (16)	n-alkanols, acids, diols ($n = 23$)	$\log P_w = -2.24 + 0.81 \log K_{ow} - 0.013\ MW$	17	$18 < MW < 160$ $-1.4 < \log K_{ow} < 3.0$
15	Potts and Guy (16)	Flynn Database ($n \approx 90$)	$\log P_w = -2.72 + 0.71 \log K_{ow} - 0.0061\ MW$	1	$18 < MW < 765$ $-3 < \log K_{ow} < 6$
16	Siddiqui et al. (18)	Steroids ($n = 7$)	$\log P_w = -6.66 + 1.05 \log K_{ow}$	18	$288 < MW < 476$ $1.0 < \log K_{ow} < 3.5$
17	Fiserova-Bergerova et al. (19)	Misc. ($n = ?$)	$\log P_w = \log(0.00253 + 0.0102\ K_{ow})$ $- 0.00695\ MW$	NS[i]	NS[i]
A	Wilschut et al. (20) based on Model 4	Misc. ($n = 123$)	$\log P_w = -2.12 + 0.502 \log K_{ow}$ $- \log(14.0 + K_{ow}^{0.50})$	20	$18 < MW < 765$ $-3 < \log K_{ow} < 6$
B	Wilschut et al. (20) based on Model 17	Misc. ($n = 123$)	$\log P_w = \log(0.00284 + 0.000256\ K_{ow})$ $- 0.00591\ MW$	20	$18 < MW < 765$ $-3 < \log K_{ow} < 6$
C	Wilschut et al. (20) based on Model 11	Misc. ($n = 123$)	$\log P_w = 0.6 - 2.15 \log MW$ $+ \log(0.28 + 1.7\ K_{ow}^{0.82})$ $- \log(1 + 0.0051\ K_{ow}^{0.82})$	20	$18 < MW < 765$ $-3 < \log K_{ow} < 6$

| D | Wilschut et al. (20) based on Model 15 | Misc. ($n = 123$) | $\log P_w = -1.55 + 0.481 \log K_{ow} - 0.143 \sqrt{MW}$ | 20 | $18 < MW < 765$ $-3 < \log K_{ow} < 6$ |
| E | Wilschut et al. (20) based on Model 10 | Misc. ($n = 123$) | $P_w = \left[\dfrac{1}{P_{lip} + P_{pol}} + \dfrac{1}{P_{aq}} \right]^{-1}$ $\log P_{lip} = -1.326 + 0.61 \log K_{ow} - 0.179 \sqrt{MW}$ $P_{pol} = 1.52 \times 10^{-4}/\sqrt{MW}$ $P_{aq} = 2.5/\sqrt{MW}$ | 20 | $18 < MW < 765$ $-3 < \log K_{ow} < 6$ |

* Equations for Models 7 and 11 are corrections from those listed in Table 6.2 of reference (25).

† Not specified.

[a] General description of chemicals; n = the total number of data points; the number of different chemicals may be fewer.

[b] Although not directly specified, there is some indication that this data is largely composed of the alkanols, alkanoic acids, and alkanediols of (17) and the phenols reported by (8).

[c] The Bronaugh and Barton correlation (Model 3) is essentially identical to the Potts and Guy correlation (Model 15).

[d] Data adjusted to 31°C by assuming the permeability coefficient doubles for every 10°C increase in temperature.

[e] Modification of Potts and Guy equation (Model 15).

[f] This database of more than 130 permeability measurements shares some permeability measurements with the Flynn database. Most of the additional values were measured either with shed snake skin as the membrane (13 values) or with human skin from the vehicle propylene glycol (78 values).

[g] Includes many of the permeability measurements for chemicals of MW < 227 tabulated in the Flynn database. In vivo guinea pig permeability measurements for three chemicals were also included.

[h] Lipophilicity is represented with the mineral oil–water partition coefficient rather than the more conventional log K_{ow}. For unknown reasons, these two measures of lipophilicity are poorly correlated for the ten chemicals, but log K_{ow} tends to be higher on average; when log K_{ow} are substituted for the mineral oil–water partition coefficients, the permeability is overestimated.

[i] The published form of correlation 17 was written in terms of steady-state flux from a saturated aqueous solution (J_{sat}, $mg\ h^{-1}\ cm^{2}$) rather than P_w. The correlation shown here was derived from $P_w = J_{sat}/C_{sat}$ where C_{sat} ($mg\ mL^{-1}$) is the aqueous solubility limit of the penetrating solute. Fiserova-Bergerova et al. (19) cite Berner and Cooper (21) as the original source of Model 17. However, the Model 17 equation does not appear in Berner and Cooper (21,22). In a more recent publication (23), Fiserova-Bergerova state that Model 17 was "the equation by Berner-Cooper, as written by Osborn (24)." We were unable to confirm this source, because this corporate report was not readily available.

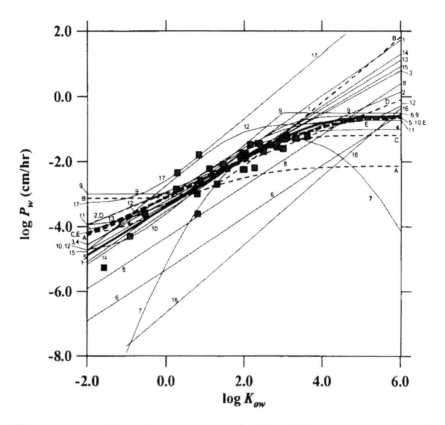

Figure 1   Permeability coefficient correlations for MW = 100 compared to experimental data from the Flynn permeability coefficient database (Ref. 1) for chemicals of 50 < MW < 150. Model 5 is designated by the solid heavy curve, Model E by the dashed heavy curve.

be explained in terms of operational variables (e.g., temperature and state of ionization) that effect the measured permeability coefficient. The remaining residual indicates either that lipophilicity (log K_{ow}) and MW are not able to model the important aspects of the skin permeability coefficient data or that experimental uncertainties and individual skin variability limit the prediction of skin permeability coefficients. Figure 2 shows analogous results, although the uncertainties with the data and the discrepancies between the model equations have increased.

At the highest MW, Fig. 3 shows that most of the model equations describe the data poorly, primarily because the data sets on which many of these equations were developed included only a few or no chemicals with a MW this high. Be-

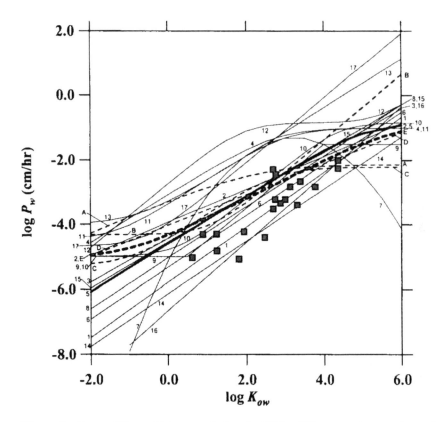

Figure 2 Permeability coefficient correlations for MW = 300 compared to experimental data from the Flynn permeability coefficient database (Ref. 1) for chemicals of 250 < MW < 350. Model 5 is designated by the solid heavy curve, Model E by the dashed heavy curve.

cause it was derived from similar chemicals with larger MW, the El Tayar equation (Model 6) based on the hydrocortisone-21-yl-esters describes these data better than do the other equations. This figure dramatically illustrates the adverse effects of extrapolating permeability coefficient equations outside the range of the training database.

Based on the results in Figs. 1 through 3, we are able to make the following observations: (a) many of the equations predict a similar dependence on K_{ow}; (b) the distinctive differences between equations are in their prediction of the effect of MW; (c) nearly all of the equations reasonably predict the experimental data for chemicals of MW $\approx$ 100 (exceptions are Models 6, 7, 8, 16, 17, and A); (d) differences between model equations and between equations and experimental

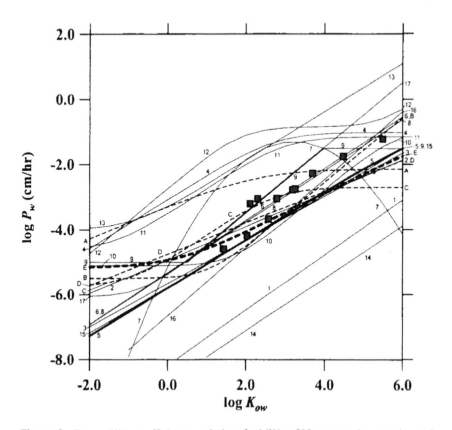

Figure 3 Permeability coefficient correlations for MW = 500 compared to experimental data from the Flynn permeability coefficient database (Ref. 1) for chemicals of 450 < MW < 550. Model 5 is designated by the solid heavy curve, Model E by the dashed heavy curve.

values increase with the MW of the penetrating chemical; (e) some of the model equations listed in Table 1 were developed from data with a limited range of properties, and not surprisingly those equations do not perform well when used outside of that range (Models 11, 14, and 16 are examples of this). Model equations 11 and 14 were developed for chemicals with low MW (and therefore, these models perform well at low MW and poorly at higher MW). Model 16 was developed for high MW chemicals (and therefore, it performs well at high MW but poorly at lower MW). Model 17 overestimates the permeability coefficient data by one to two orders of magnitude for MW values of 100 and 300, possibly owing to errors in its derivation (22).

B.   The Flynn Database

The Flynn database (1) has been used so extensively in the development of equations for estimating permeability coefficient data that many of its features are common knowledge. Flynn assembled 97 human skin permeability coefficient values for 94 compounds with a relatively broad range of properties ($18 <$ MW $<$ 765 and $-3 <$ log $K_{ow} < 6$). It is important to recognize that only a few data are available for chemicals at the extremes of log K_{ow} or high MW. The Flynn database consists of measured permeation, primarily in vitro, from (mostly) aqueous solutions. Because of its size and diversity, this database has been widely used in developing permeability coefficient equations. Five of the model equations listed in Table 1 were developed from the Flynn database (i.e., Models 3, 5, 8, 9, and 15) with subtle changes in the forms of equations or slightly different log K_{ow} values. Several equations in Table 1 (i.e., Models 6, 7, and 12 among others) are based on subsets of the Flynn database.

Though consisting of a relatively diverse set of toxic and pharmacological compounds, the Flynn database illustrates that hydrophilic chemicals have been studied much less than lipophilic chemicals, perhaps because hydrophilic chemicals penetrate more slowly than lipophilic chemicals of similar size. Of the 97 assembled permeability coefficients for 94 different chemicals, only nine compounds (2,3-butanediol, ethanol, 2-ethoxy ethanol, methanol, N-nitrosodiethanolamine, ouabain, sucrose, scopolamine, and water) have log $K_{ow} \leq 0.0$ and only four (N-nitrosodiethanolamine, ouabain, sucrose, and water) of those are at log $K_{ow} \leq -1$. In addition, the Flynn database includes only a few highly lipophilic compounds (i.e., log $K_{ow} > 4$). More data are needed so that penetration of ionic, hydrophilic, and highly lipophilic chemicals can be better quantified.

The appendix lists experimental permeability coefficient values included in the Flynn database along with MW, log K_{ow}, the temperature at which the permeability coefficient was determined, the fraction of the penetrating compound that is in its nonionized form, and the permeability coefficient as predicted by Model 15. Two sets of log K_{ow} values are listed (Set A and Set B). The Set A values of log K_{ow}, most of which were tabulated by Flynn (1) and reprinted exactly by the USEPA (13), were used in developing several of the permeability coefficient equations based on the Flynn database. Flynn (1) did not list log K_{ow} values for four chemicals (chlorpheniramine, diethylcarbamazine, N-nitrosodiethanolamine, and ouabain), and so Set B values were used for these chemicals in analyses described later, unless specified otherwise. To indicate that these values were not provided in the original Flynn database we enclose the log K_{ow}

values for these four chemicals in parentheses [e.g., for chlorpheniramine (3.39)].
The Set B log K_{ow} values are recommended *star* (★) values from Hansch et al.
(26) unless contained in brackets (e.g., for chloroxylenol [3.48]), in which case
they were calculated using Daylight (27). According to Hansch et al., the *star*
designates preferred measurements of log K_{ow} as or converted to the neutral (nonionized) form of compounds (26). Neither a star value nor a calculated value
were available for 17-hydroxypregnenolone.

The fraction of the penetrating compound that was nonionized was computed by SPARC (28). The program SPARC (SPARC performs automated reasoning in chemistry) is an expert system for the estimation of chemical and physical reactivity. Its computational algorithms are based on considerations of
molecular structure that are arrived at using the reasoning process that an expert
chemist might apply in reactivity analysis. The computational approaches in
SPARC are a blending of conventional linear free-energy theory (LFET) and
perturbed molecular orbital (PMO) methods. In general, SPARC utilizes LFET
to compute thermodynamic properties and PMO theory to describe quantum effects such as delocalization energies or polarizabilities of π-electrons. SPARC-
calculated and IUPAC pK, values for more than 3,500 different compounds have
been calculated and compared. For this statistical comparison, the regression coefficient r^2 was 0.994 and the root-mean-square error (RMSE) was 0.37 (29).
These statistics indicate a high level of predictive power.

Figure 4 compares experimental permeability coefficients from the Flynn
database (P_{exp}) with predictions (P_{pred}) from the Potts and Guy equation (Model
15). The dashed upper and lower horizontal lines distinguish those permeability
coefficients that are underestimated or overestimated by more than an order of
magnitude. The Potts and Guy correlation was developed by regression to the
Flynn database, and thus the P_{exp}/P_{pred} is distributed more or less symmetrically
around $P_{exp}/P_{pred} = 1$ (i.e., log $P_{exp}/P_{pred} = 0$). This plot does not indicate a great
representation of the experimental data (i.e., there is approximately a one to two
orders of magnitude discrepancy between data and prediction), but it is an improvement over the discrepancy between the model equations (approximately six
orders of magnitude) shown in Fig. 3.

The Flynn database includes three in vivo measurements [ethylbenzene,
styrene, and toluene from Dutkiewicz and Tyras (30,31)] which are designated
in Fig. 4 by asterisks. These measurements consistently exceed the predictions.
The fact that the measurements are high is expected from the experimental protocol and the procedure for calculating permeability from the absorption data. The
measurements were made on hands, and palmar skin is more permeable than
abdominal, forearm, back, or forehead skin (32). Permeability coefficients were
calculated from loss of chemical in the vehicle before a steady state was established. This also should overestimate the steady-state permeability coefficient,
which equations like Model 15 are meant to approximate. Finally, the permeabil-

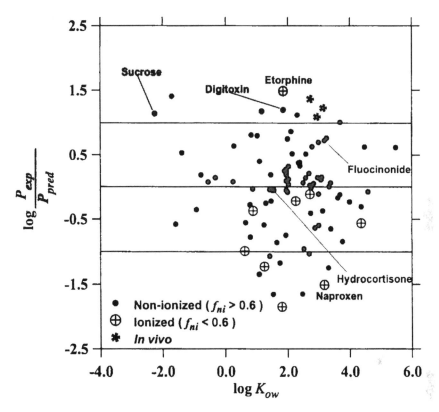

Figure 4 The effect of ionization on experimental permeability coefficients from the Flynn database (1) compared with predictions from the Potts and Guy equation (Model 15 in Table 1).

ity coefficients for these three compounds were calculated using the vehicle concentration at the end of the experiment, which was smaller than the average of the exposure period.

Permeability coefficients of several individual compounds are identified in Fig. 4. The measured permeability coefficient of hydrocortisone (33), determined from a 5% ethanol vehicle [ethanol can enhance penetration (34)], and the permeability coefficient of naproxen (35), determined from an aqueous gel vehicle, may not be representative of exclusively aqueous vehicles. The permeability coefficient of etorphine (36) is questionable, since it is much larger than permeability coefficients measured for compounds with very similar structure and size [see (36)]. Also, the permeability coefficient measured for etorphine using hairless mouse skin was much more consistent with the data from these related chemicals, suggesting that the human skin measurement may not be reliable. The permeability coefficient of digitoxin (9) is distinguished because digitoxin is the largest molecule in the database (MW = 765) and is expected to require the longest time to reach a steady state. Unfortunately, the exposure time for the digitoxin diffusion cell experiment was not reported, and we cannot assess whether a steady state was attained.

Finally, the permeability coefficients for fluocinonide and sucrose (12) were excluded on the recommendation of Anderson (37). Anderson noted a discrepancy between the permeability coefficient of fluocinonide in the experimental notebook (1.7×10^{-2} cm/h) and that reported in their paper (1.7×10^{-3} cm/h) (12). The correct value has not been decisively resolved. Anderson also suggested excluding their reported permeability coefficient for sucrose. He indicated that subsequent studies by Peck (38) explored the permeability coefficient of polar penetrants in much greater depth and solved some of the problems leading to variability in permeability coefficients for these compounds.

Figure 4 also compares the permeability coefficients of ionized and nonionized compounds. It has been repeatedly verified [e.g., (4,9,39,40) and Chapter 3 of this book among others] that ionized forms of chemicals penetrate through skin at a much slower rate than the corresponding nonionized form. Frequently, nonionized species penetrate between one and two orders of magnitude faster than ionized species of the same compounds. Consequently, the fraction of nonionized chemical available for penetration should greatly influence the magnitude of the observed permeability coefficient. When the ionized species penetrates approximately 100 times more slowly through skin than the nonionized species, and a significant fraction of the compound exists in a nonionized form (greater than 10%), it is approximately correct to disregard the ionized species concentration and calculate permeability coefficients from the concentration of only the nonionized species (4).

Permeability coefficient values reported for nine compounds in the Flynn database (atropine, codeine, etorphine, fentanyl, hydromorphone, meperidine, morphine, naproxen, and salicylic acid) appear to have been calculated using the total concentration (i.e., including the concentration of the ionized and nonionized species), although the experimental pH was such that at least 40% of the chemical existed in an ionized state. Six compounds (codeine, etorphine, hydromorphone, meperidine, morphine, and naproxen) were more than 90% ionized in the vehicle. As indicated in Fig. 4, except for etorphine, permeability coefficients of partially ionized compounds are lower on average than predictions developed with the entire Flynn database (e.g., Model 15). Clearly, chemical ionization is one source of uncertainty in the Flynn database.

Figure 5 examines the effect of temperature on permeability coefficients in the Flynn database. Permeability coefficients measured at lower temperatures (i.e., 25–26°C) are smaller on average than other permeability coefficients (and

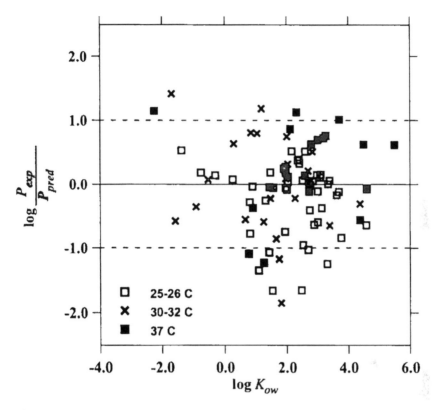

Figure 5 The effect of temperature on experimental permeability coefficients from the Flynn database (1) compared to those predicted by the Potts and Guy equation (Model 15 in Table 1). In vivo data (ethylbenzene, styrene, and toluene) and measurements made at unknown temperature (naproxen and benzyl alcohol) are not shown.

are overestimated on average by Model 15), while permeability coefficients measured at higher temperatures (i.e., 37°C) are larger on average than other permeability coefficients (and are underestimated on average by Model 15). To quantify this observation, the log residuals (defined as $\log P_{exp} - \log P_{pred}$) were calculated for data collected at different temperatures. At 25–26°C, the mean value of $\log(P_{exp}/P_{pred})$ is −0.195 (i.e., the experimental value is low compared to the prediction); at 30–32°C, the mean value of $\log(P_{exp}/P_{pred})$ is 0.005 (i.e., the experimental value is almost the same as the prediction); and at 37°C, the mean value of $\log(P_{exp}/P_{pred})$ is not 0.330 (i.e., the experimental value is high compared to the prediction). These residuals indicate that the permeability coefficient roughly doubles when the temperature increases by 5–7°C. These observations are in approximate agreement with Scheuplein and Blank, who reported that the permeability coefficient can more than double over the 10–12°C temperature range common to dermal absorption experiments (32). Consequently, differences in temperature used to measure the permeability coefficients cause some of the uncertainty in predicting measurements in the Flynn database.

C. Other Databases

Other databases exhibit similar problems. The database compiled by Wilschut and others (20) is a more extensive compilation than that of Flynn, but many of the added measurements are not of the same quality. Permeability coefficients for the alkanols studied by Scheuplein and Blank were duplicated. Measurements from Barry et al. (41) were included although the vehicle was 50% ethanol, which would likely enhance penetration relative to a water-only vehicle (34). Additionally, measurements for ionized compounds were included. For example, Wilschut et al. (20) chose to include the permeability coefficient values measured for the 100% ionized form rather than the nonionized form for six drugs (diclofenac, diethylamine salicylate, indomethacin, naproxen, piroxicam, and salicylic acid), although Singh and Roberts (42) reported experimental values for both. The relevance of log K_{ow} determined for the nonionized compound with permeability coefficients determined for the ionized compound is questionable.

Kasting and others (2,43) compiled a database of more than 130 permeability coefficient measurements. Many of these measurements overlap with those in the Flynn database. However, other measurements were determined with shed snakeskin as the membrane (13 values), or with human skin but from propylene glycol vehicles (78 values). Permeability coefficients in snakeskin have been reported to mimic human skin (44), but the relationship has not been precisely demonstrated (4). Permeability coefficients measured from propylene glycol solution were adjusted to account for partition coefficient differences (using the propylene glycol–water partition coefficient), but other effects (e.g., propylene glycol induced damage or penetration enhancement) were not considered.

McKone and Howd (14) formed a database consisting mostly of permeability coefficient measurements for chemicals in the Flynn database (1) with MW < 227. In addition, they included permeability coefficient measurements for three chemicals measured in vivo in hairless guinea pigs. The database assembled by Morimoto et al. (15) to develop Model 13 was independent of the Flynn database. However, several of the compounds in the Morimoto database were ionized (4).

D. The Predictors log K_{ow} and MW

Permeability coefficients and log K_{ow} are both affected by chemical ionization. Permeability coefficient values measured for ionized compounds are lower than permeability coefficients measured for nonionized compounds (4,9,39,40). log K_{ow} values measured with both ionized and nonionized species present, and without correction for ionization, are lower than log K_{ow} values measured when only nonionized chemical is present, because the ionized form has a higher water solubility. Permeability coefficients measured for nonionized chemicals are commonly related to log K_{ow} also measured for the nonionized chemical.

Different approaches are used to relate permeability coefficients determined for partially ionized chemicals (i.e., the permeability coefficient value observed when nonionized and ionized species are simultaneously penetrating skin) with log K_{ow}. For example, the data regressions to develop Models 9 and 15 related permeability coefficients for several chemicals that were > 40% ionized (e.g., atropine, naproxen, salicylic acid) with log K_{ow} measured for the nonionized penetrant. In another approach, permeability coefficients for partially ionized chemicals have been regressed with log K_{ow} values measured at the same pH (i.e., with the same proportions of ionized and nonionized species). This method assumes that the partitioning of ionized and nonionized species into octanol is representative of their partitioning into stratum corneum. Model 13 was developed using this method (15). Chapter 3 and Vecchia (4,40) provide a more detailed discussion of the relationship between permeability coefficients for partially ionized chemicals and log K_{ow}.

All equations of permeability coefficients that use MW and log K_{ow} as predictors include correlation among these two parameters. Figure 6 examines the relationship between log K_{ow} (Set A from Appendix 4A) and MW for all 94 chemicals in the Flynn permeability coefficient database. There is a slight positive correlation between log K_{ow} and MW (i.e., log K_{ow} increases with increasing MW), although the correlation coefficient is quite low ($r^2 = 0.07$). In addition, at least two chemicals (sucrose and ouabain) are clear exceptions to this trend. The poor correlation coefficient of the log K_{ow} with MW regression does not demonstrate that log K_{ow} and MW are independent. Indeed, closer examination reveals that some homologous series of chemicals within the database (for example, the normal alkanols) exhibit a strong correlation between log K_{ow} and MW as shown in Fig. 6. Alternatively, the steroids exhibit almost no correlation between these two parameters, because the MW is nearly constant over a wide range of log K_{ow}. We are led to conclude that the poor correlation between MW and log K_{ow} for the entire database arises from combining several groups of chemicals with different degrees of correlation between log K_{ow} and MW. Equations developed to estimate permeability coefficient values from a chemical series (e.g., the normal alkanols), with a strong log K_{ow}–MW correlation, are most affected by the correlation between log K_{ow} and MW and least able to make predictions for chemicals with different correlation between log K_{ow} and MW. Equations developed from a database containing many different compounds and chemical series (each with a different type of log K_{ow}–MW correlation) are best able to make predictions for chemicals outside of the database.

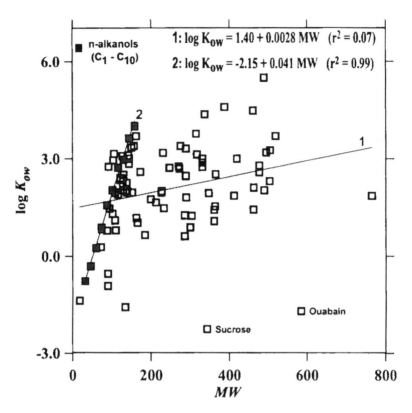

Figure 6 Correlation between log K_{ow} and MW for all compounds (line 1, $\square$, $\blacksquare$) and for n-alkanols (line 2, $\blacksquare$) in the Flynn database.

E. Reanalysis of the Flynn Database

Several good equations for estimating skin permeability coefficients from aqueous solution are based on the Flynn database. However, permeability coefficient values for compounds in this database were measured at different temperatures and different levels of ionization (from nonionized to completely ionized). The model equations do not account for these differences and therefore do not make optimal use of the permeability coefficient measurements in the database. In this section, we explore the Flynn database in several ways: (a) using log K_{ow} that are tabulated by Flynn (1) (i.e., the log K_{ow} values from Set A) and comparing results to analysis when log K_{ow} are recommended by Hansch et al. (26) (i.e., the log K_{ow} values from Set B), (b) making an adjustment for ionization and comparing results to analysis with no adjustment for ionization, (c) incorporating the temperature of the permeability coefficient measurement into analysis and comparing results with the analysis results when no temperature incorporation was made, and (d) analyzing a critically validated portion of the Flynn database and comparing the results to an analysis of the entire Flynn database.

First, we compare results when the Flynn database is analyzed using log K_{ow} from the Flynn tabulation (1) and using the recommended values of Hansch et al. (26). Using a standard multiple linear regression program, JMP (45), and the Set A list of log K_{ow} values (from Appendix A), we obtained the following fit to the entire Flynn database without adjustment for ionization:

$$\log P_w \text{ [cm/h]} = -2.77(0.16) + 0.677(0.06) \log K_{ow} \tag{2}$$
$$- 0.0057(0.0005) \text{ MW}$$

($n = 97$, $r^2 = 0.698$, $r^2_{adj} = 0.691$, RMSE = 0.730, F-ratio = 108.4). Uncertainties in parentheses are standard errors of the regression coefficients. Equation (2) shows that approximately 69.8% of the variability in log P_{cw} is explained by variation in log K_{ow} and MW. The r^2_{adj} statistic is analogous to r^2 but allows for more relevant comparisons between regressions with different numbers of fitted parameters [JMP User's Guide (45)]. Specifically, $(1 - r^2) =$ error sum of squares/total sum of squares and $(1 - r^2_{adj}) = (1 - r^2)(n - 1)/(n - p)$, where n is the number of data points and p is the number of parameters. RMSE is the root-mean-square error of the regression; RMSE is zero when the regression perfectly correlates the data. The F-ratio statistic is defined as the ratio of the sum of squares for the regression divided by the degrees of freedom for the regression and the sum of squares for the regression divided by the degrees of freedom for the error. The F-ratio = 1 when there is no correlation with the parameters and is large for correlations with good predictive power. Because the number of fitted parameters is in the denominator of the F-ratio, changes in the regression F-ratio with an increase in the number of parameters should reflect the effect on predictive power relative to the number of fitted parameters. Thus an equation with a larger number of parameters might give a higher r^2 (or r^2_{adj}) but a lower F-ratio than an equation with fewer parameters. This would indicate that the improvement in predictive power (as indicated by a larger r^2) was not as large per parameter as for the equation with fewer parameters.

Alternatively, using the Set B list of log K_{ow} values from the appendix, we obtained the following fit to the entire Flynn database without adjustment for ionization:

$$\log P_w \text{ [cm/h]} = -2.69(0.20) + 0.53(0.06) \log K_{ow} \tag{3}$$
$$- 0.0045(0.0006) \text{ MW}$$

($n = 97$, $r^2 = 0.568$, $r^2_{adj} = 0.559$, RMSE = 0.873, F-ratio = 61.7). Equation (3) shows that approximately 56.8% of the variability in log P_{cw} is explained by variation in log K_{ow} and MW.

The fit obtained with log K_{ow} values tabulated by Flynn ($r^2 = 0.698$) is superior to that obtained with log K_{ow} values reported by Hansch et al. ($r^2 = 0.568$). Similarly, the Set A values of log K_{ow} will also provide better fits than the Set B values of log K_{ow} in regressions discussed later. This may occur because for some compounds the log K_{ow} values tabulated by Flynn were measured at the same conditions by the same investigators as P_{cw}. Specifically, the log K_{ow} values from Set A and Set B are quite different for the set of hydrocortisone esters (12). The Set B values were calculated using Daylight (27) and are not the Hansch star values. Generally, we favor the use of the Hansch et al. recommended star values, despite these statistics, because they have been validated and are fully documented.

Next, we compare results when the Flynn database is adjusted for ionization to results when there is no adjustment for ionization. Specifically, permeability coefficients measured for partially ionized penetrants were divided by the fraction nonionized when no more than 90% of the compound was ionized, and permeability coefficients for six chemicals (i.e., codeine, etorphine, hydromorphone, meperidine, morphine, and naproxen) were excluded because more than 90% of the compound was ionized. Using multiple linear regression with the Set A values of log K_{ow}, we obtained the following fit to the ionization-adjusted Flynn database:

$$\log P_w \text{ [cm/h]} = -2.72(0.15) + 0.68(0.05) \log K_{ow} \qquad (4)$$
$$- 0.0057(0.0005) \text{ MW}$$

($n = 91$, $r^2 = 0.724$, $r^2_{adj} = 0.717$, RMSE = 0.691, F-ratio = 115.2). Equation (4) shows that approximately 72.4% of the variability in log P_{cw} is explained by variation in log K_{ow} and MW. Alternatively, using multiple linear regression and the Set B values of log K_{ow}, we obtained the following fit to the ionization-adjusted Flynn database:

$$\log P_w \text{ [cm/h]} = -2.63(0.19) + 0.53(0.06) \log K_{ow} \qquad (5)$$
$$- 0.0045(0.0006) \text{ MW}$$

($n = 91$, $r^2 = 0.586$, $r^2_{adj} = 0.577$, RMSE = 0.845, F-ratio = 62.4). Equation (5) shows that approximately 58.6% of the variability in log P_{cw} is explained by variation in log K_{ow} and MW.

If we compare Eqs. (2) and (4) and Eqs. (3) and (5), we see that the adjustment for ionization reduces the variability of permeability coefficients in the Flynn database and improves predictability. The statistics change only slightly because there are other sources of variability, and only a few compounds in the Flynn database are partially ionized.

Thirdly, we compare results when temperature is and is not considered.

Assuming that temperature alters the diffusion coefficient through an Arrhenius process, temperature can be used to modify the MW term in Eq. (5). Using the Set B values of log K_{ow}, we obtain the following fit to the ionization-adjusted Flynn database:

$$\log P_w \text{ [cm/h]} = -2.60(0.19) + 0.532(0.06) \log K_{ow}$$
$$- 1.42(0.18) \text{ (MW/}T)$$
(6)

($n = 91$, $r^2 = 0.597$, $r_{adj}^2 = 0.588$, RMSE = 0.833, F-ratio = 65.3). Equation (6) shows that approximately 59.7% of the variability in log P_{cw} is explained by variation in log K_{ow} and (MW/T). Equation (6) incorporates temperature in a physically realistic way without creating an additional adjustable parameter. If we compare Eqs. (5) and (6) we see that the temperature-inclusive regression reduced the regression uncertainty compared to a model that does not incorporate temperature. While temperature differences are a small contribution to uncertainty, temperature affects many measurements.

Finally, we compare results for the entire Flynn database to results for a fraction of the Flynn database that met certain quality criteria. Several measurements from the Flynn database were withheld from this analysis for various reasons. Permeability coefficients for ionizable penetrants were adjusted (by dividing the observed permeability coefficient by the fraction ionized) or excluded (when more than 90% is ionized), as mentioned previously. The three in vivo measurements for ethylbenzene, styrene, and toluene (30,31) were excluded from the analysis for reasons described previously. The hydrocortisone measurement (33) and the naproxen measurement (35) were excluded because they were measured from nonaqueous vehicles. The permeability coefficient of digitoxin was excluded because it is likely that it was not measured at a steady state. The permeability coefficients of fluocinonide and sucrose (12) were excluded as recommended by one of the authors (37).

Using the Set B values of log K_{ow}, we obtained the following fit to this validated database:

$$\log P_w \text{ [cm/h]} = -2.76(0.20) + 0.52(0.06) \log K_{ow}$$
$$- 0.0041(0.0006) \text{ MW}$$
(7)

($n = 84$, $r^2 = 0.537$, $r_{adj}^2 = 0.526$, RMSE = 0.820, F-ratio = 47.0). Equation (7) shows that approximately 53.7% of the variability in log P_{cw} is explained by variation in log K_{ow} and MW. A temperature-inclusive regression of the form described above for this same group of data produced the following equation:

$$\log P_w \text{ [cm/h]} = -2.72(0.20) + 0.53(0.06) \log K_{ow}$$
$$- 1.32(0.20) \text{ (MW/}T)$$
(8)

($n = 84$, $r^2 = 0.549$, $r_{adj}^2 = 0.538$, RMSE = 0.809, F-ratio = 49.4). Equation

(8) shows that approximately 54.9% of the variability in log P_{cw} is explained by variation in log K_{ow} and (MW/T).

Interestingly, the removal of questionable data did not reduce the variability in the P_{cw} regression. The fit of Eq. (7) is poorer than the fit obtained when the entire unadjusted Flynn database is analyzed [i.e., Eq. (3)] using the same values of log K_{ow}. Goodness-of-fit statistics usually, but not always, increase when measurements with questionable validity are removed. However, reduction of variability is not the only criterion for discriminating between regressions. As much as possible, regressions should be based on data that are physically relevant and have been consistently measured and analyzed. Permeability coefficients that have been determined or calculated inconsistently (e.g., ionized and nonionized chemicals) and permeability coefficients from different physical situations (e.g., nonaqueous vehicles, animal skin, etc.) should be excluded from analysis whether the goodness-of-fit statistic improves, remains the same, or worsens.

Like other equations developed from the Flynn database, Eqs. (7) and (8) are trained with permeability coefficients for compounds of diverse structure and properties (as measured by MW and log K_{ow}), and so should be useful for predicting permeability coefficients for a wide range of organic compounds. We recommend Eqs. (7) and (8) over other equations based on the entire Flynn database because the effects of ionization [and also temperature in Eq. (8)] have been included and the training data set has been more thoroughly reviewed.

The analysis described here begins the validation of permeability coefficient data by considering some operational variables that are most easily assessed. Some of the variability that cannot be explained by the criteria proposed here may be explained by differences in experimental protocols, and these effects are addressed more fully in Chapter 3 and elsewhere (4,40). Through data validation, more certain methods of estimating human skin permeability coefficients should be possible.

IV. CONCLUSIONS

Skin permeability coefficients can be estimated with semitheoretical equations developed from a large and diverse set of valid permeability coefficient measurements. We have compared 22 published equations with a large and well-known permeability coefficient database (the Flynn database with 97 measurements for 94 different chemicals). Model equations developed from large and diverse databases are best able to predict this database; equations developed from small or nondiverse databases may have excellent fit statistics but have poor predictive utility beyond the database from which they were developed.

Permeability coefficient data should be critically evaluated before being included in the development of predictive equations. A portion of the variance in permeability coefficients in the Flynn and other databases can be explained by different experimental factors (e.g., temperature, fraction of chemical that is ionized, and vehicle type). Based on reasonable quality criteria, some permeability coefficients from the Flynn database should not be used to develop correlations.

ACKNOWLEDGMENTS

This work was supported in part by the United States Environmental Protection Agency under Assistance Agreement Nos. CR817451 and CR822757, by the U.S. Air Force Office of Scientific Research under agreement F49620-95-1-0021, and by the National Institute of Environmental Health Sciences under grant No. R01-ES06825. We thank Dr. Richard Guy for his helpful comments.

NOTATION

f_{ni}	=	Fraction of the total chemical dose that is nonionized in the vehicle
F-ratio	=	Statistic measuring the predictive power of the regression
K_{ow}	=	Octanol–water partition coefficient of the penetrating chemical
K_{mw}	=	Mineral oil–water partition coefficient of the penetrating chemical
MW	=	Molecular weight of the absorbing chemical
n	=	Number of data points included in the regression
p	=	Number of fitted parameters in the regression
P_{cw}	=	Steady-state permeability coefficient of the SC from water
P_{ew}	=	Steady-state permeability coefficient of the VE from water
P_{exp}	=	Experimental steady-state permeability coefficient from water
P_{pred}	=	Predicted steady-state permeability coefficient from water
P_w	=	Steady-state permeability coefficient of the SC–VE composite membrane from water
r^2	=	Goodness of fit parameter
r^2_{adj}	=	Goodness of fit parameter adjusted for the number of fitted parameters and the number of data points
RMSE	=	Root-mean-square error of the regression
SC	=	Stratum corneum
T	=	Absolute temperature (Kelvin)
VE	=	Viable epidermis

APPENDIX

Table A   Information Relevant to Interpretation of the Permeability Coefficients in the Flynn Database

Compound	MW	Set A $\log K_{ow}^{a}$	Set B $\log K_{ow}^{b}$	T (°C)	f_{ni}	P_{exp} (cm/h)	P_{prod}^{c} (cm/h)	Reference No.
Aldosterone	360.4	1.08	1.08	26	1	3.0E-6	6.7E-5	46
Amobarbital	226.3	1.96	2.07	30	1	2.3E-3	1.8E-3	33
Atropine	289.4	1.81	1.83	30	0.55	8.6E-6	6.0E-4	9
Barbital	184.2	0.65	0.65	30	1	1.1E-4	3.9E-4	33
Benzyl alcohol	108.1	1.1	1.10	25	1	6.0E-3	2.4E-3	47
4-Bromophenol	173.0	2.59	2.59	25	1	3.6E-2	1.1E-2	8
2,3-Butanediol	90.1	-0.92	-0.92	30	1	<5.0E-5	1.1E-4	32
Butanoic acid	88.1	0.79	0.79	25	1	1.0E-3	1.9E-3	32
n-Butanol	74.1	0.88	0.88	25	1	2.5E-3	2.7E-3	32
2-Butanone	72.1	0.28	0.29	30	1	4.5E-3	1.0E-3	32
Butobarbital	212.2	1.65	1.73	30	1	1.9E-4	1.4E-3	33
4-Chlorocresol	142.6	3.10	3.10	25	1	5.5E-2	3.9E-2	8
2-Chlorophenol	128.6	2.15	2.15	25	1	3.3E-2	1.0E-2	8
4-Chlorophenol	128.6	2.39	2.39	25	1	3.6E-2	1.5E-2	8
Chloroxylenol	156.6	3.39	[3.48]	25	1	5.9E-2	5.1E-2	8
Chlorpheniramine	274.8	(3.39)	3.39	30	0.96	2.2E-3	9.7E-3	9
Codeine	299.3	0.89	1.14	37	<0.1	4.9E-5	1.2E-4	39
Cortexolone	364.5	2.52	2.52	26	1	7.5E-5	6.6E-4	46
Cortexone	330.5	2.88	2.88	26	1	4.5E-4	1.9E-3	46
Corticosterone	346.5	1.94	1.94	26	1	6.0E-5	3.3E-4	46
Cortisone	360.5	1.42	1.47	26	1	1.0E-5	1.2E-4	46
o-Cresol	108.1	1.95	1.95	25	1	1.6E-2	9.6E-3	8
m-Cresol	108.1	1.96	1.96	25	1	1.5E-2	9.7E-3	8

p-Cresol	108.1	1.95	1.94	25	1	1.8E-2	9.6E-3	8
n-Decanol	158.3	4.0^d	4.57	25	1	8.0E-2	1.4E-1	7
2,4-Dichlorophenol	127.6	3.08	3.06	25	1	6.0E-2	4.6E-2	8
Diethylcarbamazine	199.3	(1.75)	[1.75]	30	1	1.3E-4	1.9E-3	9
Digitoxin	764.9	1.86^c	2.83	30	1	1.3E-5	8.1E-7	9
Ephedrine	165.2	1.03	0.93	30	0.92	6.0E-3	9.5E-4	9
β-Estradiol	272.4	2.69	4.01	26	1	3.0E-4	3.2E-3	46
β-Estradiol (2)	272.4	2.69	4.01	30	1	5.2E-3	3.2E-3	9
Estriol	288.4	2.47	2.45	26	1	4.0E-5	1.8E-3	46
Estrone	270.4	2.76	3.13	26	1	3.6E-3	3.7E-3	46
Ethanol	46.1	-0.31	-0.31	25	1	8.0E-4	5.7E-4	32
2-Ethoxy ethanol	90.1	-0.54	-0.32	30	1	2.5E-4	2.1E-4	32
Ethyl benzene	106.2	3.15	3.15	24	1	1.2	7.0E-2	30
Ethyl ether	74.1	0.83	0.89	30	1	1.6E-2	2.5E-3	32
4-Ethylphenol	122.2	2.4	2.58	25	1	3.5E-2	1.6E-2	8
Etorphine	411.5	1.86	[1.41]	37	<0.1	3.6E-3	1.2E-4	36
Fentanyl	336.5	4.37	4.05	37	0.59	5.6E-3	2.0E-2	39
Fentanyl (2)	336.5	4.37	4.05	30	0.85	1.0E-2	2.0E-2	9
Fluocinonide	494.6	3.19	3.19	37	1	1.7E-3	3.2E-4	12
Heptanoic acid	130.2	2.5	[2.41]	25	1	2.0E-2	1.7E-2	17
n-Heptanol	116.2	2.72	2.72	25	1	3.2E-2	3.0E-2	32
Hexanoic acid	116.2	1.9	1.92	25	1	1.4E-2	7.9E-3	17
n-Hexanol	102.2	2.03	2.03	25	1	1.3E-2	1.2E-2	32
Hydrocortisone (HC)	362.5	1.53	1.61	26	1	3.0E-6	1.4E-4	46
Hydrocortisone (HC) (2)	362.5	1.53	1.61	30	1	1.2E-4	1.4E-4	33
[HC-21-yl]-N,N dimethyl succinamate	489.6	2.03	[0.88]	37	1	6.7E-5	5.1E-5	12
[HC-21-yl]-hemipimelate	504.6	3.26	[1.82]	37	0.80	1.8E-3	3.1E-4	12
[HC-21-yl]-hemisuccinate	462.5	2.11	[0.91]	37	0.78	6.3E-4	8.6E-5	12

Table A Continued

Compound	MW	Set A $\log K_{ow}$[a]	Set B $\log K_{ow}$[b]	T (°C)	f_{ni}	P_{exp} (cm/h)	P_{pred}[c] (cm/h)	Reference No.
[HC-21-yl]-hexanoate	460.6	4.48	[3.28]	37	1	1.8E-2	4.2E-3	12
[HC-21-yl]-hydroxy hexanoate	476.6	2.79	[1.29]	37	1	9.1E-4	2.1E-4	12
[HC-21-yl]-octanoate	488.7	5.49	[4.34]	37	1	6.2E-2	1.5E-2	12
[HC-21-yl]-pimelamate	503.6	2.31	[0.82]	37	1	8.9E-4	6.7E-5	12
[HC-21-yl]-proprionate	418.5	3.0	[1.69]	37	1	3.4E-3	6.8E-4	12
[HC-21-yl]-succinamate	461.6	1.43	[0.17]	37	1	2.6E-5	2.9E-5	12
Hydromorphone	285.3	1.25	[0.55]	37	<0.1	1.5E-5	2.5E-4	39
17-Hydroxypregnenolone	330.5	3.0	(3.0)	26	1	6.0E-4	2.3E-3	46
17-Hydroxyprogesterone	330.5	2.74	3.17	26	1	6.0E-4	1.5E-3	46
Isoquinoline	129.2	2.03	2.08	30	0.99	1.7E-2	8.1E-3	33
Meperidine	247.0	2.72	2.45	37	<0.1	3.7E-3	4.8E-3	39
Methanol	32.0	-0.77	-0.77	25	1	5.0E-4	3.3E-4	32
Methyl-[HC-21-yl]-succinate	476.6	2.58	[1.38]	37	1	2.1E-4	1.5E-4	12
Methyl-[HC-21-yl]-pimelate	518.6	3.7	[2.20]	37	1	5.4E-3	5.2E-4	12
Methyl-4-hydroxy benzoate	152.1	1.96	1.96	25	1	9.1E-3	5.2E-3	8
Morphine	285.3	0.62	0.76	37	<0.1	9.3E-6	9.0E-5	39
2-Naphthol	144.2	2.84	2.70	25	1	2.8E-2	2.5E-2	8
Naproxen	230.3	3.18	3.34	NR	<0.1	4.0E-4[f]	1.3E-2	35
Nicotine	162.2	1.17	1.17	30	0.95	1.9E-2	1.3E-3	33
Nitroglycerine	227.1	2.0[g]	[0.98]	30	1	1.1E-2	2.0E-3	9
3-Nitrophenol	139.1	2.0	2.00	25	1	5.6E-3	6.7E-3	8
4-Nitrophenol	139.1	1.96	1.91	25	1	5.6E-3	6.3E-3	8
N-nitrosodiethanolamine	134.1	(-1.58)	[-1.58]	32	1	5.5E-6	2.1E-5	48
n-Nonanol	144.3	3.62	4.26	25	1	6.0E-2	8.8E-2	32
Octanoic acid	144.2	3.0	3.05	25	1	2.5E-2	3.2E-2	17
n-Octanol	130.2	2.97	3.00	25	1	5.2E-2	3.7E-2	32
Ouabain	584.6	(-1.70)	-1.70	30	1	7.8E-7	3.0E-8	9

Pentanoic acid	102.1	1.3	1.39	25	1	2.0E-3	3.6E-3	17
n-Pentanol	88.2	1.56	1.56	25	1	6.0E-3	6.7E-3	32
Phenobarbital	232.2	1.47	1.47	30	1	4.5E-4	7.6E-4	33
Phenol	94.1	1.46	1.46	25	1	8.2E-3	5.2E-3	8
Pregnenolone	316.5	3.13	4.22	26	1	1.5E-3	3.5E-3	46
Progesterone	314.5	3.77	3.87	26	1	1.5E-3	1.0E-2	46
n-Propanol	60.1	0.25	0.25	25	1	1.4E-3	1.2E-3	32
Resorcinol	110.1	0.80	0.80	25	1	2.4E-4	1.4E-3	8
Salicylic acid	138.1	2.26	2.26	30	0.49	6.3E-3	1.0E-2	33
Scopolamine	303.4	1.24	[−0.20]	30	0.76	5.0E-5	1.9E-4	9
Styrene	104.1	2.95	2.95	24	1	6.4E-1	5.2E-2	31
Sucrose	342.3	−2.25	−3.70	37	1	5.2E-6	3.7E-7	12
Sufentanyl	387.5	4.59	3.95	37	0.90	1.2E-2	1.4E-2	39
Testosterone	288.4	3.31	3.32	26	1	4.0E-4	7.0E-3	46
Thymol	150.2	3.34	3.30	25	1	5.3E-2	5.1E-2	8
Toluene	92.1	2.75	2.73	24	1	1.0	4.4E-2	31
2,4,6-Trichlorophenol	162.0	3.69	3.69	25	1	5.9E-2	7.7E-2	8
Water	18.01	−1.38	−1.38	25	1	5.0E-4	1.5E-4	32
3,4-Xylenol	122.2	2.35	2.23	25	1	3.6E-2	1.5E-2	8

[a] log K_{ow} tabulated by Flynn (1) and reprinted exactly by US EPA (13). Values for four chemicals (chlorpheniramine, diethylcarbamazine, N-nitrosodiethanol-amine, and ouabain) contained within parentheses are from the Set B list. According to Flynn (1): (i) log K_{ow} were taken from the permeability coefficient references where these data were provided in addition to permeability coefficients, and (ii) the actual source of many of these data and the source of values not co-listed with permeability coefficients is the invaluable, extensive compilation of Hansch and Leo.

[b] log K_{ow} taken from the recommended star values of Hansch et al. (26), unless contained within brackets (e.g., for chloroxylenol [3.48]), in which case it was calculated using Daylight (27). The log K_{ow} for 17-hydroxypregnenolone is contained in parentheses to designate that Set A values were used for this chemical.

[c] Predicted by the Potts and Guy permeability coefficient equation (Model 15).

[d] Interpreted from other values in homologous series.

[e] Value reported for log $K_{butanol/w}$.

[f] Interpreted from data obtained with gels containing naproxen in solution.

[g] Value reported for partitioning into "oils."

REFERENCES

1. GL Flynn. Physicochemical determinants of skin absorption. In: TR Gerrity, CJ Henry, eds. Principles of Route-to-Route Extrapolation for Risk Assessment. New York: Elsevier, 1990, pp 93–127.
2. GB Kasting, RL Smith, BD Anderson. Prodrugs for dermal delivery: solubility, molecular size, and functional group effects. In: KB Sloan, ed. Prodrugs: Topical and Ocular Drug Delivery. New York: Marcel Dekker, 1992, pp 117–161.
3. MH Abraham, HS Chadha, RC Mitchell. The factors that influence skin penetration of solutes. J Pharm Pharmacol 47:8–16, 1995.
4. BE Vecchia. Estimating the dermally absorbed dose from chemical exposure: data analysis, parameter estimation, and sensitivity to parameter uncertainties. M.S. thesis, Colorado School of Mines, Golden, CO, 1997.
5. RL Bronaugh, CN Barton. Prediction of human percutaneous absorption with physiochemical data. In: RG Wang and HI Maibach, eds. Health Risk Assessment through Dermal and Inhalation Exposure and Absorption of Toxicants. Boca Raton, FL: CRC Press, 1991.
6. SL Brown, JE Rossi. A simple method for estimating dermal absorption of chemicals in water. Chemosphere 19:1989–2001, 1989.
7. RJ Scheuplein, IH Blank. Mechanism of percutaneous absorption. IV. Penetration of nonelectrolytes (alcohols) from aqueous solutions and from pure liquids. J Invest Dermatol 60:286–296, 1973.
8. MS Roberts, RA Anderson, J Swarbrick. Permeability of human epidermis to phenolic compounds. J Pharm Pharmacol 29:677–683, 1977.
9. AS Michaels, SK Chandrasekaran, JE Shaw. Drug permeation through human skin: theory and in vitro experimental measurement. AIChE J 21:985–996, 1975.
10. RL Cleek, AL Bunge. A new method for estimating dermal absorption from chemical exposure. 1. General approach. Pharm Res 10:497–506, 1993.
11. N El Tayar, RS Tsai, B Testa, PA Carrupt, C Hansch, A Leo. Percutaneous penetration of drugs: a quantitative structure–permeability relationship study. J Pharm Sci 80:744–749, 1991.
12. BD Anderson, WI Higuchi, PV Raykar. Heterogeneity effects on permeability–partition coefficient relationships in human stratum corneum. Pharm Res 5:566–573, 1988.
13. US EPA. Dermal exposure assessment: principles and applications, EPA/600/8-91/011B, Exposure Assessment Group, Office of Health and Environmental Assessment, Office of Research and Development, Washington, DC, 1992.

14. TE McKone, RA Howd. Estimating dermal uptake of nonionic organic chemicals from water and soil. I. Unified fugacity-based models for risk assessments. Risk Analysis 12:543–557, 1992.
15. Y Morimoto, T Hatanaka, K Sugibayashi, H Omiya. Prediction of skin permeability of drugs: comparison of human and hairless rat skin. J Pharm Pharmacol 44:634–639, 1992.
16. RO Potts, RH Guy. Predicting skin permeability. Pharm Res 9:663–669, 1992.
17. RJ Scheuplein. Molecular structure and diffusional processes across intact epidermis. Contract No. DA18-108-AMC-148(A), U.S. Army Chemical Research and Development Laboratories, Edgewood Arsenal, MD, 1967.
18. O Siddiqui, MS Roberts, AE Polack. Percutaneous absorption of steroids: relative contributions of epidermal penetration and dermal clearance. J Pharmacokinet Biopharm 17:405–424, 1989.
19. V Fiserova-Bergerova, JT Pierce, PO Droz. Dermal absorption potential of industrial chemicals: criteria for skin notation. Am J Ind Med 17:617–635, 1990.
20. A Wilschut, WF ten Berge, PJ Robinson, TE McKone. Estimating skin permeation. The validation of five mathematical skin permeation models. Chemosphere 30:1275–1296, 1995.
21. B Berner, ER Cooper. Models of skin permeability. In: AF Kydonieus, B Berner, eds. Transdermal Delivery of Drugs. Boca Raton, FL: CRC Press, 1987, pp 41–56.
22. AL Bunge. Re: "Dermal absorption potential of industrial chemicals: criteria for skin notation." Am J Ind Med 34:89–90, 1998.
23. V Fiserova-Bergerova. Re: Response to Bunge's letter to the editor. Am J Ind Med 34:91, 1998.
24. DW Osborne. Computation methods for predicting skin permeability in pharmaceutical manufacturing. Technical update, Upjohn Co., Kalamazoo, MI, 1986.
25. AL Bunge, JN McDougal. Dermal uptake. In: S Olin, ed. Exposure to Contaminants in Drinking Water: Estimating Uptake Through the Skin and by Inhalation. Boca Raton, FL: CRC Press, 1999, pp 137–181.
26. C Hansch, A Leo, D Hoekman. Exploring QSAR: Hydrophobic, Electronic, and Steric Constants. Washington, DC: American Chemical Society, 1995.
27. PCModel. Ver. 4.2, Daylight Chemical Information Systems, Inc., Mission Viejo, CA, 1995.
28. SPARC (SPARC performs automated reasoning in chemistry): an expert system for estimating physical and chemical reactivity. Ver. Windows Prototype Version 1.1, US EPA (Ecosystem Research Division) and University of Georgia, Athens, GA, 1995.

29. SH Hilal, SW Karickhoff, LA Carreira. A rigorous test for SPARC's chemical reactivity models: estimation of more than 4300 ionization pK,'s. QSARDI 14:348–355, 1995.

30. T Dutkiewicz, H Tyras. A study of the skin absorption of ethylbenzene in man. Br J Ind Med 24:330–332, 1967.

31. T Dutkiewicz, H Tyras. Skin absorption of toluene, styrene, and xylene by man. Br J Ind Med 25:243, 1968.

32. RJ Scheuplein, IH Blank. Permeability of the skin. Physiol Rev 51:702–747, 1971.

33. J Hadgraft, G Ridout. Development of model membranes for percutaneous absorption measurements. I. Isopropyl myristate. Int J Pharm 39:149–156, 1987.

34. MS Roberts, RA Anderson. The percutaneous absorption of phenolic compounds: the effect of vehicles on the penetration of phenol. J Pharm Pharmacol 27:599–605, 1975.

35. ZT Chowhan, R Pritchard. Effect of surfactants on percutaneous absorption of naproxen. I. Comparisons of rabbit, rat, and human excised skin. J Pharm Sci 67:1272–1274, 1978.

36. LM Jolicoeur, MR Nassiri, C Shipman, HK Choi, GL Flynn. Etorphine is an opiate analgesic physicochemically suited to transdermal delivery. Pharm Res 9:963–965, 1992.

37. BD Anderson. Department of Pharmaceutics and Pharmaceutical Chemistry, University of Utah, personal communication, 1995.

38. KD Peck. Ph.D. thesis, University of Utah, Salt Lake City, 1996.

39. SD Roy, GL Flynn. Transdermal delivery of narcotic analgesics: comparative permeabilities of narcotic analgesics through human cadaver skin. Pharm Res 6:825–832, 1989.

40. GE Parry, AL Bunge, GD Silcox, LK Pershing, DW Pershing. Percutaneous absorption of benzoic acid across human skin. I. In vitro experiments and mathematical modeling. Pharm Res 7:230–236, 1990.

41. BW Barry, SM Harrison, PH Dugard. Vapour and liquid diffusion of model penetrants through human skin; correlation with thermodynamic activity. J Pharm Pharmacol 37:226–236, 1985.

42. P Singh, MS Roberts. Skin permeability and local tissue concentrations of nonsteroidal anti-inflammatory drugs after topical application. J Pharmacol Exp Ther 268:144–151, 1994.

43. GB Kasting, RL Smith, ER Cooper. Effect of lipid solubility and molecular size on percutaneous absorption. In B Shroot, H Schaefer, eds. Skin Pharmacokinetics. Basel: Karger, 1987, pp 138–153.

44. T Itoh, J Xia, R Magavi, T Nishihata, JH Rytting. Use of shed snake skin as a model membrane for in vitro percutaneous penetration studies: comparison with human skin. Pharm Res 7:1042–1047, 1990.

45. JMP Statistical Discovery Software. Ver. 3.1. Cary, North Carolina; SAS Institute, 1995.
46. RJ Scheuplein, IH Blank, GJ Brauner, DJ MacFarlane. Percutaneous absorption of steroids. J Invest Dermatol 52:63–70, 1969.
47. MS Roberts. Percutaneous absorption of phenolic compounds. Ph.D. thesis, University of Sydney, Sydney, Australia, 1976.
48. RL Bronaugh, ER Congdon, RJ Scheuplein. The effect of cosmetic vehicles on the penetration of N-nitrosodiethanolamine through excised human skin. J Invest Dermatol 76:94–96, 1981.

3

Skin Absorption Databases and Predictive Equations

Brent E. Vecchia* and Annette L. Bunge
Colorado School of Mines, Golden, Colorado, U.S.A.

I. INTRODUCTION

Obtaining a viable estimate of the amount of compound absorbed into the skin is important for reasons related to medicinal therapy or to protect human health during exposure to environmental pollutants and toxic compounds. Experimentally determined permeability coefficients are useful for quantifying the dermal absorption of compounds through the skin, and they have been determined for many compounds in many different exposure scenarios. However, measurements do not exist for many compounds with the potential for dermal absorption. Hence the motivation for developing equations to estimate permeability coefficient values based on readily available input parameters.

Several equations for estimating the permeability of skin to aqueous organic compounds in terms of the penetrating chemical's molecular weight (MW) and octanol–water partition coefficient (K_{ow}) were reviewed in Chapter 2 along with the databases upon which these equations were developed. We identified numerous measurements in the most commonly utilized databases that are inconsistent with the database in various ways. For example, some measurements were made at an unsteady state; others were measured from vehicles other than water alone, and still other measurements were from animal and not human skin. Also the different penetration rates of ionized and nonionized penetrants were not usually

---

* *Current affiliation*: Blakely Sokoloff Taylor & Zafman LLP, Denver, Colorado, U.S.A.

considered. Permeability coefficients for ionogenic compounds were frequently calculated using the total vehicle concentration, rather than the concentration of the nonionized species, although ionized species penetrate more slowly through stratum corneum (SC) than nonionized species. Using the Flynn database, we showed that permeability coefficients for ionized compounds were consistently lower than expected from the nonionized species alone. Finally, temperature was usually not incorporated into the analysis of permeability coefficients, although permeability coefficients in the Flynn database measured at 37°C were larger (on average) than permeability coefficients measured at 25–26°C.

Permeability coefficients can be more accurately predicted if more of the many skin penetration studies are critically evaluated and utilized. As shown in Chapter 2, chemical ionization, temperature, and certain model-conformity criteria (e.g., a steady state) influence permeability. Additional criteria pertaining to the predictor parameters used in the permeability correlation (e.g., MW and K_{ow}) will also be discussed. In this work we develop and apply quality criteria to a large database of experimentally determined permeability coefficients. These data provide the means to develop mechanistically relevant predictive models of percutaneous absorption. Finally, equations representing these data are presented and discussed.

II. DATA VALIDATION CRITERIA

All permeability coefficient data examined in this chapter were measured in vitro through human skin from aqueous vehicles. Prior to developing regressed equations, each data point was critically reviewed with respect to several criteria deemed important to permeability coefficient values. Three collectively exhaustive (taken together they contain all permeability measurements we have considered) and mutually exclusive (measurements appearing in one database do not appear in others) databases are presented: (a) a fully validated (FV) database that meets all of the quality criteria (containing the measurements used to develop predictive equations), (b) a provisional database reserved for subsequent analysis (containing measurements that are missing pertinent information with respect to the quality criteria), and (c) an excluded database (containing measurements that do not meet one or more of the validation criteria). These three databases are listed in Appendix A. Details about how the permeability coefficients were extracted from the original references and their validation are contained in Appendix B.

Data in the FV database were required to meet five criteria: (a) the temperature must be known and be between 20 and 40°C, (b) more than 10% of the penetrating compound must be in a nonionized form, (c) a valid log K_{ow} [either a recommended star (★) value from Hansch et al. (1) or a value calculated using

Daylight software, which was developed from these recommended K_{ow} values (2)] must represent the penetrating molecule (usually the nonionized compound), (d) the measurement must have been determined at a steady state, and (e) the donor and receptor fluids do not compromise (more than water does) the barrier of the skin. Steady-state permeability coefficients require either constant vehicle concentration and sink conditions in the receptor or adjustment of the data to account for changing vehicle and/or receptor concentrations.

The excluded database includes a few measurements of ionic compounds that are valid in all respects except that a suitable log K_{ow} is not available. Although these measurements cannot be used to develop predictive equations involving log K_{ow}, information can be obtained about the penetration of these ions (cations, anions, zwitterions) without attempting to correlate them with log K_{ow}.

A. Adjustment for Ionization

Permeability coefficients have been measured for chemicals that exhibit diverse ionization behavior. Many compounds are essentially nonionized at the pH of the experiment (i.e., typically $2 < pH < 10$). However, many others (e.g., aniline, caffeine, codeine, isoquinoline, ibuprofen, and nicotine among others) exist in equilibrium with a charged species (frequently protonated amines or dissociated carboxylic acid). Some compounds are always completely ionized (e.g., paraquat or tetraethylammonium bromide). Others are zwitterionic (i.e., net neutral) but never nonionized (e.g., 5-fluorouracil). Still others coexist as a complex mixture of zwitterionic, charged, and nonionized species (e.g., 2-amino-4-nitrophenol, dopamine, hydromorphone, isoprenaline, levodopa, morphine, and nicotinic acid). The relative rates of penetration of anionic, cationic, and zwitterionic species are not precisely known. Sznitowska and colleagues have measured the penetration of net anionic, cationic, and zwitterionic forms of several amino acids and claim that permeability coefficients are essentially the same (within experimental error) when measured for the different ionic forms (3). Despite this, intuition suggests that zwitterionic (net neutral) species might penetrate through skin more rapidly than species carrying a charge.

Ionized organic compounds do not penetrate through the SC as rapidly as nonionized compounds. For example, Fig. 1a demonstrates that the observed permeability coefficients ($P_{cw,obs}$) of fentanyl and sufentanil (4) at different extents of ionization is correlated reasonably with the fraction of chemical that is nonionized (f_{ni}). Fentanyl and sufentanil are weak bases that ionize more as the pH decreases. In Fig. 1a the nonionized permeability coefficients ($P_{cw} = 0.0339$ cm/h for fentanyl and $P_{cw} = 0.0327$ cm/h for sufentanil) were estimated by averaging the observed permeability coefficients that were measured when the fraction nonionized [determined from calculated pK_a values for fentanyl (7.0 at 37°C) and sufentanil (6.2 at 37°C)], was more than 0.99 (as described in Appendix B).

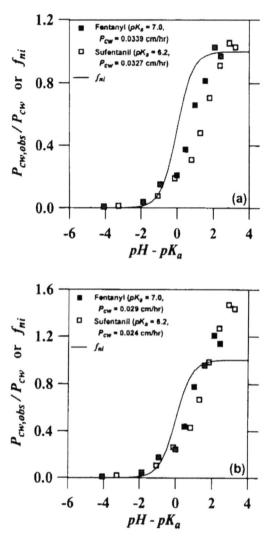

Figure 1   Observed permeability coefficients ($P_{cw,obs}$) normalized by P_{cw} for weak bases fentanyl and sufentanil at a range of pH (4) compared to f_{ni}: (a) P_{cw} was calculated by averaging all $P_{cw,obs}$ values at pH where $f_{ni} > 0.99$; (b) P_{cw} was calculated by averaging all $P_{cw,obs}/f_{ni}$ values at pH where $f_{ni} \geq 0.1$.

Note that these values for P_{cw} for fentanyl and sufentanil, derived from measurements through skin of a single subject, differ from those listed in Appendix A, which were calculated from measurements through skin from several subjects.

As illustrated in Fig. 1a, the permeability coefficient for an ionized molecule is substantially less than for its nonionized form. The SC permeability coefficients for nonionized compounds are frequently one to two orders of magnitude larger than permeability coefficients for ionized forms of the same compound. The exact relationship between the nonionized and ionized forms should depend upon the compound and the lipophilicity of the nonionized chemical, in particular. Specifically, we expect that the penetration rates for nonionized and ionized forms of the same chemical should be more similar when the nonionized species is hydrophilic and less similar when it is lipophilic.

If penetration can be attributed to the nonionized species alone, then

$$P_{cw} = \frac{P_{cw,obs}}{f_{ni}} \tag{1}$$

Assuming the penetration rate of the nonionized species is two orders of magnitude faster than for the ionized form of the same compound, then penetration rate of the ionized species will contribute negligibly (i.e., <10% of the penetration rate) as long as <90% of the compound is ionized (i.e., f_{ni} > 0.1). In Fig. 1b, P_{cw} (0.029 cm/h for fentanyl and 0.024 cm/h for sufentanil) was calculated by averaging $P_{cw,obs}/f_{ni}$ for all pH values at which at least 10% of the compound is nonionized. (See discussion in Appendix B for more details.) As shown in Fig. 1b, the estimated value of P_{cw} is less than $P_{cw,obs}$ at high pH (i.e., $P_{cw,obs}/P_{cw}$ > 1 at high pH), probably because the pK_a values are not correct. Roy and Flynn (4,5) reported pK_a values (8.9 for fentanyl and 8.5 for sufentanil at 37°C) without reference, which are quite different from the calculated pK_a values we used (using SPARC, which will be discussed later). However, the pK_a values calculated using SPARC appear to be more consistent with the data than the pK_a values reported by Roy and Flynn (4,5). Although theoretically and experimentally supported, the accuracy of estimates for nonionized permeability coefficients, made using permeability coefficients measured on partially ionized compounds, will depend strongly on accurate values for pK_a.

Unless specified otherwise, permeability coefficients in the FV database are for the nonionized species calculated using Eq. (1) from the observed permeability coefficient (calculated using the total concentration) for the partially but not significantly ionized species (i.e., f_{ni} > 0.1). When more than 90% of the compound is ionized, the rate of penetration of the ionized species cannot be neglected, and the simple calculation described by Eq. (1) is no longer applicable. Consequently, permeability coefficient measurements for compounds with f_{ni} < 0.1 are listed in the excluded database.

The fraction of nonionized compound in the vehicle, f_{ni}, for compounds with one dominant acid–base reaction, can be calculated from the acid–base dissociation constant (pK_a) and the vehicle pH (6):

$$f_{ni} = \frac{1}{(1 + 10^g)} \tag{2}$$

where $g = (pH - pK_a)$ for acids and $g = (pK_a - pH)$ for bases. We have used the vehicle pH rather than the skin pH (typically about pH = 4) because the skin has low buffer capacity to balance a vehicle with large volume.

The program SPARC [SPARC performs automated reasoning in chemistry] (7) was used to calculate pK_a values at 25°C. SPARC is an expert system for the estimation of chemical and physical reactivity. Its computational algorithms are based on considerations of molecular structure that are arrived at using the reasoning process that an expert chemist might apply in reactivity analysis. The computational approaches in SPARC blend conventional linear free energy theory (LFET) and perturbed molecular orbital (PMO) methods. In general, SPARC utilizes LFET to compute thermodynamic properties and PMO theory to describe quantum effects such as delocalization energies or polarizabilities of π-electrons. SPARC-calculated and IUPAC pK_a values for more than 4000 different compounds have been compared. For this statistical comparison, the regression coefficient r^2 was 0.994 and the root-mean-square error was 0.37 (8).

Frequently, the pK_a is not known precisely enough for small differences in temperature to affect it (6). Nevertheless, we adjusted all pK_a values to the reported experimental temperature by using the following integrated form of the van't Hoff equation (9):

$$\ln \frac{(K_a)_2}{(K_a)_1} = \frac{-\Delta H}{R} \left(\frac{1}{T_2} - \frac{1}{T_1} \right) \tag{3}$$

Enthalpies of ionization ΔH (given for the acid dissociation reaction with the equilibrium constant $K_a = 10^{-pK_a}$), used in adjusting pK_a values for temperature, were estimated from enthalpies of protonation presented elsewhere (10): 0.0 for carboxylic acids, 5.0 for phenolic compounds, 10 for amines (primary, secondary, or tertiary), 7.5 for aniline (or amines attached directly to an aromatic ring system), and 5 for aromatic nitrogen (pyridine derivatives and isoquinoline). We assume that the enthalpy of ionization is constant over a small temperature range (25–37°C) and disregard the temperature dependence of heat capacities. Multiple pK_a values for the same molecule were adjusted independently for effects of temperature.

When buffer solutions were not used and the pH was not reported, we calculated the expected natural pH using the solution concentration and all pK_a values, assuming neutral water prior to chemical addition. A general treatment of simultaneous equilibrium is required, involving equations for all linearly inde-pendent reactions, the water dissociation reaction ($K_s = 1.0 \times 10^{-14}$), a molecular balance on the active species, and an equation requiring solution electroneutrality (11).

B. Selection of K_{ow}

There are three different approaches for relating permeability coefficients for ionogenic compounds with log K_{ow}. If only nonionized species penetrate, the best approach is to use both P_{cw} and K_{ow} for the nonionized species. The next best approach is to disregard ionization effects and use the observed (for all ionized and nonionized species) permeability coefficient $P_{cw,obs}$ and the observed (for all ionized and nonionized species) $K_{ow,obs}$ measured at the same conditions (i.e., the same pH, concentration, f_{ni}, and temperature). This approach assumes that the ability of nonionized, ionized, and zwitterionic species to partition into octanol from water is representative of their ability to partition into and permeate through SC from water. The least appropriate method and a flaw of several existing equa-tions relating P_{cw} and K_{ow} (e.g., those based on the Flynn database as discussed in Chapter 2) is to use $P_{cw,obs}$ (based on the total concentration of ionized and nonionized species) and K_{ow} for the nonionized chemical. The three approaches are equivalent for compounds that are nonionized or that exist in only one state of ionization (e.g., paraquat, tetraethylammonium ion, and 5-fluorouracil at low pH).

In the data analyses that follow, we assume that only nonionized species significantly penetrate the SC. Therefore we calculate the permeability of the nonionized species and relate it to the preferred log K_{ow} star ($\star$) values reported by Hansch and colleagues (1), which these authors state were "measured as or converted to the neutral form." When these recommended values were not avail-able, Daylight software (2) was used to calculate surrogate values for the nonion-ized species. These calculated surrogates are generated using algorithms trained to the database of preferred values. Values of K_{ow} for the neutral species are larger than would be measured for a partly ionized chemical because the ionized species is more water soluble.

III. PERMEABILITY COEFFICIENT MODELS

The steady-state permeability across the SC from an aqueous vehicle (P_{cw}) into an infinite sink depends on the effective diffusivity of the chemical in the SC (D_c), the SC thickness (L_c), and the equilibrium partition coefficient between the SC and the water vehicle (K_{cw}) as given below (12):

$$P_{cw} = \frac{K_{cw}D_c}{L_c} \tag{4}$$

where K_{cw} is defined as the concentration of chemical in the SC (mass/volume of SC at absorbing conditions) divided by the equilibrium concentration in the vehicle (mass/volume) (13). As defined here, D_c is based on the SC thickness rather than the true diffusivity based on the actual molecular diffusion path length, which is not known.

The barrier resistance provided by the SC is $1/P_{cw}$. However, for intact skin the barrier includes the viable epidermis (VE) in addition to the SC, and the steady-state resistance for dermal absorption $(1/P_w)$ is the sum of these two resistances:

$$\frac{1}{P_w} = \frac{1}{P_{cw}} + \frac{1}{P_{ew}} \tag{5}$$

where the steady-state permeability through the VE from an aqueous vehicle is defined as

$$P_{ew} = \frac{K_{ew}D_e}{L_e} \tag{6}$$

in which K_{ew} is the VE–water partition coefficient, and D_e and L_e are the effective diffusivity of chemical in and thickness of the VE, respectively. If the vehicle itself does not alter the thermodynamic character of the SC or VE,

$$K_{ew} = \frac{K_{cw}}{K_{ce}} \tag{7}$$

and the resistance of the SC–VE composite barrier (i.e., the intact epidermis) is then

$$\frac{1}{P_w} = \frac{1}{P_{cw}} \left[1 + \frac{K_{ce}D_cL_e}{D_eL_c} \right] = \frac{1}{P_{cw}}(1 + B) \tag{8}$$

where the parameter B, defined as

$$B = \frac{D_cL_eK_{ce}}{D_eL_c} = \frac{P_{cw}}{P_{ew}} \tag{9}$$

measures the relative permeability of the SC to the VE. B is independent of the vehicle, provided that the vehicle has not altered physicochemically the SC or VE. Consequently, B values from one vehicle, such as water, can be used in calculations for other vehicles.

Because diffusivity in the VE is much larger than in the SC, B is often small, and the permeability across the SC–VE barrier nearly equals the permeability of the SC alone. However, the VE is much more hydrophilic than the SC, and K_{ce} will be similar to partitioning between lipophilic and hydrophilic solvents, such as octanol and water. Consequently, the VE resistance to chemicals with large log K_{ow} could be similar to or even larger than the resistance presented by the SC (i.e., $B \geq 1$). These highly lipophilic compounds enter the relatively hydrophilic VE with difficulty, thereby causing the total permeability of the combined SC–VE barrier to be less than the permeability across the SC alone. Since many chemicals of environmental interest are highly lipophilic, dermal absorption estimates should include effects of the SC–VE combined barrier.

Most permeability coefficients have been measured with SC–VE composite membranes (i.e., P_w) rather than with isolated SC membranes (i.e., P_{cw}), but the database still predicts the SC permeability coefficient, because for most compounds the VE is unimportant. Only a few permeability coefficients have been measured for compounds lipophilic enough that the VE may provide a significant resistance to penetration. Most regressed, including the regressed models developed in this chapter, do not adequately incorporate the VE resistance and are actually models for the SC alone rather than for the SC–VE composite.

A. The Conventional Model

Equation (4) is the basis of several equations for estimating SC permeability coefficients [e.g., (14)]. In these it is assumed that the SC–water partition coefficient is related to the octanol–water partition coefficient through a power function of the general form

$$\log K_{cw} = \log \hat{a} + \hat{b} \log K_{ow} \tag{10}$$

in which the parameter $\hat{b}$ accounts for differences in lipophilic character of the SC lipids compared to octanol (14). The SC–water partition coefficient data in Chapter 4 indicate that the intercept (i.e., log $\hat{a}$) predicted by Eq. (10) is small. The diffusion of small molecules in rubbery polymers is generally considered to be an activated process that varies exponentially with the size of the penetrant (15):

$$D_c = D_0 \exp(-\gamma_1^* \, MV) \tag{11}$$

where D_0 is the diffusion coefficient of a hypothetical molecule having zero molecular volume (MV), and γ_1^* is a constant. Equations (10) and (11) can then be combined as indicated by Eq. (4). Potts and Guy (14) showed no significant degradation in predictive power when MV was replaced by MW (i.e., $D_c = D_0 \exp(-\gamma_1 \, MW)$) and recommended the following functional form:

$$\log P_{cw} = a + b \log K_{ow} + d_1 \, MW \tag{12}$$

in which $a = \log \hat{a} + \log(D_0/L_c)$ and $d_1 = \gamma_1 \log e = 0.424 \gamma_1$. However, their data set consisted primarily of hydrocarbons. One would expect that MV would be better than MW for describing a chemically more heterogeneous data set (e.g., including halogenated hydrocarbons). Analysis of skin permeability measurements with Eq. (12) will provide values for a, b, and d_1 that have attributable physicochemical meaning.

B. Effect of Temperature

The effect of temperature on P_{cw} has been explored experimentally by Blank and colleagues (16), who showed that P_{cw}, measured for the normal alcohols, increased approximately 2.9-fold for a temperature increase of 10°C. Likewise, in Chapter 2 we observed that residuals (measurement–model prediction for a model developed from the data in Chapter 2) were positive (on average) for measurements made at temperatures greater than 30°C and negative (on average) for measurements made at temperatures less than 30°C. We now analyze the data with an equation that explicitly incorporates the temperature effect in the hope of reducing unexplained variance.

According to free-volume theory, the transport of small molecules in rubbery polymers is viewed as an activated process in which the diffusion coefficient obeys an Arrhenius type relationship (15):

$$D_c = D_0 \exp\left(\frac{-E_a}{RT}\right) = D_0 \exp\left(-\gamma_2^{\sharp} \cdot \frac{MV}{T}\right) \tag{13}$$

where E_a is a size-dependent activation energy required for the penetrant to make a diffusional jump, $\gamma_2^{\sharp}$ is a constant, and T is the absolute temperature at which the permeability coefficient was determined.

Assuming that MW is an appropriate substitute for MV [i.e., $D_c = D_0 \exp(-\gamma_2 \, MW/T)$], Eqs. (13) and (10) can be substituted into Eq. (4) to obtain an expression for the SC permeability coefficient of the form

$$\log P_{cw} = a + b \log K_{ow} + d_2 \frac{MW}{T} \tag{14}$$

in which $a = \log \hat{a} + \log (D_0/L_c)$ and $d_2 = \gamma_2 \log e = 0.424 \gamma_2$. In Eq. (14) temperature is assumed to reduce the activation energy for transport of penetrants through the SC, primarily acting through the MW dependence. Notably, Eq. (14) has the same number of adjustable parameters as the conventional model, Eq. (12).

C. Liquid Density (LD) Corrections and LD-Temperature Corrections

According to free-volume theory [Eq. (11)], the SC diffusion coefficient is influenced by the MV of the penetrating compound. As already described, MV is often replaced by MW. Since the MW of most hydrocarbons are related similarly to MV, this is a reasonable assumption for a group of hydrocarbons (i.e., $\sigma \approx$ MV/MW is approximately constant). However, this assumption will introduce systematic errors if the database includes chemicals composed of heavy elements such as halogens, for which σ is significantly smaller than for hydrocarbons. For example, based on data in the Merck Index, $\sigma = 0.63$ cm^{3}g^{-1} for carbon tetrachloride and 0.68 for chloroform compared to 1.29 for cyclohexane and 1.43 for n-octane. Unfortunately, MV data are not readily available. By correcting the MW with an experimental LD for each compound, it might be possible to reduce these errors without having to know or calculate the MV.

We will incorporate a pseudo-LD correction to make MW more closely resemble MV using two approaches. In the first approach, we assume that the coefficient of thermal expansion (K_T) is the same for all compounds. K_T is a thermodynamic property of materials [i.e., $K_T \equiv (\partial(MV)/\partial T)/(MV)$ at constant pressure] that to a first approximation is similar for all organic liquids. Based on values reported by Welty and colleagues (17) for several liquids [water, aniline, ammonia, dichlorodibromomethane (Freon-12), n-butyl alcohol, benzene, and glycerin] in the temperature range of 80–100°F, K_T varies between $8.3 \times 10^{-5}/K$ for water to $9.6 \times 10^{-4}/K$ for dichlorodibromomethane. The K_T for these compounds differ by only a factor of 11.5. An experimental liquid density, ρ_{ref}, which was measured at a temperature T_{ref}, is adjusted to the temperature of the permeability coefficient measurement (T) using K_T as follows:

$$\rho(\text{at } T) = \rho_{ref} \exp[K_T(T_{ref} - T)] \tag{15}$$

The density ρ is a pseudo–liquid density calculated through a hypothetical adjustment of experimental density (ρ_{ref}) measured at temperature T_{ref} without accounting for phase changes. Using the fact that MV is related approximately to MW normalized by this pseudo–liquid density (i.e., MV $\approx$ MW/ρ), and using Eq. (12), we derive the following expression for log P_{cw}:

$$\log P_{cw} = a + b \log K_{ow} + d_1 \frac{MW}{\rho_{ref}} \exp[-K_T(T_{ref} - T)] \tag{16}$$

If, in addition, the effect of temperature on the permeability coefficient is included as given in Eq. (14), we obtain an expression of the form

$$\log P_{cw} = a + b \log K_{ow} + d_2 \frac{MW}{T\rho_{ref}} \exp[-K_T(T_{ref} - T)] \tag{17}$$

Equations (16) and (17) are equivalent except for T in the denominator of the last term of Eq. (17).

In the second approach, we assume that the product of K_T and the critical temperature (T_c) is constant. T_c is the highest temperature, at any pressure, that a pure material can exist in vapor/liquid equilibrium. Starting with the definition of K_T and the theorem of corresponding states (which suggests that properties of materials are similar when related to their critical point temperature, pressure, and volume), it is easy to show that the product $K_T T_c$ is approximately the same for all liquids (i.e., $K_T T_c$ should be more constant than K_T). Using tabulated K_T values (17) for several liquids [water, aniline, ammonia, dichlorodibromomethane (Freon-12), n-butyl alcohol, benzene, and glycerin] in the temperature range 80–100°F, and the T_c for these same compounds (18), we found that the product $K_T T_c$ varies from a low of 0.054 for water to a high of 0.368 for dichlorodibromomethane. Thus $K_T T_c$ differs by only a factor of 6.8 for these compounds. Equation (18) is analogous to Eq. (15) but uses the product $K_T T_c$ to adjust densities for temperature as follows:

$$\rho(\text{at } T) = \rho_{\text{ref}} \exp\left[K_T T_c \left(\frac{T_{\text{ref}}}{T_c} - \frac{T}{T_c} \right) \right] \tag{18}$$

Assuming MV can be estimated by adjusting MW by this liquid density (i.e., $MV = MW/\rho$) and using Eq. (12), we obtain an expression for $\log P_{cw}$ that is analogous to Eq. (16):

$$\log P_{cw} = a + b \log K_{ow} + d_1 \frac{MW}{\rho_{\text{ref}}} \exp\left[-K_T T_c \left(\frac{T_{\text{ref}}}{T_c} - \frac{T}{T_c} \right) \right] \tag{19}$$

As before, the separate effect of temperature on the permeability coefficient can also be included to produce an expression that is analogous in form to Eq. (17):

$$\log P_{cw} = a + b \log K_{ow} + d_2 \frac{MW}{T\rho_{\text{ref}}} \exp\left[-K_T T_c \left(\frac{T_{\text{ref}}}{T_c} - \frac{T}{T_c} \right) \right] \tag{20}$$

In summary, Eqs. (16) and (19) adjust MW using LD alone (i.e., temperature in these equations only adjusts the LD value), while Eqs. (17) and (20) also adjust for temperature of the permeability coefficient measurement.

Eqs. (16) through (20) require values for K_T, which are not always known. If $K_T(T_{\text{ref}} - T)$ is small, then the following simple linear equation accurately represents the exponential expression in these equations:

$$\exp[-K_T(T_{\text{ref}} - T)] \approx 1 - K_T(T_{\text{ref}} - T) = 1 - K_T T_c \left(\frac{T_{\text{ref}}}{T_c} - \frac{T}{T_c} \right) \tag{21}$$

With this simplification, Eqs. (16) and (17) become

$$\log P_{cw} = a + b \log K_{ow} + d_1 \frac{MW}{\rho_{ref}} - e_1 MW \frac{T_{ref} - T}{\rho_{ref}} \tag{22}$$

$$\log P_{cw} = a + b \log K_{ow} + d_2 \frac{MW}{T\rho_{ref}} - e_2 MW \frac{T_{ref} - T}{T\rho_{ref}} \tag{23}$$

in which $e_j = d_j K_T$ (for $j = 1$ or 2) can be treated as an additional regression parameter if K_T is approximately constant. Similarly, incorporating Eq. (21) into equations that assume that $K_T T_c$ is approximately constant [i.e., Eqs. (19) and (20)] yields

$$\log P_{cw} = a + b \log K_{ow} + d_1 \frac{MW}{\rho_{ref}} - e_1 MW \frac{T_{ref} - T}{T_c \rho_{ref}} \tag{24}$$

$$\log P_{cw} = a + b \log K_{ow} + d_2 \frac{MW}{T\rho_{ref}} - e_2 MW \frac{T_{ref} - T}{TT_c \rho_{ref}} \tag{25}$$

In this case, $e_j = d_j K_T T_c$ (for $j = 1$ or 2) can be treated as additional regression parameters if $K_T T_c$ is approximately constant.

D. The Two-Pathway Model

Until now we have essentially ignored the fact that the SC is a heterogeneous membrane composed of lipophilic domains (the intercellular lipids) and hydrophilic domains (the cellular protein). The protein and lipid domains in the SC are histologically revealed as a mosaic of cornified cells containing cross-linked keratin filaments and intercellular lipid-containing regions (19). The cellular protein is not itself a homogeneous domain, but the differences within this domain are small when compared with the differences between the lipids and this protein phase. Many researchers believe that lipophilic compounds penetrate the SC by a transcellular pathway through the lipid domains (20). The mechanism for penetration of hydrophilic compounds is subject to more controversy.

Many authors have suggested that dermal absorption of hydrophilic and lipophilic compounds is governed by separate mechanisms (21–25). Potts and Guy performed an analysis of the Flynn database and they concluded that one mechanism was sufficient to describe all data in that database (14). In their investigation, partially ionized and nonionized compounds were all related to log K_{ow} values for the nonionized compound.

Here we revisit the question of separate pathways for penetration of lipophilic and hydrophilic compounds with a larger set of data, including several more measurements for hydrophilic compounds. Furthermore, analysis of this database will be more meaningful because (1) steps have been taken to avoid unequal comparisons between partially ionized and nonionized species, and (2) partially ionized species are not included in the data regressions that use log K_{ow} values for only nonionized compounds.

To test the two-pathway hypothesis, we have developed a mathematical model with a different linear dependence on log K_{ow} for hydrophilic (P_{cw}^H) and lipophilic compounds (P_{cw}^L):

$$\log P_{cw}^H = a + c \log K_{ow} + d_1 MW \tag{26}$$

$$\log P_{cw}^L = a - (b - c) \log K_{ow}^* + b \log K_{ow} + d_1 MW \tag{27}$$

Equations (26) and (27) can be combined to represent hydrophilic and lipophilic chemicals:

$$\begin{aligned}
\log P_{cw} = {}& a + c \log K_{ow} \\
& + (b - c)(\log K_{ow} - \log K_{ow}^*)U(K_{ow} - K_{ow}^*) \\
& + d_1 MW
\end{aligned} \tag{28}$$

where K_{ow}^*, representing the transition between hydrophilic and lipophilic mechanisms, is determined in the regression. The unit step function, $U(K_{ow} - K_{ow}^*)$, is defined as

$$\begin{aligned}
U(K_{ow} - K_{ow}^*) &= 0 \quad \text{if} \quad K_{ow} < K_{ow}^* \\
&= 1 \quad \text{if} \quad K_{ow} \geq K_{ow}^*
\end{aligned} \tag{29}$$

In this development, we have assumed that the effect of MW is not a function of the pathway. If, in addition, the permeability coefficients for hydrophilic compounds do not vary with K_{ow}, then $c = 0$ and Eq. (28) simplifies to

$$\log P_{cw} = a + b(\log K_{ow} - \log K_{ow}^*)U(K_{ow} - K_{ow}^*) + d_1 MW \tag{30}$$

In Eq. (30) the parameter a represents log P_{cw}^H.

E. Linear Solvation-Energy Relationships Model

The theoretical foundations of linear solvation-energy relationships (LSERs) have been described elsewhere (26–28), and LSER models have been developed to interpret and predict SC permeability coefficients (29,30). Appendix C provides additional information on LSERs. Briefly, LSER models have been successful at representing free energy–dependent properties exhibited by a class of solutes in the same solvent. In this application, the parameters of the LSER model (i.e., the solvatochromic parameters α, β, π, and V_x) correspond to the solutes rather than the solvent. Thus equations of the general form

$$\log \lambda = \log \lambda_0 + a\alpha + b\beta + c\pi + dV_x \tag{31}$$

were found for various solvent-dependent free-energy-based properties, λ. The solute solvatochromic parameters have the following physicochemical interpretation: α is the effective hydrogen-bond acidity, β is the effective hydrogen-bond basicity, π is the solute dipolarity/polarizability, and V_x (with units of cm^3/mol/ 100) is the characteristic volume of McGowan (31), which is calculated simply from molecular structure and is independent of intermolecular forces such as hydrogen bonding. For a given property and set of compounds, the coefficients $\log(\lambda_0)$ and the parameters a through d are determined using multilinear regression analysis of data.

SC permeability coefficients can be modeled with LSER parameters using an equation of the form

$$\log P_{cw} = \log P_{cw}^0 + a\alpha + b\beta + c\pi + dV_x \tag{32}$$

This equation assumes that the difference in the chemical potential (between solute in the vehicle and in the skin) driving the solute across the SC results from differences in van Der Waals forces, size effects, polarizability, and the preference of a solute to act as an electron donor and an electron acceptor in hydrogen bonds.

IV. RESULTS AND DISCUSSION

Appendix A summarizes all of the permeability coefficient values collected in this study. Table A1 contains the FV database of permeability coefficients. The provisional database is contained in Table A2. Permeability coefficients that were excluded from further analysis are listed in Table A3. The adjustment of pK$_a$ values for temperature and the calculation of f_{ni} and the natural pH for unbuffered solutions at known solute concentration are summarized in Table A4.

Figure 2 shows the FV database and several excluded and provisional measurements plotted as a function of log K_{ow}. Several of the measurements in Fig. 2 are labeled and deserve further discussion. Digitoxin, with the largest MW in the database (MW = 764.9), was excluded from the FV database because the exposure time was not specified and very long times are expected to establish a steady state. Likewise, ouabain (MW = 584.6) was made provisional because an exposure time was not specified. Other excluded data include the measurements of fluocinonide and sucrose reported by Anderson et al. (20). According to Anderson (32), an irresolvable discrepancy exists between the original notebook and the published value for fluocinonide. Anderson recommended that the permeability coefficient for sucrose be excluded because it was not measured with a technique that is necessary to measure permeability coefficients of hydrophilic compounds (see further documentation in Appendix B). Etorphine was excluded because the permeability coefficient reported was unusually large relative to per-

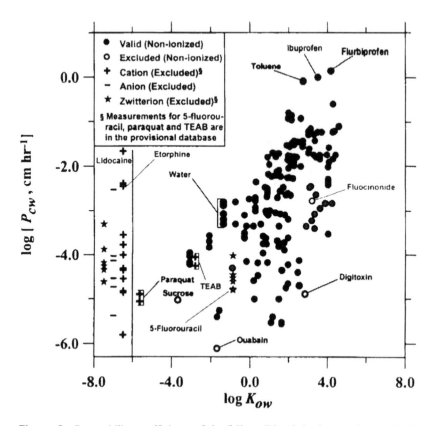

Figure 2 Permeability coefficients of the fully validated database and several values from the excluded and provisional databases plotted as a function of K_{ow}. Ionic species, without an appropriate log K_{ow}, are plotted to the left of all other measurements at an assigned log K_{ow} (cations at log $K_{ow} = -6.5$, anions at log $K_{ow} = -7.0$, and zwitterions at log $K_{ow} = -7.5$). Paraquat, tetraethylammonium (TEAB), and 5-fluorouracil are plotted at log K_{ow} reported for these ionized species.

meability coefficients for compounds of similar size and structure. In studies on hairless mouse skin, Jolicoeur et al. determined permeability coefficient values for etorphine that were comparable to other compounds of similar size and structure in their study (33). Perhaps the human skin used to study etorphine was damaged or otherwise compromised.

     The compounds producing the three largest permeability coefficient values (toluene, ibuprofen, and flurbiprofen) are also identified. These permeability coefficients are much larger than expected, particularly after adjustment for ionization, but we could not identify an explanation other than unusually large random experimental error, so they were not excluded. Permeability coefficients for water are separately labeled to serve as a benchmark for assessing the penetration of other compounds. Ionic species, without an appropriate log K_{ow}, are plotted to the left of all other measurements. Cations are plotted at log $K_{ow} = -6.5$, anions at log $K_{ow} = -7.0$, and zwitterions at log $K_{ow} = -7.5$. Paraquat, tetraethylammonium bromide (TEAB), and 5-fluorouracil are shown and separately labeled because valid log K_{ow} are available for these ions. Other excluded measurements, which we believe should not follow even the qualitative trends of the data (due to nonaqueous vehicle effects, vehicle depletion, and so on), are not shown, but are listed in the excluded database (Table A3) for reference. The measurement for hydroxypregnenolone, from the provisional database, is not shown because a suitable log K_{ow} value was not available for this compound.

As indicated in Fig. 2, permeability coefficients in the FV database vary by approximately six orders of magnitude. These permeability coefficients are clearly correlated with log K_{ow} and exhibit a nearly linear dependence over the entire range shown. A portion of the difference among measurements that is not explained by log K_{ow} can be explained by differences in molecular size, as will be shown later. The permeability coefficient values for ionic species, although apparently quite variable, are generally lower than for nonionized lipophilic species. In this database, permeability coefficients for cations are larger than permeability coefficients for either anions or zwitterions. However, many more measurements are necessary to determine whether this trend persists.

A. Comparison with the Flynn Permeability Coefficient Database

The FV database contains 170 permeability coefficient values for 127 compounds, encompassing many different organic compound structural classes. The entire FV database is broad in the sense of log K_{ow} (-3.10 [mannitol] $<$ log $K_{ow} < 4.57$ [decanol]; mean $= 1.66$, median $= 1.94$, standard deviation $= 1.75$) and MW (18.0 [water] $<$ MW 584.6 [ouabain]; mean $= 201.9$, median $= 160.25$, standard deviation $= 129.11$) and should be useful for predicting permeability coefficient values for a diverse set of organic compounds. Permeability coefficients for large MW compounds (MW > 600) are not represented in the database and cannot be accurately estimated. The FV database has measurements for only 13 hydrophilic compounds (defined as log $K_{ow} < 0.0$): 2,3-butanediol, caffeine, ethanol, 2-ethoxy ethanol, α-hydroxyphenyl acetamide, mannitol, methanol, N-nitrosodiethanolamine, ouabain, p-phenylenediamine, scopolamine, urea, and water.

By contrast, Flynn (21) assembled 97 human skin permeability coefficient values for 94 compounds with a relatively broad range of log K_{ow} (-2.25 [sucrose] $<$ log $K_{ow} < 5.49$ [HC-21-yl octanoate]; mean $= 2.05$, median $= 2.03$,

standard deviation = 1.40) and MW (18 [water] < MW < 765 [digitoxin];
mean = 238.4, median = 184.2, standard deviation = 148.8). These statistics are
based on the values of log K_{ow} reported by Flynn (21) for all but four compounds
(chlorpheniramine, diethylcarbamazine, N-nitrosodiethanolamine, and ouabain).
Values of K_{ow} for these compounds were either obtained from Hansch et al. (1)
(chlorpheniramine and ouabain) or calculated using Daylight software (2) (diethylcarbamazine and N-nitrosodiethanolamine). The Flynn database includes three
in vivo measurements (for benzene, styrene, and toluene). There are nine hydrophilic (log K_{ow} < 0.0) compounds in the Flynn database (2,3-butanediol, ethanol,
2-ethoxy ethanol, methanol, N-nitrosodiethanolamine, ouabain, sucrose, scopolamine, and water). Only four (N-nitrosodiethanolamine, ouabain, sucrose, and
water) of those compounds have a log $K_{ow} \leq -1$.

Names of compounds listed in the FV database (Table A1) with permeability coefficient values in the Flynn database are contained within brackets. Up-
dated values (2) have replaced the n-alcohol permeability coefficients appearing
in the Flynn database (34). (The only practical difference is that a lower perme-
ability coefficient for propanol [1.2×10^{-3} cm h^{-1}] replaced the value [1.4×10^{-3} cm h^{-1}] included in the Flynn database.) In vivo measurements included in
the Flynn database [ethylbenzene, styrene, and toluene from (35,36)] were ex-
cluded from the current analysis. The measured permeability coefficient of hydrocortisone (37), determined from a 5% ethanol vehicle, and the permeability of
naproxen, determined from an aqueous gel vehicle, were placed in the excluded
database. Likewise, etorphine was excluded since Jolicoeur and others (33) identified it as being very different from quite similar compounds, and much different
in hairless mouse skin. Digitoxin was excluded since the measurement was probably not at steady state. Subsequent investigations for morphine (38) and fentanyl
and sufentanil (4) were considered in addition to the prior investigation (5) cited
in the Flynn database. Various compounds from the Flynn database were excluded because more than 90% of the compound was ionized in the experiments.

B. Data Regression Analysis

Table 1 summarizes the results of linear regressions of the permeability coefficient data to the various model equations developed above. These regressions
were performed by standard procedures using JMP (39). Uncertainties in the
coefficients of the input parameters, given as the standard errors of the coefficients, are listed in parentheses. Table 1 also lists regression statistics: r^2, r^2_{adj},
RMSE, and the F-ratio. The r^2_{adj} statistic is analogous to r^2 but allows for more
relevant comparisons between models with different numbers of fitted parameters
(JMP User's Guide, Ref. 39). Specifically, $(1 - r^2)$ = error in sum of squares/
total sum of squares and $(1 - r^2_{adj}) = (1 - r^2)(n - 1)/(n - p)$, where n is the
number of data points and p is the number of parameters. RMSE is the root-

mean-square error of the model, which is zero if the model fits the data perfectly. The F-ratio is defined as the ratio of the sum of squares for the model divided by the degrees of freedom for the model and the sum of squares for the error divided by the degrees of freedom for the error. The F-ratio $= 1$ when there is zero correlation with the parameters, and it is large for equations with good predictive power. Because the number of fitted parameters is in the denominator of the F-ratio, changes in the model F-ratio with an increase in the number of parameters reflect the effect of the number of fitted parameters on the predictive power. Thus an equation with a larger number of parameters might give a higher r^2 but a lower F-ratio than an equation with fewer parameters. This would indicate that the improvement in predictive power (as indicated by a larger r^2) was not as large per parameter as for the equation with fewer parameters.

C. Analysis with the Conventional Model

Equation T1 in Table 1 is the regression of permeability data in the FV database to Eq. (12). This equation is based on almost twice as many measurements as are contained in the Flynn database, with roughly half again as many new compounds. Equation T1 accounts for approximately 55% of the variability in the 170 permeability measurements by variation in log K_{ow} and MW. Equation T1 should reasonably estimate the skin permeability coefficient for a wide range of organic compounds.

One test of validity, for a quantitative structure–activity relationship, is the ability to predict similar regression coefficients when different subsets of data are analyzed. Equation T2 in Table 1 is the equation developed in Chapter 2 from the regression to Eq. (12) of the data in the Flynn database that satisfy the validation criteria presented in this chapter. Significantly, the regression coefficients in Eqs. T1 and T2 are quite similar when uncertainty is considered. It is not surprising that r^2 is slightly smaller for Eq. T1 than for Eq. T2, since, with more data collected by different research groups, there are more potential sources of variation. Without eliminating these sources of variability, it is likely that equations derived from a larger (but equivalently validated) set of data will not produce more meaningful coefficients.

Importantly, regression equations should not be used for estimating permeability coefficient values for compounds that are very different from those used to develop the database. That is, appropriate bounds on log K_{ow} and MW need to be set for equations used to estimate permeability coefficient values. In the FV database, 95% of the chemicals have $-2.83 \leq \log K_{ow} \leq 4.27$ with an equal percentage of compounds (i.e., 2.5%) with higher and lower values than these bounds. Likewise, 80% of the chemicals have $-1.3 \leq \log K_{ow} \leq 3.85$. The more conservative lower bound (i.e., log $K_{ow} \geq -1.3$) and the less conservative upper bound (i.e., log $K_{ow} \geq 4.27$) are chosen. A reasonable lower bound for MW is that of water (i.e., the lowest MW chemical in the database, MW = 18). Permeability coefficients for water are of high quality, and there are many permeability coefficients for chemicals with slightly larger MW. In the FV database, 10% of chemicals have MW ≥ 392 and 2.5% have MW ≥ 500. The less conservative bound (i.e., MW ≤ 500) is chosen. Generally, the regressions developed from the entire FV database will provide a reasonable estimate of P_{cw} for aqueous organic compounds with log K_{ow} in the range −1.3 to 4.3 and MW in the range 18 to 500.

Equation T1, based on the conventional analysis of permeability data, will be useful for examining effects in the database that are not accounted for by log K_{ow} and MW. In Figs. 3–10 we examine the factors that influence permeability coefficient values by comparing permeability coefficient data from the FV database with values calculated by Eq. T1 in various ways.

D.   Trends in the FV Database

Permeability measurements in the FV database but not in the Flynn database, and those in both the validated Flynn database and the FV database, are compared with predictions using Eq. T1 in Fig. 3 plotted as a function of log K_{ow}. The log K_{ow} values in Table A1, which were only different from the log K_{ow} reported by Flynn (21) for a few chemicals (see Table A in the Appendix to Chapter 2 for details), were used to make this plot. Measurements appearing above the upper dashed line are underestimated by more than an order of magnitude, and those below the lower dashed line are overestimated by more than an order of magnitude (these measurements are identified by ↑ and ↓, respectively, in Table A1). The overall uncertainty appears similar in both sets of data. Figure 3 shows that more valid hydrophilic measurements are available in the FV database than in the validated portion of the Flynn database.

Figure 4 compares the FV permeability coefficient measurements and predictions against the year that the measurements were collected. There is no systematic trend over the period of time shown. The results in Fig. 4 indicate that uncertainty in measuring permeability coefficients is nearly the same in recent measurements as in those measured as early as 1964. This plot also shows that measurements of dermal absorption have been made for several decades and still continue until today. The last decade has been a particularly active period of research in dermal absorption, and many more measurements have been published since this study was suspended in 1997.

Figure 5 identifies the effects of temperature and ionization on the ratio of FV measurements to predictions made using Eq. T1. As observed in the analysis of the Flynn database in Chapter 2, measurements at temperatures less than 30°C are overestimated (on average), and measurements made at temperatures greater than 30°C are underestimated (on average). This effect is anticipated in most theories, including free-volume theory. To quantify better the effect of tempera-

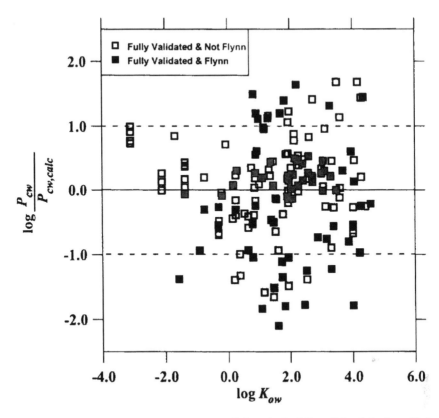

Figure 3 A comparison of permeability coefficients in the fully validated database (P_{cw}) with calculations from Eq. T1 ($P_{cw,calc}$).

ture, average residuals ($\log P_{cw} - \log P_{cw,calc}$) have been calculated. The average residuals are -0.256 and $+0.366$ when $T \leq 30°C$ and $T > 30°C$, respectively. When $T \leq 30°C$ (the average is 26.7°C), permeability coefficients are overestimated by an average factor of 1.8. When $T > 30°C$ (the average is 36.1°C), permeability coefficients are underestimated by an average factor of 2.3. The average temperature for all permeability coefficient measurements shown in Fig. 4 is 30.6°C. The increase of SC permeability coefficients by a factor of two resulting from approximately a 5°C temperature change is in agreement with experimental evidence of the effect of temperature (34). These results provide encouraging evidence that a large database of SC permeability coefficients can predict physically realistic trends that are difficult to determine from individual measurements.

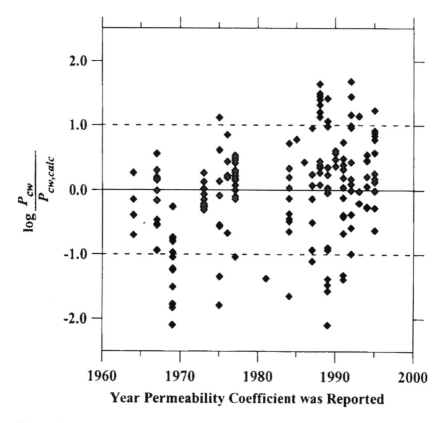

Figure 4 A comparison of permeability coefficients in the fully validated database (P_{cw}) to those calculated by Eq. T1 ($P_{cw,calc}$) as a function of the year in which the data were published.

Figure 5 also shows ionization-unadjusted (dashes) and adjusted (solid or open squares) permeability coefficients, for partly ionized compounds, connected to one another by vertical line segments. These results show that adjusting for ionization always increases the value of the permeability coefficient. Several adjustments were inconsequential. The three most under-estimated permeability coefficients (flurbiprofen, ibuprofen, and indomethacin), measured by Morimoto and colleagues (25), are in better agreement with Eq. T1 before adjustment for f_{ni}. This caused us to question whether these authors had already adjusted the measurement for ionization. However, there is no indication in the publication that they did, and we were unable to obtain an answer from them.

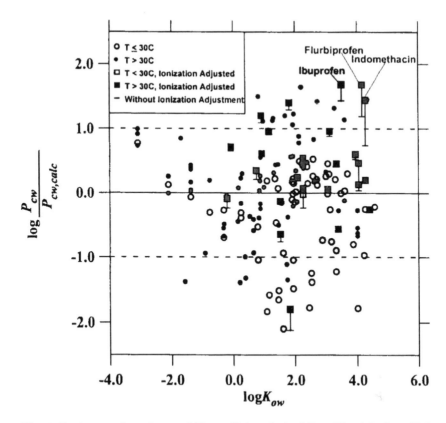

Figure 5 A comparison of permeability coefficients in the fully validated database (P_{cw}) to those calculated by Eq. T1 ($P_{cw,calc}$) with different temperatures and levels of ionization designated.

Figure 6 identifies the measurements in the database from different investigations of the same compound. Multiple measurements of a single compound are indicated with dashes that are connected by a vertical line. Because p-cresol and hydrocortisone have similar K_{ow}, the replicated measurements for these two compounds overlap, and the p-cresol data are identified separately. The FV database contains 66 replicate measurements for 23 different compounds. Several coefficients for the same compound disagree by more than an order of magnitude, and some by two orders of magnitude. These results should change our understanding of equations for predicting permeability coefficient values in two important ways: (1) we must learn to accept that very accurate predictions cannot be

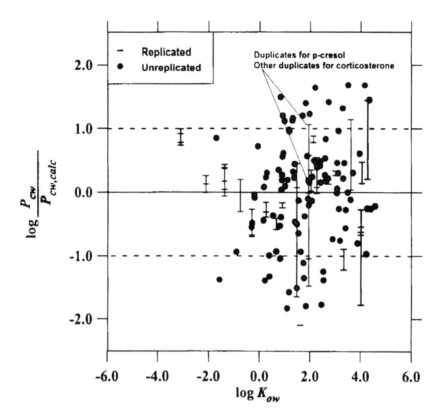

Figure 6   A comparison of permeability coefficients in the fully validated database (P_{cw}) to those calculated by Eq. T1 ($P_{cw,calc}$) with replicated (multiple permeability coefficients measured for the same compound) measurements designated.

made if measurements as variable as these appear to be are used in the equation development, and (2) regressions will provide much more certain estimates on average than individual measurements (i.e., when there are differences between measurements and a judiciously developed predictive equation, the weight of the evidence strongly favors the regression equation). The multiple permeability coefficients of water are in close agreement. This may indicate that skin is consistently permeable to water or that researchers select experimental results that are consistent with expected permeability values for water.

As shown in Fig. 7, relative to an equation representing the entire database (Eq. T1), measurements that were made in some laboratories can vary systematically from measurements made in other laboratories. The permeability coefficients shown have been adjusted for the effects of ionization. One would expect,

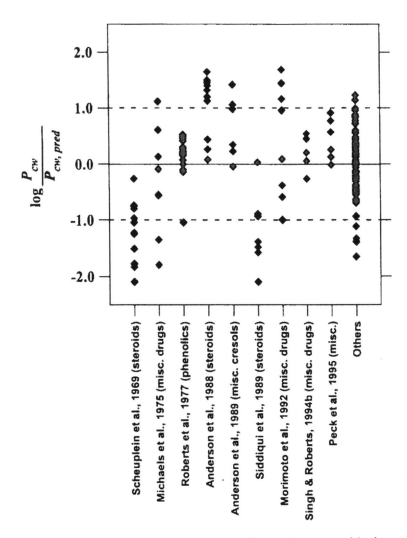

Figure 7 A comparison of permeability coefficients (P_{cw}) reported in the prominent investigations within the fully validated database to those calculated by Eq. T1 ($P_{cw,calc}$).

assuming that MW and log K_{ow} effects are dominant, that permeability coefficients measured in all studies would be randomly scattered about the line of perfect estimation (i.e., log $P_{cw}/P_{cw,calc} = 0.0$). Permeability coefficient values for the steroids measured by Scheuplein and others (40) are always overestimated by Eq. T1. Likewise, for a similar group of chemicals, Siddiqui et al. (41) report permeability coefficient values that are smaller than calculated by Eq. T1 for six of the seven chemicals studied. In contrast, the penetration of slightly modified hydrocortisone steroids measured by Anderson and colleagues (20) are systematically underestimated by Eq. T1. All of the compounds in these three studies are structurally similar. Anderson and Raykar also report higher than expected permeability coefficient values for several cresols (42) compared to other FV measurements. In light of these two studies, it is possible that the skin used in the Anderson investigations was uncommonly permeable, or that the experimental procedure used by Anderson and colleagues lead to higher than normal values for the permeability coefficient.

The investigation of Morimoto and colleagues (25) also had some poorly estimated measurements, although not systematically high or low. After extensive study of their data and procedures we have not been able to identify a cause. Permeability coefficients for the hydrophilic compounds measured by Peck and others (43) are always overestimated, but hydrophilic compounds may not be optimally represented by Eq. T1 (which is based predominantly on compounds with log $K_{ow} > 0$). The remaining permeability coefficient values (labeled others) show the anticipated scatter around log $P_{cw}/P_{cw,calc} = 0.0$. This plot shows that at least some of the variability not resolved by regressions such as Eq. T1 may arise from systematic differences in measurement between laboratories. Whether this is due to skin properties or to an unidentified experimental protocol is unknown. In Chapter 4 we show that systematic differences also occur in the measurement of SC–water partition coefficients (K_{cw}). Interestingly, when permeability coefficients and partition coefficients were measured by the same research group, systematic differences were usually in the same direction. Elucidation of potential experimental causes for these results would be useful.

Figure 8 shows the prediction of the FV permeability coefficient measurements determined using isolated SC, whole epidermis (EPID, which includes SC and VE), and split or full-thickness skin (SPLIT/FULL) as a function of log K_{ow}. The skin type for some measurements was not specified, and these are designated as N/A. More of the permeability coefficient measurements determined with isolated SC (21 values have log $P_{cw}/P_{cw,calc} > 0.0$) are underestimated by Eq. T1 than are overestimated (9 values have log $P_{cw}/P_{cw,calc} < 0.0$). The average residual, log $P_{cw}/P_{cw,calc}$, for these 30 measurements is 0.30 (i.e., $P_{cw}/P_{cw,calc} \approx 10^{0.3} = 2$). This would be expected if the extra barriers presented by the viable epidermis (EPID and SPLIT/FULL measurements) and dermis (for SPLIT/FULL measurements) were significant relative to other uncertainties and effects, or if the isolated SC membranes were susceptible to damage during preparation. The number of measurements for which $P_{cw} > P_{cw,calc}$ was nearly the same as the number of measurements for which $P_{cw} < P_{cw,calc}$ in experiments that used whole epidermis (EPID) and also in experiments that used tissues with dermis (SPLIT/FULL). The average residual, log $P_{cw}/P_{cw,calc}$, was determined to be slightly negative

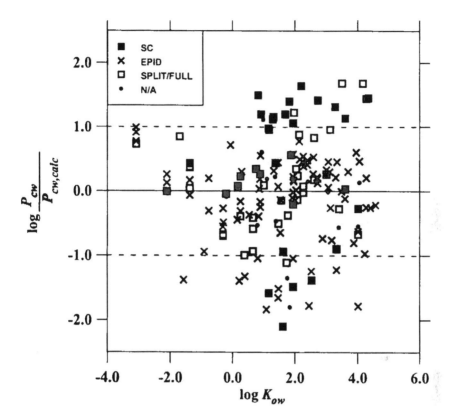

Figure 8 A comparison of permeability coefficients in the fully validated database (P_{cw}) to those calculated by Eq. T1 ($P_{cw,calc}$) with the skin layers present designated: isolated stratum corneum (SC), intact epidermis (EPID), split or full-thickness skin (SPLIT/FULL), or not specified (N/A).

(-0.15) for 89 EPID measurements and slightly positive (0.19) for 36 SPLIT/FULL measurements. Based on these observations, it seems likely that the presence of dermis had almost no effect compared to the VE itself. Also, given the variability in the database as a whole, the difference between isolated SC and other skin samples is relatively small.

Figure 9 shows the ratio of measured to calculated permeability coefficient values for skin surgically removed from live humans (labeled patient) and skin removed postmortem (cadaver) as a function of MW of the penetrating compound. Most commonly, patient skin was from the abdomen or breast (female), although this variable has not been noted. Usually, patient skin was used fresh, or stored frozen for varying periods of time. Cadaver skin may be subject to harsh

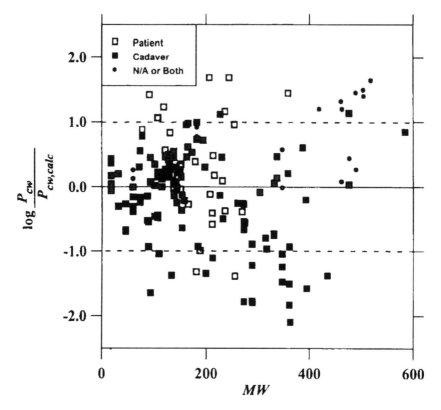

Figure 9   A comparison of permeability coefficients in the fully validated database (P_{cw}) to those calculated by Eq. T1 ($P_{cw,calc}$) as a function of MW with the skin source designated: excised from living donors (patient), excised from cadavers (cadaver), and not specified (N/A) or an average permeability coefficient from measurements with both patient and cadaver skin (both).

cleaning procedures prior to removal, which might compromise the integrity of the transport barrier. However, the data shown in Fig. 9 indicate that on average cadaver skin provided a more resistive barrier to absorption. The average values of $\log(P_{cw}/P_{cw,calc})$ calculated for the 119 measurements using cadaver skin is -0.206, and the average value for 32 measurements using skin excised from live humans is 0.311. It is not clear whether these results can be separated from other factors, such as whether the skin was frozen and laboratory effects. For example, many of the cadaver results are for steroid measurements by Scheuplein et al. (40) and Siddiqui et al. (41) that were identified in Fig. 7 and discussed earlier. The skin type for some measurements either was not specified (and labeled N/A) or was an average of data from both patient and cadaver and are identified as such on Fig. 9.

Figure 10 shows the ratio of measured to calculated permeability coefficient values as a function of the solution pH. Permeability coefficients for partially ionized chemicals have been adjusted (i.e., divided by the fraction nonionized when $f_{ni} > 0.1$). Consequently, all data shown in Fig. 10 should represent the permeability coefficient for the nonionized species. Measurements for chemicals that do not dissociate (labeled ND in Table A1) are grouped at pH = 7 when the pH was not specified. One group was treated differently. Our understanding is that the permeability coefficient values for several carboxylic acids (44) were measured at a low enough pH to be nonionized, but the actual pH is unknown.

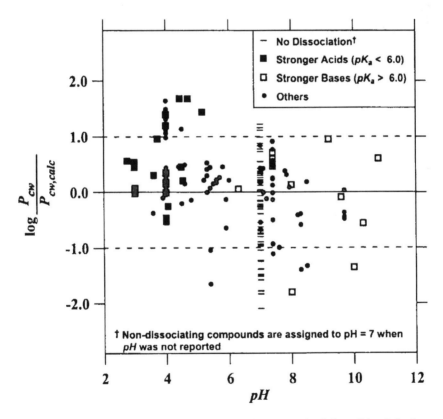

Figure 10 A comparison of permeability coefficients in the fully validated database (P_{cw}) to those calculated by Eq. T1 ($P_{cw,calc}$) as a function of the pH of the vehicle. Measurements were plotted at pH = 7 when pH was not reported for compounds that do not dissociate.

These measurements are placed at pH = 4 in Fig. 10. Singh and Roberts (45) reported that their permeability coefficient measurements were determined at the pH at which $f_{ni} = 0.5$ (i.e., pH = pK_a if there is one dominant acid–base reaction). However, they did not list pH or pK_a values for any of the chemicals they studied. Lacking these values, pK_a values for each chemical were calculated using SPARC. Permeability coefficient values for compounds determined using a range of pH measurements (i.e., ethyl nicotinate, methyl nicotinate, and salicylic acid) were placed at the pH corresponding to the midpoint of the pH interval given in Table A1.

The data in Fig. 10 appear to scatter uniformly around log $P_{cw}/P_{cw,calc}$ = 0. However, there is some indication that permeability coefficient measurements determined from acidic solutions are larger (on average) than permeability coefficient measurements determined from basic solutions. The compounds studied from acidic solutions were primarily carboxylic acids, and the compounds studied from basic solutions are primarily amines. To explore the possibility that acid–base interactions between the skin and the penetrating molecule cause this weak trend, we distinguished the stronger acids ($pK_a < 6.0$) and the stronger bases ($pK_a > 6.0$) from nondissociating compounds or weak acids ($pK_a > 6.0$) and weak bases ($pK_a < 6.0$). It appears that the pH more than a chemical acid or basic strength is responsible for the trend in Fig. 10. We neglected the measurements arbitrarily plotted at pH = 7 and regressed the residual log $P_{cw}/P_{cw,calc}$ on pH for the 105 remaining measurements. The best-fit line, log $P_{cw}/P_{cw,calc} = 1.06 - 0.146$ pH, has a slope that was statistically different from zero at the 95% level of confidence, although the goodness-of-fit statistics were poor ($r^2 = 0.17$, $r^2_{adj} = 0.16$, RMSE = 0.66, F-ratio = 20). The data in Fig. 10 suggest that acidic conditions may weakly enhance the permeability of the SC in some way. However, many more data are needed to confirm this conclusion.

In summary, the results presented in Figs. 3–10 indicate that several factors do and do not affect uncertainty in SC permeability coefficient measurements. As shown in Fig. 3, the 84 validated Flynn database permeability coefficients and the 86 newly assembled permeability coefficients have nearly the same amount of uncertainty. Based on Figs. 4–10, we conclude that several factors have no effect on the permeability coefficient: (a) permeability coefficient measurements as early as 1960 appear to be of equal quality to more recent measurements, (b) additional resistances to penetration by the viable epidermis and dermis appear to have minimal effect on the permeability coefficient measurements for chemicals in the FV database, and (c) there is no consistent difference between skin from living and dead subjects. Temperature, ionization, and solution acidity do affect permeability coefficient values, and variations in these may explain a portion of the overall uncertainty in the database. Also, permeability coefficient measurements determined using isolated SC membranes are modestly larger than measurements determined using whole epidermis, split thickness skin, or whole skin, suggesting that isolated SC membranes may be difficult to prepare without introducing damage. However, these effects do not explain the larger differences between replicate permeability coefficients measured for the same compound, which often surpass an order of magnitude.

In the next section, we use the FV database to examine quantitatively (a) the effects of temperature, (b) the use of MW as a representation for molecular volume, (c) the existence of a different mechanism of absorption for hydrophilic chemicals, and (d) the use of LSER parameters to model SC permeability coefficients.

E. Effect of Temperature

In Chapter 2 we found that permeability coefficients from the Flynn database (21) increased on average with increasing temperature. Model equations that incorporated temperature were able to explain more of the variability in permeability coefficients (see Chapter 2). In this section, we investigate the FV database for the effects of temperature to explain the apparent trend in Fig. 5.

The absolute temperature at which P_{cw} was measured was used to reduce the variability of SC permeability coefficients from the FV database according to Eq. (14). The resulting equation, Eq. T3 in Table 1, can explain 56.4% of the variability in log P_{cw} by variation in log K_{ow} and MW/T compared to 55.1% when T was not included (i.e., Eq. T1). Consequently, adding T to the equation for estimating P_{cw} provides only a minor improvement. This may arise because temperature weakly influences individual permeability coefficient values, altering them by a factor somewhere between 2 and 5. However, many measurements are influenced by temperature, so incorporation of temperature does improves the model.

F. Effect of Penetrant Liquid Density

Experimental reference liquid densities (ρ_{ref}) and the temperatures at which they were measured (T_{ref}), melting point temperatures (T_m), and critical temperatures (T_c) for some compounds from the FV permeability coefficient database are summarized in Table A6 (18,46). Critical temperatures are not known for some of the compounds for which reference liquid densities, reference temperatures, and melting point temperatures are known. Eqs. T4–T6 were developed using chemicals from the FV database for which reference liquid densities, reference temperatures, and melting point temperatures were available. Eqs. T7–T11 were a subset of these data for which T_c was also known.

First, we assume that K_T is approximately constant for all organic liquids. Eq. T4 was derived from Eq. (22), which only adjusts the liquid density value for temperature. Eq. T5, based on Eq. (23), includes the effect of temperature

Table 1 Comparison of Data Regressions to Various Model Equations

No.	Data set[a]	n/m[b]	Model Eq.	$\log P_{cw}$ [cm h^{-1}][c,d]	r^2	r^2_{adj}	RMSE	F-ratio
T1	FV	170/127	12	$= -2.44(0.12) + 0.514(0.04) \log K_{ow}$ $- 0.0050(0.0005)MW$	0.551	0.546	0.80	102.6
T2	MF	80/79	12	$= -2.76(0.20) + 0.52(0.06)\log K_{ow}$ $- 0.0041(0.0006)\ MW$	0.537	0.526	0.82	47.0
T3	FV	170/127	14	$= -2.41(0.12) + 0.52(0.04)\log K_{ow}$ $- 1.58(0.16)\ MW/T$	0.564	0.559	0.79	108.2
T4	A	53/32	22	$= -1.74(0.25) + 0.85(0.10)\log K_{ow}$ $- 0.016(0.004)\ MW/\rho_{ref}$	0.755	0.740	0.47	50.3
T5	A	53/32	23	$- 9.0(4.0) \times 10^{-5}\ MW\ (T_{ref} - T)/\rho_{ref}$ $= -1.75(0.24) + 0.84(0.10)\log K_{ow}$ $- 4.74(1.06)\ MW/(T\ \rho_{ref})$	0.761	0.746	0.47	51.9
T6	A	53/32	12	$- 0.024(0.012)\ MW\ (T_{ref} - T)/(T\ \rho_{ref})$ $= -1.89(0.26) + 0.81(0.11)\log K_{ow}$ $- 0.014(0.004)\ MW$	0.716	0.704	0.51	62.9
T7	B	44/22	24	$= -1.93(0.33) + 0.75(0.15)\log K_{ow}$ $- 0.013(0.005)\ MW/\rho_{ref}$ $- 0.17(0.091)\ MW\ (T_{ref} - T)/(T_c\ \rho_{ref})$	0.720	0.700	0.50	34.4

T8	B	44/22	25	$= -1.91(0.31) + 0.76(0.14)\log K_{ow}$ $- 4.01(1.47)\ MW/(T\ \rho_{ref})$	0.727	0.707	0.50	35.5
T9	B	44/22	12	$= -47.4(27.4)\ MW\ (T_{ref} - T)/(TT_c\ \rho_{ref})$ $- 1.74(0.41) + 0.86(0.19)\log K_{ow}$ $- 0.017(0.007)\ MW$	0.675	0.659	0.54	42.5
T10	B	44/22	22	$= -1.90(0.32) + 0.76(0.15)\log K_{ow}$ $- 0.013(0.005)\ MW/\rho_{ref}$	0.721	0.700	0.50	34.4
T11	B	44/22	23	$= -0.00018(0.0001)\ MW\ (T_{ref} - T)/\rho_{ref}$ $- 1.89(0.30) + 0.76(0.14)\log K_{ow}$ $- 4.09(1.45)\ MW/(T\ \rho_{ref})$	0.727	0.707	0.50	35.6
T12	FV	170/127	28	$= -0.052(0.030)\ MW\ (T_{ref} - T)/(T\ \rho_{ref})$ $- 2.9(0.2) + *0.11(0.14)\log K_{ow}$ $+ 0.5(0.17)(\log K_{ow} + 0.5)U(K_{ow} - 0.32)$ $- 0.0052(0.0005)\ MW$	0.573	0.566	0.79	74.4
T13	FV	170/127	30	$= -3.01(0.13)$ $+ 0.635(0.05)\ (\log K_{ow} + 0.5)U(K_{ow} - 0.32)$ $- 0.0052(0.0005)\ MW$	0.572	0.567	0.79	111.5

[a] FV = Fully validated database; MF = Modified Flynn database; A = Subset of FV database; B = Subset of FV database.

[b] n = number of data points, m = number of different compounds.

[c] The uncertainties expressed within parenthesis are reported as standard error in the coefficients.

[d] Coefficients indicated with an asterisk (*) are not meaningfully different from zero at the 95% confidence level.

on the permeability coefficient. For comparison, Eq. T6 shows regression of the same data set using the conventional model, Eq. (12). Although the predictive power of Eq. T4 is greater than that of Eq. T6, as indicated by larger r^2, the predictive power per parameter is not as large (F-ratio = 50.3 compared to 62.9). When temperature is included (i.e., Eq. T5), the overall predictive power and predictive power per parameter both increase, although only slightly.

Next we analyze the permeability coefficient data assuming that the product $K_T T_c$ is constant. As described earlier, for the compounds considered here, $K_T T_c$ has approximately half the variability of K_T. However, equations derived assuming that $K_T T_c$ is approximately constant [Eqs. (24) and Eqs. (25)] require that T_c be known for each compound. Unfortunately, T_c was not found for a few chemicals included in the last regression, and consequently the database for regression to Eq. (24) and Eq. (25) is slightly smaller (i.e., $n = 44$) than that for regressions T4 and T5 ($n = 53$). Eq. T7 adjusts for liquid density alone, while Eq. T8 incorporates temperature effects on the permeability coefficient. For comparison, Eq. T9 is the regression of the same set of data to the conventional model, Eq. (12). Here too, the predictive power of Eq. T7 is greater than that of Eq. T9, as indicated by larger r^2, but the predictive power per parameter, as measured by the F-ratio, is not as large. When temperature is included (i.e., Eq. T8), the overall predictive power and predictive power per parameter both increase, although only slightly.

In a limited test to determine whether the data are better represented by equations that assume constant K_T or constant $K_T T_c$, the same set of 44 data points were regressed to Eqs. (22) and (23), which assumed that K_T was constant. Eq. T10 adjusts for liquid density alone, while Eq. T11 incorporates temperature effects on the permeability coefficient. The values of r^2 for regressing the same set of 44 data points to the constant K_T equations are 0.721 when liquid density alone is included (Eq. T10) and 0.727 when liquid density and T are both included (Eq. T11). These values are not statistically different from Eqs. T7 (i.e., $r^2 = 0.720$) and T8 (i.e., $r^2 = 0.727$). The same conclusion is reached by comparing the F-ratios for these fits. Analysis with this limited database indicates that introducing the factor T_c provides no significant improvement in the predictive power. Since T_c are not available for many compounds, it is more practical to assume that K_T is constant rather than assuming that $K_T T_c$ is constant.

The correction of MW with a pseudo–liquid density has been relatively easy to accomplish and has removed some of the uncertainty in the data. The reduction of unexplained uncertainty is in agreement with results reported by Kasting et al. when they substituted MW for MV in an analysis with fewer data and observed that r^2 was reduced by about 0.05 (22). We have attained essentially the same change in regression statistics without having to calculate MV.

Figure 11 illustrates the inadequacy of MW to represent MV in equations for predicting the permeability coefficient values for organic compounds that have both low and high density. Experimental permeability coefficients for chem-

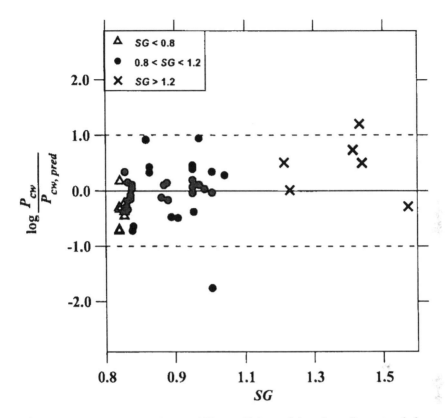

Figure 11 A comparison of permeability coefficients of the subset of compounds from the fully validated database (P_{cw}) for which experimental liquid densities are available to those calculated by Eq. T1 ($P_{cw,calc}$).

icals listed in Table A6 are compared with predictions using the conventional model (i.e., Eq. T1) as a function of specific gravity (SG) based on density of liquid water at 4°C (i.e., 1.000 g cm^{-3}). This figure shows that permeability coefficient values for low-density compounds (SG < 0.8, designated with triangles) are overestimated on average and permeability coefficient values for high density compounds (SG > 1.2, designated with ✖) are underestimated on average. Clearly, MW does not sufficiently represent the MV for compounds with significantly different liquid densities.

Figure 12 illustrates the effect of modifying MW by liquid density using Eq. T5 (i.e., assuming K_T is constant and accounting for temperature effects on the permeability coefficient) for the same experimental permeability coefficient values as are shown in Fig. 11. It is easy to see the improvement in prediction

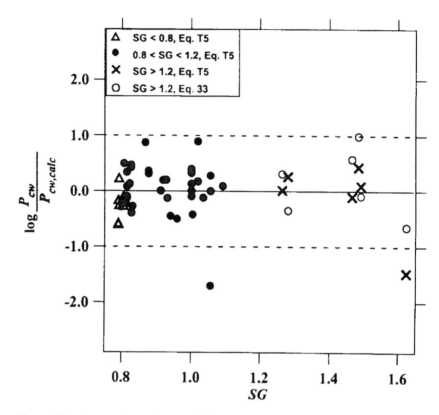

Figure 12 A comparison of permeability coefficients of the subset of compounds from the fully validated database (P_{cw}) for which experimental liquid densities are available to those calculated ($P_{cw,calc}$) by Eqs. T5 or Ref. 33.

of the permeability coefficients for low- and high-density organic compounds. The effect is most pronounced for chemicals with SG > 1.2, designated with ✖. However, the number of data points is quite small, and hence it is not possible to demonstrate conclusively that this observation is generally true. Nevertheless, these results suggest that a better estimate of molecular size than MW should improve prediction of permeability coefficient values.

Based on these results, we propose that permeability coefficient estimates for dense chemicals (i.e., SG > 1.2) can be improved by using an effective MW (i.e., MW_{eff}) defined as

$$MW_{eff} = MW \frac{SG_{DB}}{SG} \qquad (33)$$

in the conventional model. SG_{DB} is the average SG of the compounds in the database used to derive the conventional model equation. Strictly, SG and SG_{DB} should be at a uniform set of conditions (e.g., for liquid state at the same convenient T such as 25°C). However, given that conventional model estimates have errors of at least 1 order of magnitude, a reasonable estimate for SG_{DB} may be sufficient. $SG_{DB} \approx 0.9$ for the data set used to derive Eq. T6. This is probably typical for many of the conventional models reported in the literature. Thus we argue that trichloroethylene (MW = 131, SG = 1.464) will penetrate the SC like a compound with a MW of 81 (= 131 × 0.9/1.464). This adjustment becomes significant for compounds like dibromomethane (MW = 174, SG = 2.497), which based on Eq. (33) is estimated to penetrate the SC like a compound with a MW of 63. The open circles in Fig. 12 compare the experimental values for compounds in the database with $\rho_{ref} > 1.2$ g mL^{-1} to calculations made using MW_{eff} in Eq. T1. This simple adjustment does improve estimated permeability coefficient values, though not as much as Eq. T5.

G. Analysis with the Two-Pathway Model

Several authors have argued that hydrophilic chemicals may penetrate skin by a different mechanism than more lipophilic chemicals. Specifically, based on data like those shown in Fig. 2, they have hypothesized that permeability coefficient values for hydrophilic chemicals depend less strongly on K_{ow} than permeability coefficient values for more lipophilic chemicals. Here we statistically analyze the data in Fig. 2 to determine if a two-mechanism model is supported by the data. This analysis faces two difficulties: (1) permeability coefficient measurements of hydrophilic chemicals are often more variable than for lipophilic chemicals of similar size, and (2) there are very few measured permeability coefficients for hydrophilic compounds.

First, we arbitrarily divide the data in the FV database into a group of lipophilic (log $K_{ow} > 0.0$) and hydrophilic (log $K_{ow} \leq 0.0$) chemicals. Each of these two groups was separately regressed to an equation of the form given by Eq. (12). The resulting equations are provided in Table 2 along with a description of the database analyzed (n = number of data points and m = number of different chemicals) and regression statistics. Uncertainties listed in parenthesis are reported as the standard error of the coefficients. Coefficients that are not meaningfully different from zero, at the 95% confidence level, are indicated with an asterisk. When it was determined that a coefficient was not statistically different from zero, the regression was repeated with this term deleted. The hydrophilic fraction of the FV database was analyzed with and without the inclusion of permeability coefficients for three chemicals (5-fluorouracil, paraquat, and TEAB) from the provisional database.

The equations listed in Table 2 do support the hypothesis that penetration of

Table 2 Analysis of the Hydrophilic (log $K_{ow} \leq 0.0$)) and Lipophilic (log $K_{ow} > 0.0$) Fractions of the Fully-Validated (FV) and Provisional Permeability Coefficient Data Sets with the Conventional Model

Data set[a]	n/m[b]	log P_{cw} [cm h^{-1}][c,d]	r^2	r^2_{adj}	RMSE	F-ratio
FVL	142/114	$= -2.59(0.18) + 0.62(0.062)$ log $K_{ow} - 0.00538(0.00059)$ MW	0.51	0.51	0.83	74
FVH	28/13	$= -2.92(0.17) + *0.18(0.097)$ log $K_{ow} - 0.0039(0.00077)$ MW	0.57	0.53	0.48	17
FVH	28/13	$= -3.13(0.13) - 0.0041(0.0008)$ MW	0.51	0.49	0.50	27
FVH+	38/16	$= -3.13(0.15) + *0.110(0.072)$ log $K_{ow} - 0.0043(0.0009)$ MW	0.51	0.49	0.55	18
FVH+	38/16	$= -3.24(0.14) - 0.0048(0.0008)$ MW	0.48	0.47	0.56	33
FV	170/127	$= -2.9(0.2) + *0.11(0.14)$log K_{ow} $+ 0.5(0.17)(\log K_{ow} + 0.5)U(K_{ow} - 0.32) - 0.0052(0.0005)$ MW	0.57	0.57	0.79	74
FV	170/127	$= -3.0(0.1) + 0.64(0.05)$ (log $K_{ow} + 0.5)U(K_{ow} - 0.32)$ $- 0.0052(0.0005)$ MW	0.57	0.57	0.79	112

[a] FVL = Lipophilic fraction of the FV database; FVH = Hydrophilic fraction of the FV database; FVH+ = Hydrophilic fraction of the FV database and TEAB, paraquat and 5-fluorouracil (i.e., hydrophilic fraction of the provisional database).

[b] n = number of data points, m = number of different compounds.

[c] The uncertainties expressed within parenthesis are reported as standard error in the coefficients.

[d] Coefficients indicated with an asterisk (*) are not meaningfully different from zero at the 95% confidence level.

hydrophilic and lipophilic compounds may be mechanistically different. Indeed, regressions of the permeability coefficient data for hydrophilic and lipophilic compounds apparently do depend differently upon K_{ow}. Lipophilic compounds have a stronger dependence upon log K_{ow} (around 0.615) than hydrophilic compounds (around 0.11 to 0.18, and frequently, not meaningfully different from zero at the 95% confidence level). Permeability coefficient values for hydrophilic compounds do depend upon MW, although somewhat less strongly than for the lipophilic fraction (the absolute value of the coefficient multiplying MW is approximately 0.004–0.0045 for the hydrophilic chemicals compared to about 0.0054 for lipophilic chemicals). The leading coefficient of the regressions is significantly smaller (i.e., coefficients are more negative) for hydrophilic compounds (approximately -3) compared to that for lipophilic chemicals (approximately -2.6). Since this coefficient physically represents D_c/L_c for a molecule of MW = 0, it would seem that hydrophilic chemicals may diffuse more slowly than lipophilic chemicals. This could arise if hydrophilic species were confined to the polar regions of the lipid bilayer material, meaning that the area available for their penetration is smaller than that for lipophilic compounds which penetrate through the larger nonpolar region of the lipid bilayer. However, the hydrophilic fraction in this database is small, and the hypothesis that hydrophilic chemicals diffuse more slowly than lipophilic chemicals needs to be tested using a larger number of measurements for hydrophilic chemicals.

While the regression equations in Table 1 are informative, these models do not force a smooth transition in P_{cw} for hydrophilic and lipophilic compounds. Also, in the analysis summarized in Table 2 it was arbitrarily assumed that the criterion log K_{ow} = 0.0 divided chemicals penetrating by hydrophilic and lipophilic mechanisms. We also analyzed the FV database using Eqs. (28) and (30), which require the equations for hydrophilic and lipophilic chemicals to intersect at K_{ow}^*, defined as the transition between hydrophilic and lipophilic mechanisms. In using these equations, we assume that the only difference between hydrophilic and lipophilic chemicals is in the effect of K_{ow}. That is, the effect of molecular size (i.e., MW) is the same for both hydrophilic and lipophilic chemicals.

Eq. T12 in Table 1 was developed by regressing the FV database to Eq. (28) for different values of log K_{ow}^* (i.e., log K_{ow}^* = -1, -0.75, -0.5, -0.25, 0.0, 0.25, 0.5, 0.75, and 1.0) to produce the best fit at log K_{ow}^* = -0.5 (i.e., K_{ow} = $10^{-0.5}$ = 0.32). Perhaps it is significant that the coefficient multiplying log K_{ow} in Eq. T12 (i.e., 0.11) is not meaningfully different from zero at the 95% level of confidence, although it must be remembered that this result is based on only 20 measurements for six chemicals with log $K_{ow}^* \leq -0.5$. Comparing Eq. T12 with Eqs. (26) and (27), we determine that the coefficient multiplying log K_{ow} is 0.11 (= c) for hydrophilic chemicals compared to 0.61 [= b = ($b - c$) + c = 0.5 + 0.11] for lipophilic chemicals. The coefficient multiplying MW (d_1) is -0.0052 (assumed to be the same for both pathways), and -2.90 (= a) and

$-2.65[= (a - (b - c) \cdot \log K_{ow}^s) = -2.90 - 0.5\,(-0.5)]$ are the intercepts for hydrophilic and lipophilic compounds, respectively.

Eq. T13 is the result when we analyzed the FV database assuming that permeability coefficient values for hydrophilic chemicals do not depend upon K_{ow} at all [i.e., using Eq. (30) in which $c = 0$]. Comparing Eqs. T13 and (30), $a = -3.01$ is the average $\log P_{cw}$ for the hydrophilic chemicals (i.e., $\sim 10^{-3}$), and the slope is 0.64 $(= c)$ and the intercept $-2.69 = [a - c \log K_{ow}^s = -3.01 - 0.64 \times (-0.5)]$ for lipophilic compounds.

As indicated in Eq. T12, hydrophilic chemicals do not have a meaningful dependence on $\log K_{ow}$, so the statistics decreased only slightly when K_{ow} dependence was removed for the hydrophilic chemicals (i.e., Eq. T13). This result is consistent with the premise that hydrophilic chemicals penetrate by a different mechanism, but the data are limited and these findings must be substantiated by more measurements that are of better quality.

H.   Analysis with the LSER Model

The LSER model, Eq. (32), has been used to analyze chemicals from the FV database for which solvatochromic parameters could be found. These chemicals, identified in Table A1 with an asterisk to the left of the chemical name, represent a low MW fraction of the FV database. The solvatochromic parameters, all from Abraham et al. (28), were calculated by averaging multiple normalized solvent effects on a variety of chemical properties involving many varied types of indicators (28). Table A5 contains the LSER parameters used in this analysis.

The LSER database consists of 65 FV permeability coefficient values for 43 different chemicals and is larger than other known databases analyzed with the LSER model [e.g., (30)]. For purposes of analysis, the entire LSER database is divided into lipophilic ($\log K_{ow} > 0.0$) and hydrophilic ($\log K_{ow} \leq 0.0$) fractions. Various regressions of the entire LSER database, the lipophilic fraction, and the hydrophilic fraction of the FV database are listed in Table 3, along with a description of the database analyzed (n = number of data points and m = number of different chemicals) and the r^2 statistic. Generally, each of the three data sets was analyzed with the conventional model first [(i.e., Eq. (12)], followed by analysis with the full LSER model [i.e., Eq. (32)] and then by a reduced LSER model which includes only the parameters that were meaningful at the 95% level of confidence. Coefficients that are not meaningfully different from zero, at the 95% confidence level, are indicated with an asterisk. Uncertainties are listed in parentheses as the standard error of the coefficients.

Analysis of the entire LSER database is presented first in Table 3. The regression to all LSER parameters gives a better fit to the data ($r^2 = 0.76$) than a fit to the conventional model ($r^2 = 0.68$). However, for this set of data, only the hydrogen bond basicity (β) and molecular volume (V_x) are statistically significant

Table 3 Regressions of a Subset of the Fully Validated (FV) Database with LSER Parameters and with log K_{ow} and MW

Data set	n/m[a]	Model Eq.	log P_{cw} [cm h^{-1}][b,c]	r^2
LSER[d]	65/43	12	= −2.03(0.24) + 0.74(0.11)log K_{ow} − 0.011(0.004) MW	0.68
LSER	65/43	32	= −1.67(0.23) − *0.46(0.24)π − *0.46(0.27)α − 3.44(0.41)β + 1.76(0.19)V_x	0.76
LSER	65/43	modified 32	= −2.22(0.20) − 3.28(0.41)β + 1.73(0.17) V_x	0.70
LSER	65/43	modified 32 with K_{ow}	= −2.00(0.19) − 1.69(0.41)β + 0.41(0.04) log K_{ow}	0.72
LSER	65/43	modified 32 with MW	= −2.43(0.28) − 2.07(0.50)β + 0.012(0.002) MW	0.56
LSER-L[e]	51/39	12	= −2.23(0.33) + 0.76(0.12) log K_{ow} − 0.0098(0.004) MW	0.54
LSER-L	51/39	32	= −2.04(0.32) − *0.12(0.28)π − 1.08(0.37) α − 3.01(0.50) β + 2.01(0.29)V_x	0.66
LSER-L	51/39	modified 32	= −2.35(0.32) − 3.23(0.53) β + 1.85(0.33)V_x	0.52
LSER-H[f]	14/4	12	= −3.09(0.39) − *0.18(0.23) log K_{ow} − *0.0064(0.006) MW	0.52
LSER-H	14/4	modified 32	= −3.13(0.42) − *1.55(1.75) β − *2.20(1.24)V_x	0.55
LSER-H	14/4	modified 32	= −2.77(0.11) − 1.15(0.32)V_x	0.51

[a] n = number of data points, m = number of different compounds.
[b] The uncertainties expressed within parenthesis are reported as standard error in the coefficients.
[c] Coefficients indicated with an asterisk (*) are not meaningfully different from zero at the 95% confidence level.
[d] Measurements in the fully validated database for which LSER parameters are available (primarily low MW compounds).
[e] The lipophilic fraction of the LSER data set is arbitrarily defined as log K_{ow} > 0.0.
[f] The hydrophilic fraction of the LSER data set arbitrarily defined as log K_{ow} ≤ 0.0.

predictors of permeability. Regression on only these two parameters gave a comparable fit ($r^2 = 0.70$) to the conventional model ($r^2 = 0.68$). Interestingly, the fit is much poorer when the LSER database is analyzed with β and MW ($r^2 = 0.56$) than when analyzed with β and V_x ($r^2 = 0.70$), indicating that for the LSER database MW is not a good substitute for V_x. Regression on the two most significant parameters, β and log K_{ow}, produced a minimal improvement of fit ($r^2 = 0.72$) compared to when β and V_x are used ($r^2 = 0.70$), indicating that V_x and log K_{ow} probably contain similar chemical information.

Regressions of the lipophilic fraction of the LSER database are listed next in Table 3. The regressions to the entire LSER database and the lipophilic fraction are not meaningfully different, but this is probably because the lipophilic fraction dominates the LSER database. The hydrogen bond basicity, β, and calculated volume, V_x, are meaningful, and the hydrogen bond acidity, α, is marginally significant at the 95% confidence level in the analysis of the lipophilic chemicals. The fit on β and V_x alone give coefficients that are not statistically different from the fit of the entire LSER database, although the fit is poorer ($r^2 = 0.70$ for the LSER database compared to $r^2 = 0.52$ for the lipophilic fraction). This fit on β and V_x is comparable in quality to the fit of the conventional model to the same data set ($r^2 = 0.54$).

The hydrophilic chemicals were analyzed with the conventional model, and neither MW nor log K_{ow} were found to be significant for the data set in this analysis. No terms in the LSER model were significant when at least two terms (i.e., any two of α, β, π, or V_x) were included in analysis. V_x was significant when it was included alone in the LSER model, and that fit was able to represent as much variability in permeability coefficient values for the hydrophilic chemicals ($r^2 = 0.51$) as when the conventional model was used ($r^2 = 0.52$). The leading coefficients are significantly larger negative numbers (approximately = -2.8 for the LSER model with only V_x) than the leading coefficients for the entire LSER database (approximately = -2.2 for the LSER model with β and V_x) or the lipophilic fraction of that database (approximately = -2.3 for the LSER model with β and V_x). This could be further evidence that hydrophilic species were confined to the polar regions of the lipid bilayer material, meaning that the area available for their penetration is smaller than that for lipophilic compounds that penetrate through the larger nonpolar region of the lipid bilayer.

We conclude that regression on the independent chemical descriptors (α, β, π, V_x) provides insight into the separate physicochemical effects of transport through skin but does not greatly improve the predictive ability of the model. Importantly, the chemical diversity of the solutes used in the LSER regression is by no means optimal (much less diverse than the entire FV database), which makes precise specification of the coefficients difficult. Moreover, LSER parameters are not available, and would be difficult to determine, for many of the compounds for which permeability coefficient values have been measured.

V. CONCLUSIONS

Overall, then, we have presented a sizable (170 measurements for 127 compounds) and diverse (MW ranging from 18 to 584, and log K_{ow} ranging from -3.1 to 4.6) database of pharmacological and toxic compounds that have been examined for certain quality criteria. Ionization can dramatically impact the magnitude of the permeability coefficient values, so permeability coefficients measured when the conditions at which chemical was partially ionized were adjusted for ionization. Analysis of these data have shown that temperature variation in the data (between 25 and 37°C) may introduce 2–5-fold differences in the permeability measurements. Despite the application of a set of quality criteria, repeated measurements from different labs can differ by more than one order of magnitude. Identifying the cause of differences between replicate measurements from different laboratories is required for development of more reliable equations for estimating permeability coefficient values.

The model based on only K_{ow} and MW is the simplest equation to provide reasonable estimates for the SC permeability coefficient. Other factors can be considered with some improvement in predictability. For example, Eq. T5 includes temperature and liquid density, which better represents compounds with relatively low or high densities. The results of Eq. T5 indicate that estimates of permeability coefficients for low or high MW compounds can be improved if the MW is adjusted by liquid density or by using a method that includes MV instead of MW. Based on the subset of the FV database, LSER models do not appear to be more useful than conventional models, especially since the parameters are less available. More permeability coefficient data for hydrophilic compounds is required to decide conclusively whether hydrophilic chemicals penetrate skin by a different mechanism than lipophilic chemicals.

ACKNOWLEDGMENTS

This work was supported in part by the United States Environmental Protection Agency under Assistance Agreement Nos. CR817451 and CR822757, by the U.S. Air Force Office of Scientific Research under agreement F49620-95-1-0021, and by the National Institute of Environmental Health Sciences under grant No. R01-ES06825. We thank Dr. Richard Guy for his helpful comments and suggestions.

APPENDIX A. PERMEABILITY COEFFICIENTS AND INPUT PARAMETERS

Table A1 Fully Validated Database of Human Skin Permeability Coefficients from Water

Compound[a]	log K_{ow}[b]	MW	T(°C)	$P_{cw,obs}$[c] cm h^{-1}	P_{cw} cm h^{-1}	↑↓[d]	f_{ai}[c]	pH[f]	Skin[g]	Ref. No.
[Aldosterone]	1.08	360.4	26	3.00E-06	3.00E-06	→	1	ND	EPID	40
2-Amino-4-nitrophenol	1.53	154.1	32	6.60E-04	8.59E-04		0.78	5.9	EPID	47
4-Amino-2-nitrophenol	1.53	154.1	32	2.80E-03	2.80E-03		0.98	5.9	EPID	47
Aminopyrine	1.00	231.3	37	[1.02E-03]	1.02E-03		1	7.94	FULL	25
[Amylobarbital]	2.07	226.3	30	2.27E-03	2.27E-03		1	7.4	FULL	37
Antipyrine	0.38	188.2	37	[6.58E-05]	6.58E-05		1	7.6	FULL	25
Atrazine	2.61	215.7	37	1.00E-02	1.00E-02		1	7[h]	SPLIT	48
[Atropine]	1.83	289.4	30	8.60E-06	1.39E-05	→	0.62	8	N/A	24
[Barbital]	0.65	184.2	30	1.10E-04	1.10E-04		1	7.4	FULL	37
* Benzene	2.13	78.1	31	1.11E-01	1.11E-01		1	ND	EPID	49
* Benzene	2.13	78.1	37	1.40E-01	1.40E-01		1	ND	SPLIT	48
* Benzoic acid	1.87	122.1	35	3.00E-02	3.00E-02		1	2.75	SC	13
* [Benzyl alcohol]	1.10	108.1	25[h]	6.00E-03	6.00E-03		1	ND	N/A	50
Betamethasone	1.94	392.5	37	2.44E-04	2.44E-04		1	4.5	SC[i]	51
Betamethasone-17-valerate	3.60	476.0	37	1.46E-02	1.46E-02	←	1	4.5	SC[i]	51
Betamethasone-17-valerate	3.60	476.0	25[h]	1.15E-03	1.15E-03		1	ND	SC[h]	41
* [p-Bromophenol]	2.59	173.0	25	3.61E-02	3.61E-02		1	[5.3]	EPID	52
[2,3-Butanediol]	-0.92	90.1	30	5.00E-05[j]	5.00E-05		1	ND	EPID	16
* [Butanoic acid]	0.79	88.1	N/A	1.00E-03	1.00E-03		1[k]	N/A	N/A	44
* [Butanol]	0.88	74.1	25	2.50E-03	2.50E-03		1	ND	EPID	53
* Butanol	0.88	74.1	30	3.00E-03	3.00E-03		1	ND	EPID	16
* [2-Butanone]	0.29	72.1	30	4.50E-03	4.50E-03		1	ND	EPID	16
[Butobarbitone]	1.73	212.2	30	1.90E-04	1.90E-04	→	1	7.4	FULL	37

Compound										
Caffeine	-0.07	194.2	30	1.60E-03	1.80E-03		0.89	7.4[b]	EPID[h]	54
4-Chloro-m-phenylenediamine	0.85	142.6	32	2.10E-03	2.10E-03		1	9.7	EPID	47
* [Chlorocresol]	3.10	142.6	25	5.50E-02	5.50E-02		1	[5.3]	EPID	52
** [Chloroform]	1.97	119.4	37	1.60E-01	1.60E-01	←	1	ND	SPLIT	48
* [o-Chlorophenol]	2.15	128.6	25	3.31E-02	3.31E-02		1	[4.6]	EPID	52
* [p-Chlorophenol]	2.39	128.6	25	3.63E-02	3.63E-02		1	[5.3]	EPID	52
[Chloroxylenol]	[3.48]	115.5	25	5.90E-02	5.90E-02		1	[5.3]	EPID	52
[Chlorpheniramine]	3.39	274.8	30	2.20E-03	2.27E-03	→	0.97	10.3	N/A	24
[Cortexolone]	2.52	346.5	26	7.50E-05	7.50E-05		1	ND	EPID	40
[Cortexone]	2.88	330.5	26	4.50E-04	4.50E-04		1	ND	EPID	40
Corticosterone	1.94	346.5	27	[6.47E-04]	6.47E-04	→	1	7.4	EPID	43
Corticosterone	1.94	346.5	39	[2.49E-03]	2.49E-03	→	1	7.4	EPID	43
[Corticosterone]	1.94	346.5	26	6.00E-05	6.00E-05	→	1	ND	EPID	40
Corticosterone	1.94	346.5	25[b]	2.24E-05	2.24E-05		1	ND	SC[b]	41
[Cortisone]	1.47	360.5	26	1.00E-05	1.00E-05		1	ND	EPID	40
* [m-Cresol]	1.96	108.1	25	1.52E-02	1.52E-02		1	[5.5]	EPID	52
** [o-Cresol]	1.95	108.1	25	1.57E-02	1.57E-02		1	[5.6]	EPID	52
* [p-Cresol]	1.94	108.1	37	1.20E-01	1.20E-01	←	1	4	SC	42
** [p-Cresol]	1.94	108.1	25	1.75E-02	1.75E-02		1	[5.6]	EPID	52
Cyclobarbitone	1.77	236.3	37	[8.14E-04]	8.14E-04		1	3.58	FULL	25
* [Decanol]	4.57	158.3	25	8.00E-02	8.00E-02		1	ND	EPID	53
[2,4-Dichlorophenol]	3.06	163.0	25	6.01E-02	6.01E-02		1	[4.4]	EPID	52
Diclofenac	4.40	260.7	37	1.82E-02	1.82E-02	→	0.50	pK_a[1]	EPID	45
[Diethylcarbamazine]	[1.75]	199.3	30	1.30E-04	1.30E-04		1	10	N/A	24
[Ephedrine]	0.93	165.2	30	6.00E-03	6.38E-03	→	0.94	10.8	N/A	24
β-Estradiol	4.01	272.4	30	3.89E-03	3.89E-03		1	7	FULL	55
[β-Estradiol]	4.01	272.4	30	5.20E-03	5.20E-03		1	7	N/A	24
[β-Estradiol]	4.01	272.4	26	3.00E-04	3.00E-04		1	ND	EPID	40
β-Estradiol	4.01	272.4	37	9.70E-03	9.70E-03		1	ND	SC[m]	56
β-Estradiol	4.01	272.4	32	4.31E-03[n]	4.31E-03		1	ND	EPID	57

Table A1 Continued

Compound[a]	log K_{ow}[b]	MW	T(°C)	$P_{cw,obs}$[c] cm h^{-1}	P_{cw} cm h^{-1}	↑↓[d]	f_{ni}[e]	pH[f]	Skin[g]	Ref. No.
[Estriol]	2.45	288.4	26	4.00E-05	4.00E-05	→	1	ND	EPID	40
[Estrone]	3.13	270.4	26	3.60E-03	3.60E-03		1	ND	EPID	40
* Ethanol	-0.31	46.0	22	3.00E-04	3.00E-04		1	ND	FULL	58
* [Ethanol]	-0.31	46.0	25	8.00E-04	8.00E-04		1	ND	EPID	53
* Ethanol	-0.31	46.0	30	3.17E-04	3.17E-04		1	ND	FULL	59
* [2-Ethoxy ethanol]	-0.32	90.1	30	2.50E-04	2.50E-04		1	ND	EPID	16
* [Ethyl ether]	0.89	74.1	30	1.60E-02	1.60E-02		1	ND	EPID	16
* [p-Ethylphenol]	2.58	122.2	25	3.49E-02	3.49E-02		1	[5.7]	EPID	52
[Fentanyl]	4.05	336.5	30	1.00E-02	1.14E-02		0.88	8	N/A	24
Fentanyl	4.05	336.5	37	1.13E-02	1.55E-02		0.73	7.4	EPID	4
* Flurbiprofen	4.16	244.3	37	[4.62E-01]	1.43E+00	↑	0.32	4.7	FULL	25
* [Heptanoic acid]	[2.41]	130.2	N/A	2.00E-02	2.00E-02[k]		1[k]	N/A	N/A	44
* Heptanol	2.72	116.0	30	3.76E-02	3.76E-02		1	ND	EPID	16
* [Heptanol]	2.72	116.0	25	3.20E-02	3.20E-02		1	ND	EPID	53
* [Hexanoic acid]	1.92	116.2	N/A	1.40E-02	1.40E-02[k]		1[k]	N/A	N/A	44
* Hexanol	2.03	102.2	31	2.77E-02	2.77E-02		1	ND	SPLIT	60
* [Hexanol]	2.03	102.2	25	1.30E-02	1.30E-02		1	ND	EPID	53
[Hydrocortisone (HC)]	1.61	362.5	26	3.00E-06	3.00E-06	→	1	ND	EPID	40
Hydrocortisone (HC)	1.61	362.5	25[h]	2.82E-06	2.82E-06	→	1	ND	SC[b]	41
[HC-yl-succinamate]	[0.17]	461.6	37	2.60E-05	2.60E-05		1	4	SC	20
[HC-yl-N,N-dimethyl succinamate]	[0.88]	489.6	37	6.70E-05	6.70E-05		1	4	SC	20
[HC-yl-hemipimelate]	[1.82]	504.6	37	1.80E-03	2.31E-03	↑	0.78	4	SC	20
[HC-yl-hemisuccinate]	[0.91]	462.5	37	6.30E-04	8.10E-04	↑	0.78	4	SC	20
[HC-yl-hexanoate]	[3.28]	460.6	37	1.80E-02	1.80E-02	↑	1	4	SC	20
[HC-yl-6-hydroxy hexanoate]	[1.29]	476.6	37	9.10E-04	9.10E-04	↑	1	4	SC	20

Compound		MW	n	K_p (1)	K_p (2)		f	pK_a	DB	n
[HC-yl-octanoate]	[4.34]	488.7	37	6.20E-02	6.20E-02	←	1	4	SC	20
[HC-yl-pimelamate]	[0.82]	503.6	37	8.90E-04	8.90E-04	←	1	4	SC	20
[HC-yl-priopionate]	[1.69]	418.5	37	3.40E-03	3.40E-03	←	1	4	SC	20
4-Hydroxybenzyl alcohol	0.25	124.1	37	2.00E-03	2.00E-03		1	4	SC	42
α-(4-Hydroxyphenyl) acetamide	[−0.21]	151.2	37	4.50E-04	4.50E-04		1	4	SC	42
4-Hydroxyphenyl acetic acid	0.75	152.1	26	2.50E-03	3.41E-03		0.73	4	SC	42
[17α-Hydroxyprogesterone]	3.17	330.5	37	6.00E-04	6.00E-04		1	ND	EPID	40
Ibuprofen	3.50	206.3	37	[5.70E-01]	1.02E+00	←	0.56	4.44	FULL	25
Indomethacin	4.27	357.8	37	[5.05E-02]	2.54E-01	←	0.20	5.15	FULL	25
Indomethacin	4.27	357.8	30	1.48E-02	1.48E-02		0.50	pK_a[l]	EPID	45
* [Isoquinoline]	2.08	129.2	37	1.67E-02	1.68E-02	←	0.99	7.4	FULL	37
Isosorbide dinitrate	1.31	236.1	37	[1.63E-02]	1.63E-02		1	ND	FULL	25
Ketoprofen	3.12	254.3	37	[5.89E-02]	7.10E-02	←	0.83	3.72	FULL	25
Mannitol	−3.10	182.2	30	[1.11E-04]	1.11E-04		1	ND	EPID	61
Mannitol	−3.10	182.2	27	6.71E-05	6.71E-05		1	7.4	EPID	43
Mannitol	−3.10	182.2	39	9.30E-05	9.30E-05		1	7.4	EPID	43
Mannitol	−3.10	182.2	30	6.10E-05	6.10E-05		1	ND	FULL	59
* [Methanol]	−0.77	32.0	25	5.00E-04	5.00E-04		1	ND	EPID	53
* [Methanol]	−0.77	32.0	30	1.60E-03	1.60E-03		1	ND	EPID[h]	54
Methyl-4-hydroxyphenylacetate	[1.15]	166.0	37	2.00E-02	2.00E-02		1	4	SC	42
[Methyl-4-hydroxybenzoate]	1.96	152.1	25	9.12E-03	9.12E-03		1	[4.6]	EPID	52
[Methyl-HC-yl-pimelate]	[2.20]	518.6	37	5.40E-03	5.40E-03	←	1	4	SC	20
[Methyl-HC-yl-succinate]	[1.38]	476.6	37	2.10E-04	2.10E-04		1	4	SC	20
* [β-Naphthol]	2.70	144.2	25	2.79E-02	2.79E-02		1	[5.2]	EPID	52
Naproxen	3.34	230.3	37	3.82E-02	3.82E-02		0.50	pK_a[l]	EPID	45
Nicorandil	[0.65]	211.2	37	2.66E-04	2.66E-04		1	[8.0]	FULL	62
Nicorandil	[0.65]	211.2	37	[1.79E-04]	1.79E-04		1	[8.0]	FULL	25
Nicotinate, benzyl	2.40	213.2	37	1.62E-02	1.62E-02		1	[7.3]	EPID	63
Nicotinate, butyl	2.27	179.1	37	1.66E-02	1.66E-02		1	[7.7]	EPID	63
Nicotinate, ethyl	1.32	151.0	37	6.34E-03	6.34E-03		1	[7.3, 8.3]	EPID	63

Table A1   Continued

Compound[a]	log K_{ow}[b]	MW	T(°C)	$P_{cw,obs}$[c] cm h⁻¹	P_{cw} cm h⁻¹	↑↓[d]	f_m[e]	pH[f]	Skin[g]	Ref. No.
Nicotinate, hexyl	3.59	207.2	37	1.79E-02	1.79E-02		1	[7.2]	EPID	63
Nicotinate, methyl	0.87	137.1	37	3.25E-03	3.25E-03		1	[8.3, 8.5]	EPID	63
Nicotinate, PG	[0.39]	181.0	37	3.40E-05	3.40E-05	→	1	[8.3]	EPID	63
Nicotinate, TEG-Me	[0.83]	269.1	37	1.77E-04	1.77E-04		1	[8.1]	EPID	63
Nicotinate, TEG-OH	[0.21]	255.1	37	9.90E-06	9.90E-06	→	1	[8.1]	EPID	63
[Nicotine]	1.17	162.2	30	1.90E-02	1.98E-02		0.96	9.2	FULL	37
2-Nitro-p-phenylenediamine	0.53	153.1	32	5.00E-04	5.00E-04		1	9.7	EPID	47
[Nitroglycerine]	[0.98]	227.1	30	1.10E-02	1.10E-02	←	1	ND	N/A	24
* [m-Nitrophenol]	2.00	139.1	25	5.64E-03	5.64E-03		1	[4.8]	EPID	52
* [p-Nitrophenol]	1.91	139.1	25	5.58E-03	5.58E-03		1	[3.9]	EPID	52
[N-Nitrosodiethanolamine]	[-1.58]	134.1	32	5.50E-06	5.50E-06	→	1	ND	EPID	64
* [Nonanol]	4.26	144.0	25	6.00E-02	6.00E-02		1	ND	EPID	53
* [Octanoic acid]	3.05	144.2	N/A	2.50E-02	2.50E-02[k]		1[k]	N/A	N/A	44
* Octanol	3.00	130.2	22	5.20E-02	5.20E-02		1	ND	FULL	58
* [Octanol]	3.00	130.2	25	5.20E-02	5.20E-02		1	ND	EPID	53
* Octanol	3.00	130.2	30	6.10E-02	6.10E-02		1	ND	EPID[h]	54
* Ouabain	-1.70	584.6	30	3.96E-06	3.96E-06		1	7	FULL	55
* [Pentanoic acid]	1.39	102.1	N/A	2.00E-03	2.00E-03[k]		1[k]	N/A	N/A	44
* Pentanol	1.56	88.0	22	6.00E-03	6.00E-03		1	ND	FULL	58
* [Pentanol]	1.56	88.0	25	6.00E-03	6.00E-03		1	ND	EPID	53
[Phenobarbitone]	1.47	232.2	30	4.50E-04	4.50E-04		1	7.4	FULL	37
* [Phenol]	1.46	94.1	25	8.22E-03	8.22E-03		1	[5.4]	EPID	52
* Phenol	1.46	94.1	37	1.95E-02	1.95E-02		1	N/A	EPID	45
* Phenol	1.46	94.1	22	1.55E-04	1.55E-04	→	1	[5.4]	EPID[h]	54
* 2-Phenylethanol	1.36	122.2	25[b]	7.50E-03	7.50E-03		1	ND	N/A	50
o-Phenylenediamine	0.15	108.1	32	4.50E-04	4.50E-04		1	9.7	EPID	47
p-Phenylenediamine	-0.30	108.1	32	2.40E-04	2.40E-04		1	9.7	EPID	47
Piroxicam	3.06	331.4	37	3.40E-03	3.40E-03°		0.50	pK,[i]	EPID	45
Prednisolone	1.62	360.4	25[b]	4.47E-05	4.47E-05		1	ND	SC[b]	41

	Compound										
	[Pregnenolone]	4.22	316.5	26	1.50E-03		1.50E-03	1	ND	EPID	40
	[Progesterone]	3.87	314.5	26	1.50E-03		1.50E-03	1	ND	EPID	40
*	Propanol	0.25	60.0	22	1.00E-03		1.00E-03	1	ND	FULL	58
*	Propanol	0.25	60.0	30	1.70E-03		1.70E-03	1	ND	EPID	16
*	[Propanol]	0.25	60.0	25	1.20E-03		1.20E-03	1	ND	EPID	53
*	[Resorcinol]	0.80	110.1	25	2.40E-04	→	2.40E-04	1	[5.4]	EPID	52
	[Salicylic acid]	2.26	138.1	30	6.26E-03		1.28E-02	0.49	3	FULL	37
	Salicylic acid	2.26	138.1	25	Range		1.03E-02	(.2–.5)	2–4	FULL	65
	Salicylic acid	2.26	138.1	37	3.04E-02		3.04E-02	0.50	pK_a	EPID	45
	Salicylic acid^p	2.26	138.1	37	3.76E-02		3.76E-02	0.50	pK_a	EPID	45
	[Scopolamine]	[-0.20]	303.4	30	5.00E-05		6.17E-05	0.81	9.6	N/A	24
	[Sufentanil]	3.95	386.5	37	1.52E-02		1.60E-02	0.95	7.4	EPID	4
	[Testosterone]	3.32	288.4	26	4.00E-04	→	4.00E-04	1	ND	EPID	40
	Testosterone	3.32	288.4	25^h	8.51E-04		8.51E-04	1	ND	SC^b	41
*	Tetrachloroethylene	3.40	165.9	37	1.60E-02		1.60E-02	1	ND	SPLIT	48
*	[Thymol]	3.30	150.2	25	5.28E-02		5.28E-02	1	[6.0]	EPID	52
*	Toluene	2.73	92.1	37	8.30E-01	←	8.30E-01	1	4	SC	42
	Triamcinolone	1.16	394.5	25^h	3.98E-06	→	3.98E-06	1	ND	SC^b	41
	Triamcinolone acetonide	2.53	434.5	25^h	2.02E-05	→	2.02E-05	1	ND	SC^b	41
*	Trichloroethylene	2.61	131.4	37	1.20E-01		1.20E-01	1	ND	SPLIT	48
	[2,4,6-Trichlorophenol]	3.69	197.5	25	5.94E-02		5.94E-02	1	[3.6]	EPID	52
	Urea	-2.11	60.1	37	1.48E-04		1.48E-04	1	7.1^h	SC^b	66
	Urea	-2.11	60.1	27	2.01E-04		2.01E-04	1	7.4	EPID	43
	Urea	-2.11	60.1	39	2.72E-04		2.72E-04	1	7.4	EPID	43
*	Water	-1.38	18.0	30^h	1.56E-03		1.56E-03	1	7.1	SC^b	66
*	Water	-1.38	18.0	31	1.40E-03		1.40E-03	1	ND	SPLIT	60
*	Water	-1.38	18.0	32	1.55E-03		1.55E-03	1	ND	EPID	67
*	Water	-1.38	18.0	30	[8.54E-04]		8.54E-04	1	ND	EPID	61
*	Water	-1.38	18.0	30	1.58E-03		1.58E-03	1	7	FULL	55

Table A1 Continued

Compound[a]	log K_{ow}[b]	MW	$T(°C)$	$P_{cw,obs}$[c] cm h^{-1}	P_{cw} cm h^{-1}	↑↓[d]	f_{ni}[e]	pH[f]	Skin[g]	Ref. No.
* Water	-1.38	18.0	31	1.34E-03	1.34E-03		1	ND	SPLIT	68
* [Water]	-1.38	18.0	25	5.00E-04	5.00E-04		1	ND	EPID	53
* Water	-1.38	18.0	30	6.39E-04	6.39E-04		1	ND	FULL	59
* [3,4-Xylenol]	2.23	122.2	25	3.60E-02	3.60E-02		1	[5.8]	EPID	52

[a] Compounds contained within brackets (e.g., [Aldosterone]) also appeared in the Flynn database. Those indicated with an asterisk to the left (e.g., benzene) were used in the LSER analysis.

[b] Reported log K_{ow} are taken from the Hansch Starlist (1), unless contained within brackets (e.g., for chloroxylenol [3.48]), in which case they were calculated using the Daylight software (2).

[c] Permeability coefficients contained within brackets are digitized from figures in the reference.

[d] ↑ indicates a prediction by Eq. T1 that is more than one order of magnitude higher than measurement. ↓ indicates a prediction by Eq. T1 that is more than one order of magnitude lower than measurement.

[e] The nonionized fraction determined from pK_a values calculated in SPARC (7) at 25°C and adjusted to the experimental temperature as listed in Table A4.

[f] Reported solution pH unless contained within brackets (e.g., for p-bromophenol [5.3]), in which case the pH was calculated from the reported concentration and calculated pK_a values (see Table A4). Chemicals that are undissociated are indicated by ND when no pH was reported.

[g] Type of skin used in the study: isolated stratum corneum (SC), epidermal membranes (EPID), split (SPLIT), or full-thickness skin (FULL).

[h] Information was obtained through personal communication with authors.

[i] SC permeability coefficients were reported for betamethasone (BMS) and BMS 17-Valerate. The SC permeability coefficient was calculated from separate permeability measurements for split thickness skin (epidermis and part of dermis) and dermis using a multilaminate model (see Appendix B).

[j] The permeability coefficient was reported as an upper bound. Value not treated differently in the analysis.

[k] Scheuplein et al. (44) did not report pH of solution. However, we assume that the permeability coefficients for carboxylic acids were measured at a pH assuring that the chemicals were 100% unionized.

[l] Permeability coefficient equals twice that measured when pH was set at the pK_a, although the pK_a was not reported. pK_a values calculated using SPARC (7) are listed in Table A4. These may be different from the pH values used experimentally.

[m] The SC permeability coefficient was calculated from the difference of the inverse permeability coefficients of the epidermis (SC+VE) and the VE alone.

[n] Calculated from the flux provided and a concentration that was determined by personal communication with Barry (69).

[o] The permeability of piroxicam (45) is taken to be 0.0034 cm/h rather than 0.034 cm/h, which is twice the 50% ionized value and is consistent with their Fig. 2.

[p] According to Singh and Roberts (45), the salt diethylamine salicylate is ionized under the experimental conditions. Consequently, the free base form, salicylic acid, was the penetrating species.

Table A2 Provisional Database of Human Skin Permeability Coefficients from Water

Compound[a]	log K_{ow}[b]	MW	T(°C)	$P_{cw,obs}$[c] cm h^{-1}	f_m[d]	pH[e]	Skin[f]	Ref. No.
5-Fluorouracil (+ − + −)	-0.89	130.1	31	9.51E-05	<0.1	4.75s	SPLIT	70
5-Fluorouracil (+ − + −)	-0.89	130.1	32	3.00E-05	<0.1	4.75s	EPID	71
5-Fluorouracil (+ − + −)	-0.89	130.1	37	[1.58E-05]	<0.1	4.66	FULL	25
5-Fluorouracil (+ − + −)	-0.89	130.1	31	3.48E-05	<0.1	4.75s	SPLIT	68
5-Fluorouracil (+ − + −)	-0.89	130.1	31	1.66E-05	<0.1	4.75s	SPLIT	68
5-Fluorouracil (+ − + −)	-0.89	130.1	32	2.46E-05	<0.1	4.75s	EPID	72
[Hydroxypregnenolone]	N/A	N/A	26	6.00E-04	1	ND	EPID	40
[Ouabain]^i	-1.70	584.6	30	7.80E-07	1	7	N/A	24
Paraquat (dichloride)[b] (++)	[-5.65]	257.3	30	[1.28E-05]	<0.1	ND	EPID	73
Paraquat dichloride (++)	[-5.65]	257.3	30	8.70E-06	<0.1	ND	FULL	59
Tetraethylammonium bromide (+)	-2.82	210.2	27	[5.58E-05]	<0.1	7.4	EPID	43
Tetraethylammonium bromide (+)	-2.82	210.2	39	[8.82E-05]	<0.1	7.4	EPID	43

[a] Compounds contained within brackets (e.g., [Hydroxypregnenolone] also appeared in the Flynn database. All positive (+) and negative (−) ionic charges (for the chemical at experimental conditions) are indicated. For example, 5-fluorouracil with two positive and two negative charges is indicated by (+ − + −).

[b] Reported log K_{ow} are taken from the Hansch Starlist (1), unless contained within brackets (e.g., for paraquat dichloride [−5.65]), in which case they were calculated (2).

[c] Permeability coefficients contained within brackets are digitized from figures in the reference.

[d] The nonionized fraction determined from pK_a values calculated in SPARC (7) at 25°C and adjusted to the experimental temperature as shown in Table A4.

[e] Reported solution pH unless contained within brackets, in which case the pH was calculated from the reported concentration and calculated pK_a values (as given in Table A4). Compounds that essentially do not dissociate are indicated by ND when no pH was reported.

[f] Type of skin used in the study: isolated stratum corneum (SC), epidermal membranes (EPID), split (SPLIT), or full-thickness skin (FULL).

[g] Information obtained through personal communication from Barry (1996).

[h] Corresponding anion was not specified, but given it was dichloride in Scott et al. (59), it is likely that paraquat was applied as the dichloride salt in this measurement as well.

[i] No information was provided to determine whether P_{cw} for ouabain (MW = 584) was measured at steady state. We have found that the P_{cw} is large for a chemical with ouabain's properties, indicating that steady state might have been attained. Based on this analysis, the permeability coefficient for ouabain could be moved to the FV database.

Table A3 Excluded Database of Human Skin Permeability Coefficients from Water

Compound[a]	log K_{ow}[b]	MW	T(°C)	$P_{cw,obs}$[c] cm h⁻¹	f_u[d]	pH[e]	Skin[f]	Ref. No.
Amphetamine (+)	1.76	135.2	30	1.40E-05	<0.1	7	FULL	55
Aniline[g]	0.90	93.1	30	2.24E-02	1	[7.6]	SPLIT	74
Anisole[g]	2.11	108.1	30	7.37E-02	1	ND	SPLIT	74
Aspartic acid (−+−)	N/A	133.1	37	9.36E-05	<0.1	7.3	FULL	3
Aspartic acid (−+ and −+−)	N/A	133.1	37	1.33E-04	<0.1	3.4	FULL	3
Benzaldehyde[g]	1.48	106.1	30	6.08E-02	1	ND	SPLIT	74
Benzyl Alcohol[g]	1.10	108.1	30	1.69E-02	1	ND	SPLIT	74
Chromone-2-carboxylic acid I (−)	N/A	288.0	37	7.30E-05	<0.1	7	EPID	75
Chromone-2-carboxylic acid II (−)	N/A	272.0	37	2.90E-05	<0.1	7	EPID	75
Chromone-2-carboxylic acid III (−)	N/A	230.0	37	1.86E-05	<0.1	7	EPID	75
Chromone-2-carboxylic acid IV (−)	N/A	306.0	37	4.20E-06	<0.1	7	EPID	75
Cobalt chloride[h]	N/A	N/A	N/A	4.00E-04	<0.1	N/A	N/A	76
[Codeine] (+)	1.14	299.4	37	4.90E-05	<0.1	7.4	EPID	5
Coumarin[i]	1.39	146.1	37	9.10E-03	1	7.4	FULL	77
Diclofenac (−)	4.40	260.7	37	[3.00E-03]	<0.1	7.96	FULL	25
[Digitoxin][j]	2.83	764.9	30	1.30E-05	1	7	N/A	24
Dopamine (+)	[−0.05]	153.2	37	[9.90E-05]	<0.1	3.26	FULL	25
[Etorphine][k] (+)	[1.41]	411.5	37	3.60E-03	<0.1	7.3	N/A	33
[Fluocinonide][l]	3.19	494.6	37	1.70E-03	1	4	SC	20
Formaldehyde[m]	0.35	30.0	30	4.51E-04	1	7.4	FULL	78
Griseofulvin[i]	2.18	352.8	37	1.30E-03	1	7.4	FULL	77
Histidine (+−+)	−3.56	155.2	37	4.46E-05	<0.1	5	FULL	3
Histidine (−+)	−3.56	155.2	37	5.54E-05	<0.1	7.3	FULL	3
Histidine (−+)	−3.56	155.2	37	4.50E-05	<0.1	8.5	FULL	3
[Hydrocortisone][n]	1.61	362.5	30	1.19E-04	1	7.4	FULL	37
[Hydromorphone] (+)	[0.55]	285.3	37	1.50E-05	<0.1	7.4	EPID	5
Isoprenaline (+)	[0.08]	211.2	37	[2.84E-05]	<0.1	2.75	FULL	25
Ketorolac acid[o] (−)	[1.77]	255.3	32	1.30E-02	<0.1	2.1	SPLIT	79
Lead Acetate[h]	N/A	N/A	N/A	4.20E-06	<0.1	N/A	N/A	80
Levodopa (−+)	−2.74	197.0	37	[6.60E-05]	<0.1	5.42	FULL	25
Lidocaine (+)	2.26	234.3	37	[2.17E-02]	<0.1	6.82	FULL	25
Lidocaine (+)	2.26	234.3	37	4.20E-03	<0.1	7.4	EPID	81
Lysine (+−+)	−3.05	146.2	37	1.69E-04	<0.1	7.3	FULL	3
Lysine (−+) and (+−+)	−3.05	146.2	37	4.97E-04	<0.1	8.9	FULL	3
[Meperidine] (+)	2.45	247.4	37	3.70E-03	<0.1	7.4	EPID	5

Compound								
Mercuric chloride[b]	N/A	N/A	N/A	9.30E-04	<0.1	N/A	N/A	76
Morphine (+)	0.76	285.3	37	[2.84E-04]	<0.1	4.22	FULL	25
[Morphine] (+)	0.76	285.3	37	1.60E-06	<0.1	7.5	EPID	38
[Naproxen][p] (−)	3.34	230.3	N/A	4.00E-04	<0.1	6.5	FULL	82
Nickel chloride[b]	N/A	N/A	N/A	1.00E-04	<0.1	N/A	N/A	83
Nickel sulfate[b]	N/A	N/A	N/A	<9.00E-06	<0.1	N/A	N/A	84
Nicotinate, PEG 350-Me[q]	N/A	N/A	37	1.14E-05	1	[8.0]	EPID	63
Nicotinate, PPG 425[q]	N/A	N/A	37	9.39E-04	1	[7.9]	EPID	63
Nicotinic acid (−+)	[0.77]	123.1	37	2.42E-05	<0.1	[3.4]	EPID	63
2-Phenyl ethanol[a]	1.36	122.2	30	1.27E-02	1	ND	SPLIT	74
Propranolol[i] (+)	2.98	259.3	37	[1.20E-03]	<0.1	7.4	FULL	77
Silver nitrate[b]	N/A	N/A	N/A	<3.50E-04	<0.1	N/A	N/A	85
Sodium chromate[b]	N/A	N/A	N/A	2.10E-03	<0.1	N/A	N/A	86
[Sucrose][j]	-3.70	342.3	37	9.40E-06	1	4	SC	20

a Compounds contained within brackets (e.g., [Codeine]) also appeared in the Flynn database. All positive (+) and negative (−) ionic charges (for the chemical at experimental conditions) are indicated. For example, aspartic acid with two negative charges and one positive charge is indicated by (−+−).

b Reported log K_{ow} are taken from the Hansch Starlist (1), unless contained within brackets (e.g., for dopamine [−0.05]), in which case they were calculated using Daylight software (2).

c Permeability coefficients contained within brackets were digitized from figures in the reference.

d The nonionized fraction determined from pK, values calculated in SPARC (7) at 25°C and adjusted to the experimental temperature as shown in Table A4.

e Reported solution pH unless contained within brackets (e.g., for aniline [7.6]), in which case the pH was calculated from the reported concentration and calculated pK, values (see Table A4). Compounds that are undissociated are indicated by ND when no pH was reported.

f Type of skin used in the study: isolated stratum corneum (SC), epidermal membranes (EPID), split-thickness (SPLIT), or full-thickness skin (FULL).

g Ethanol was in the receptor chamber, and it is likely that the skin barrier was compromised.

h Inorganic species that was not considered further.

i The concentration of this penetrant depleted during this experiment.

j It was likely that the permeability coefficient was not measured at steady state.

k Not included in the FV database because (1) f_{ni} < 0.1, and (2) the reported permeability coefficient was higher than that of similar compounds in the study (see Appendix B).

l Anderson (32) suggested that the permeability coefficient for fluocinonide and sucrose should not be included in the FV database.

m Formaldehyde was applied in a solution of formalin, containing methanol, which may have compromised the skin barrier.

n Ethanol was in the donor solution, and it likely altered the skin permeability.

o The skin may have been damaged at this pH.

p Naproxen was delivered in an aqueous gel vehicle.

q These permeability coefficients refer to penetration of several compounds from a polydisperse mixture.

Table A4 Estimates of the Nonionized Fraction of the Penetrating Compound at Experimental Temperature

Compound[a]	pK$_a$ (25°C)[a]	T(°C)	ΔH (kcal mol^{-1})[b]	pK$_a$ (T)[c]	C$_w$ (mol L^{-1})[d]	pH[c]	f$_{ni}^i$	Ref. No.
2-Amino-4-nitrophenol		32				5.9	0.78[g]	47
(N → −)	6.84		5	6.75				
(− → ±)	6.07		7.5	5.94				
(N → +)	2.7		7.5	2.57				
4-Amino-2-nitrophenol		32				5.9	0.98[g]	47
(N → −)	7.74		5	7.65				
(− → ±)	4.34		7.5	4.21				
(N → +)	2.94		7.5	2.81				
Aniline	4.8	30	7.5	4.71	3.60E-04	[7.6]	1	74
Atropine	7.91	30	10	7.79		8.0	0.62	24
p-Bromophenol	9.4	25	5	9.40	0.058[h]	[5.3]	1	52
Caffeine	6.54	30	5	6.48		7.4	0.89	54
Chlorocresol	9.55	25	5	9.55	0.070[h]	[5.3]	1	52
o-Chlorophenol	8.25	25	5	8.25	0.078[h]	[4.6]	1	52
p-Chlorophenol	9.43	25	5	9.43	0.078[h]	[5.3]	1	52
Chloroxylenol	9.68	25	5	9.68	0.086[h]	[5.3]	1	52
Chlorpheniramine	8.95	30	10	8.83		10.3	0.97	24
m-Cresol	10.13	25	5	10.13	0.092[h]	[5.5]	1	52
o-Cresol	10.33	25	5	10.33	0.092[h]	[5.6]	1	52
p-Cresol	10.31	25	5	10.31	0.092[h]	[5.6]	1	52
2,4-Dichlorophenol	7.67	25	5	7.67	0.061[h]	[4.4]	1	52
Diclofenac	4.07	37	0	4.50		pK$_a$	0.50	45
Ephedrine	9.73	30	10	9.61		10.8	0.94	24
p-Ethylphenol	10.29	25	5	10.29	0.082[h]	[5.7]	1	52
Fentanyl	7.25	30	10	7.13		8.0	0.88	24

Compound								
Fentanyl	7.25	37	10	6.96		7.4	0.73	4
Indomethacin	4.54	37	0	4.5		pKₐ	0.50	45
Isoquinoline	5.29	30	5	5.23		7.4	0.99	37
Methyl-4-hydroxy benzoate	7.93	25	5	7.93	0.066[b]	[4.6]	1	52
β-Naphthol	9.34	25	5	9.34	0.069[b]	[5.2]	1	52
Naproxen	4.49	37	0	4.5		pKₐ	0.50	45
Nicorandil	2.87	37	5	2.73	0.188	[8.0]	1	25
Nicorandil	2.87	37	5	2.73	0.188ʲ	[8.0]	1	62
Nicotinate, benzyl	2.98	37	5	2.84	0.004	[7.3]	1	63
Nicotinate, butyl	3.17	37	5	3.03	0.021	[7.7]	1	63
Nicotinate, ethyl	3.17	37	5	3.03	0.003	[7.3]	1	63
Nicotinate, ethyl	3.17	37	5	3.03	0.331	[8.3]	1	63
Nicotinate, hexyl	3.17	37	5	3.03	0.001	[7.2]	1	63
Nicotinate, methyl	3.17	37	5	3.03	0.365	[8.3]	1	63
Nicotinate, methyl	3.17	37	5	3.03	0.730	[8.5]	1	63
Nicotinate, PEG 350-Me	2.82	37	5	2.68	0.212	[8.0]	1	63
Nicotinate, PG	3.05	37	5	2.91	0.552	[8.3]	1	63
Nicotinate, PPG 425	3.05	37	5	2.91	0.063	[7.9]	1	63
Nicotinate, TEG-Me-	2.82	37	5	2.68	0.372	[8.1]	1	63
Nicotinate, TEG-OH-	2.78	37	5	2.64	0.392	[8.1]	1	63
Nicotine	7.89	30	10	7.77		9.2	0.96	37
2-Nicotinic acid		37			0.168	[3.4]	<0.1	63
(N → ⁻)	3.43		0	3.43				
(± → ⁻)	4.79		5	4.65				
(+ → N)	3.4		5	3.26				
m-Nitrophenol	8.39	25	5	8.39	0.072[b]	[4.8]	1	52
p-Nitrophenol	6.83	25	5	6.83	0.072[b]	[3.9]	1	52
Phenol	10	25	5	10.00	0.106[b]	[5.4]	1	52
Phenol	10	37	5	10		pKₐ	0.50	45

Table A4 Continued

Compound[a]	pK$_a$ (25°C)[a]	T(°C)	ΔH (kcal mol^{-1})[b]	pK$_a$ (T)[c]	C$_w$ (mol L^{-1})[d]	pH[e]	f$_{ni}$[f]	Ref. No.
Phenol	10	22	5	10.04	0.106	[5.4]	1	54
Piroxicam	N/A	37	N/A	N/A		pK$_a$	0.50	45
Resorcinol	9.86	25	5	9.86	0.091[h]	[5.4]	1	52
Salicylic acid	4.5	37	0	4.5		pK$_a$	0.50	45
Salicylic acid	4.5	37	0	4.5		pK$_a$	0.50	45
Scopolamine	9.09	30	10	8.97		9.6	0.81	24
Sufentanil	6.44	37	10	6.15		7.4	0.95	4
Thymol	10.82	25	5	10.82	0.066[h]	[6.0]	1	52
2,4,6-Trichlorophenol	5.94	25	5	5.94	0.051[h]	[3.6]	1	52
3,4-Xylenol	10.44	25	5	10.44	0.082[h]	[5.8]	1	52

[a] pK$_a$ values calculated in SPARC (7) at 25°C using methods described in text.
[b] These heats of ionization are approximate values obtained from the literature (10).
[c] Calculated using an integrated form of the van't Hoff equation, Eq. (3).
[d] Solution concentration provided only when it was needed to calculate the pH.
[e] The pH was reported in the original paper, unless contained within brackets, in which case it was calculated from pK$_a$(T) and the solution concentration, assuming that pH was 7.0 prior to chemical addition.
[f] The nonionized fraction was calculated using Eq. (1) when one pK$_a$ is dominant. Otherwise it was determined using a more rigorous solution of simultaneous equilibrium as discussed in the text.
[g] The nonionized fraction for this compound with multiple pK$_a$ was calculated in SPARC (7) at T = 25°C.
[h] The concentrations were consistently dilute [~1% (w/v)], but not reported. We have used a concentration of 1% (w/v) to calculate the pH and the nonionized fraction.
[i] The concentration of saturated nicorandil solution at 37°C was reported by Morimoto et al. (25).

Table A5 LSER Parameters[a] and Permeability Coefficient Values for Chemicals in the LSER Database

Compound	π	α	β	$V \times 10^{-2} [cm^3\ mol^{-1}]^b$	$T(°C)$	P_{cw} (cm h^{-1})[c]	Ref. No.
Benzene	0.52	0	0.14	0.716	31	1.11E-01	49
Benzene	0.52	0	0.14	0.716	37	1.40E-01	87
Benzoic Acid	0.9	0.59	0.4	0.932	35	3.00E-02	13
[Benzyl alcohol]	0.87	0.33	0.56	0.916	25	6.00E-03	50
[p-Bromophenol]	1.17	0.67	0.2	0.95	25	3.61E-02	52
[Butanoic acid]	0.62	0.6	0.45	0.747	N/A	1.00E-03	44
[Butanol]	0.42	0.37	0.48	0.731	25	2.50E-03	53
Butanol	0.42	0.37	0.48	0.731	30	3.00E-03	16
[2-Butanone]	0.67	0.03	0.48	0.69	30	4.50E-03	16
[Chlorocresol]	1.02	0.65	0.22	1.038	25	5.50E-02	52
Chloroform	0.49	0.15	0.02	0.617	37	1.60E-01	87
[o-Chlorophenol]	0.88	0.32	0.31	0.898	25	3.31E-02	52
[p-Chlorophenol]	1.08	0.67	0.2	0.898	25	3.63E-02	52
[m-Cresol]	0.88	0.57	0.34	0.916	25	1.52E-02	52
[o-Cresol]	0.86	0.52	0.3	0.916	25	1.57E-02	52
p-Cresol	0.87	0.57	0.31	0.916	37	1.20E-01	42
[p-Cresol]	0.87	0.57	0.31	0.916	25	1.75E-02	52
[Decanol]	0.42	0.37	0.48	1.576	25	8.00E-02	53
Ethanol	0.42	0.37	0.48	0.449	22	3.00E-04	58
[Ethanol]	0.42	0.37	0.48	0.449	25	8.00E-04	53
Ethanol	0.42	0.37	0.48	0.449	30	3.17E-04	59
[2-Ethoxyethanol]	0.5	0.3	0.83	0.79	30	2.50E-04	16
[Ethyl ether]	0.25	0	0.45	0.731	30	1.60E-02	16
[p-Ethylphenol]	0.9	0.55	0.36	1.057	25	3.49E-02	52
[Heptanoic acid]	0.6	0.6	0.45	1.169	N/A	2.00E-02	44
Heptanol	0.42	0.37	0.48	1.154	30	3.76E-02	16

Table A5 Continued

Compound	π	α	β	$V \times 10^{-2}$ [cm³ mol⁻¹][b]	T(°C)	P_{cw} (cm h⁻¹)[c]	Ref. No.
[Heptanol]	0.42	0.37	0.48	1.154	25	3.20E-02	53
[Hexanoic acid]	0.6	0.6	0.45	1.028	N/A	1.40E-02	44
Hexanol	0.42	0.37	0.48	1.013	31	2.77E-02	60
[Hexanol]	0.42	0.37	0.48	1.013	25	1.30E-02	53
[Isoquinoline]	0.92	0	0.44	1.044	30	1.68E-02	37
[Methanol]	0.44	0.43	0.47	0.308	25	5.00E-04	53
Methanol	0.44	0.43	0.47	0.308	30	1.60E-03	54
[Beta-Naphthol]	1.08	0.61	0.4	1.144	25	2.79E-02	52
[m-Nitrophenol]	1.57	0.79	0.23	0.949	25	5.64E-03	52
[p-Nitrophenol]	1.72	0.82	0.26	0.949	25	5.58E-03	52
[Nonanol]	0.42	0.37	0.48	1.435	25	6.00E-02	53
[Octanoic acid]	0.6	0.6	0.45	1.31	N/A	2.50E-02	44
Octanol	0.42	0.37	0.48	1.295	22	5.20E-02	58
[Octanol]	0.42	0.37	0.48	1.295	25	5.20E-02	53
Octanol	0.42	0.37	0.48	1.295	30	6.10E-02	54
[Pentanoic acid]	0.6	0.6	0.45	0.888	N/A	2.00E-03	44
Pentanol	0.42	0.37	0.48	0.872	22	6.00E-03	58
[Pentanol]	0.42	0.37	0.48	0.872	25	6.00E-03	53
[Phenol]	0.89	0.6	0.3	0.775	25	8.22E-03	52

Phenol	0.89	0.6	0.3	0.775	37	1.95E-02	45
Phenol	0.89	0.6	0.3	0.775	22	1.55E-04	54
2-Phenylethanol	0.91	0.3	0.64	1.057	25	7.50E-03	50
Propanol	0.42	0.37	0.48	0.59	22	1.00E-03	58
Propanol	0.42	0.37	0.48	0.59	30	1.70E-03	16
[Propanol]	0.42	0.37	0.48	0.59	25	1.20E-03	53
[Resorcinol]	1	1.1	0.58	0.834	25	2.40E-04	52
Tetrachloroethylene	0.44	0	0	0.837	37	1.60E-02	87
[Thymol]	0.6	0.27	0.35	1.34	25	5.28E-02	52
Toluene	0.52	0	0.14	0.857	37	8.30E-01	42
Trichloroethylene	0.37	0.08	0.03	0.524	37	1.20E-01	87
Water	0.45	0.82	0.35	0.167	30	1.56E-03	66
Water	0.45	0.82	0.35	0.167	31	1.40E-03	60
Water	0.45	0.82	0.35	0.167	32	1.55E-03	88
Water	0.45	0.82	0.35	0.167	30	8.54E-04	61
Water	0.45	0.82	0.35	0.167	30	1.58E-03	55
Water	0.45	0.82	0.35	0.167	31	1.34E-03	68
[Water]	0.45	0.82	0.35	0.167	25	5.00E-04	53
Water	0.45	0.82	0.35	0.167	30	6.39E-04	59
[3,4-Xylenol]	0.86	0.56	0.39	1.057	25	3.60E-02	52

[a] All LSER parameters reported by Abraham and colleagues (28).

[b] The reported V_x have been divided by a factor of 100 to make the values comparable in magnitude to the other LSER parameters.

[c] Permeability coefficient values are from Table A1.

Table A6 Parameters to Modify MW with an Experimental Liquid Density

Compound[a]	log K_{ow}[b]	MW	ρ_{ref}(g cm^{-3})[c]	T_{ref} (°C)[d]	T_m (°C)[e]	T_c (°C)[f]	P_{cw} (cm hr^{-1})	T (°C)[g]	Ref. No.
* Benzene	2.13	78.1	0.8765	20	6	562	1.11E-01	31	49
* Benzene	2.13	78.1	0.8765	20	6	562	1.40E-01	37	87
[2,3-Butanediol]	-0.92	90.1	1.0030	20	8	N/A	5.00E-05	30	16
* [Butanoic acid]	0.79	88.1	0.9580	20	-6	628	1.00E-03	25	44
* [Butanol]	0.88	74.1	0.8100	20	-90	563	2.50E-03	25	53
* Butanol	0.88	74.1	0.8100	20	-90	563	3.00E-03	30	16
[2-Butanone]	0.29	72.1	0.8054	20	-87	N/A	4.50E-03	30	16
Chloroform	1.97	119.4	1.4830	20	-64	536	1.60E-01	37	87
[o-Chlorophenol]	2.15	128.6	1.2634	20	10	N/A	3.31E-02	25	52
[m-Cresol]	1.96	108.1	1.0340	20	12	706	1.52E-02	25	52
p-Cresol	1.94	108.1	1.0185	40	36	705	1.20E-01	37	42
[p-Cresol]	1.94	108.1	1.0185	40	36	705	1.75E-02	25	52
[Decanol]	4.57	158.3	0.8300	20	7	687	8.00E-02	25	53
* Ethanol	-0.31	46.0	0.7890	20	-114	514	3.00E-04	22	58
* [Ethanol]	-0.31	46.0	0.7890	20	-114	514	8.00E-04	25	53
* Ethanol	-0.31	46.0	0.7890	20	-114	514	3.17E-04	30	59
* [2-Ethoxy ethanol]	-0.32	90.1	0.9300	20	-70	N/A	2.50E-04	30	16
* [Heptanoic acid]	[2.41]	130.2	0.9180	20	-8	N/A	2.00E-02	25	44
* Heptanol	2.72	116.0	0.8220	20	-34	633	3.76E-02	30	16
* [Heptanol]	2.72	116.0	0.8220	20	-34	633	3.20E-02	25	53
* [Hexanoic acid]	1.92	116.2	0.9270	20	-3	N/A	1.40E-02	25	44
* Hexanol	2.03	102.2	0.8140	20	-45	611	2.77E-02	31	60
* [Hexanol]	2.03	102.2	0.8140	20	-45	611	1.30E-02	25	53
* [Isoquinoline]	2.08	129.2	1.0910	30	27	803	1.68E-02	30	37
* [Methanol]	-0.77	32.0	0.7910	20	-98	513	5.00E-04	25	53
* Methanol	-0.77	32.0	0.7910	20	-98	513	1.60E-02	30	54
* [m-Nitrophenol]	2.00	139.1	1.2800	100	97	N/A	5.64E-03	25	52
* [Nonanol]	4.26	144.0	0.8270	20	-5	671	6.00E-02	25	53
* [Octanoic acid]	3.05	144.2	0.9110	20	16	N/A	2.50E-02	25	44

Compound	log K_{ow}	MW	Density	Temp	MP	T_c	P		
* Octanol	3.00	130.2	0.8260	25	-16	653	5.20E-02	22	58
* [Octanol]	3.00	130.2	0.8260	25	-16	653	5.20E-02	25	53
* Octanol	3.00	130.2	0.8260	25	-16	653	6.10E-02	30	54
* [Pentanoic acid]	1.39	102.1	0.9390	20	-34	651	2.00E-03	25	44
* Pentanol	1.56	88.0	0.8140	20	-79	588	6.00E-03	22	58
* [Pentanol]	1.56	88.0	0.8140	20	-79	588	6.00E-03	25	53
* [Phenol]	1.46	94.1	1.0550	45	41	694	8.22E-03	25	52
* Phenol	1.46	94.1	1.0550	45	41	694	1.95E-02	37	45
* Phenol	1.46	94.1	1.0550	45	41	694	1.55E-04	22	54
* Propanol	0.25	60.0	0.8040	20	-126	537	1.00E-03	22	58
* Propanol	0.25	60.0	0.8040	20	-126	537	1.70E-03	30	16
* [Propanol]	0.25	60.0	0.8040	20	-126	537	1.20E-03	25	53
* Tetrachloroethylene	3.40	165.9	1.6230	20	-22	620	1.60E-02	37	87
* Toluene	2.73	92.1	0.8670	20	-95	592	8.30E-01	37	42
* Trichloroethylene	2.61	131.4	1.4640	20	-85	572	1.20E-01	37	87
* [2,4,6-Trichlorophenol]	3.69	197.5	1.4900	75	69	N/A	5.94E-02	25	52
* Water	-1.38	18.0	0.9957	30	0	647	1.56E-03	30	66
* Water	-1.38	18.0	0.9957	30	0	647	1.40E-03	31	60
* Water	-1.38	18.0	0.9957	30	0	647	1.55E-03	32	67
* Water	-1.38	18.0	0.9957	30	0	647	8.54E-04	30	61
* Water	-1.38	18.0	0.9957	30	0	647	1.58E-03	30	55
* Water	-1.38	18.0	0.9957	30	0	647	1.34E-03	31	68
* [Water]	-1.38	18.0	0.9957	30	0	647	5.00E-04	25	53
* Water	-1.38	18.0	0.9957	30	0	647	6.39E-04	30	59

a Compounds contained within brackets (e.g., [Aldosterone]) also appeared in the Flynn database. Those indicated with an asterisk to the left (e.g., benzene) were used in developing the LSER correlations.

b Reported log K_{ow} are taken from the Hansch Starlist 1, unless contained within brackets (e.g., for chloroxylenol [3.48]), in which case they were calculated using the Daylight software (2).

c Liquid density of the pure compound reported in the Handbook of Chemistry and Physics (46).

d Temperature of liquid density measurement (46).

e Melting point temperature (46) used to confirm that density values are for the liquid.

f Critical temperature of chemical (18). Chemicals for which T_c was not found are designated with N/A.

g Temperature of the reported permeability coefficient.

APPENDIX B. DOCUMENTATION OF PERMEABILITY COEFFICIENT DATA

This appendix contains specific documentation about the permeability coefficient data included in the FV, provisional, and excluded databases. Details are arranged alphabetically by the last name of the first author of the investigation.

As stated previously, all data were evaluated with respect to the requirement that they be measured at steady state. Preferably, experimental evidence of steady state (e.g., demonstration that absorption is linear in time) was shown. Sometimes steady state was not demonstrated experimentally, or, if it was, the results were not provided. Approximate estimates of the time required have been reported by Cleek and Bunge (89) at several different MW values: chemicals with MW = 50 require 30 min to reach steady state; chemicals with MW = 100 require 1 hour; and chemicals with a higher MW = 330 require 24 hours to reach steady state. Polarity is also important. Roberts et al. showed that hydrophilic chemicals with low MW require longer times to reach steady state (52). Judgment was necessary in applying these trends to validation of the permeability coefficient measurements on the basis of steady state.

Anderson et al., 1988 (20)

The measured permeability coefficients of fluocinonide and sucrose through isolated SC were taken directly from Table I, while permeability coefficients, for the remaining hydrocortisone esters, also through isolated SC, were taken from Table II. Identical values were included in the Flynn database (21). All chemicals studied were at least 10% nonionized in the vehicle (pH = 4). Although the exposure time was not precisely specified, we suspect that the values are at steady state (familiarity with authors). Figure 1 indicates that permeability coefficients, for several of the chemicals studied, are constant between 20 and 100 hours, indicating that excessive exposure times were not required for steady state to be achieved. Several different permeability coefficient values are reported for the hydrocortisone esters, but Anderson (32) recommended that the Table II permeability coefficients be used. Also, Anderson (32) noted a discrepancy between the permeability coefficient of fluocinonide in the experimental notebook (1.7×10^{-2} cm/h) and that reported in their Table I (1.7×10^{-3} cm/h) (20) and recommended that the permeability coefficient of fluocinonide be excluded. Anderson also recommended excluding their reported permeability coefficients of sucrose. He suggested that subsequent studies in the laboratory by Peck et al. (43) explored the permeability of polar permeants in much greater depth and solved some of the problems leading to variability in permeability coefficients for these compounds.

Anderson and Raykar, 1989 (42)

The reported permeability coefficients were taken directly from Table II. The structure for compound 1a in Table I is correct, and the name is more correctly given as α-(4-hydroxyphenyl)acetamide (32). α-(4-hydroxyphenyl)acetamide, 4-hydroxybenzyl alcohol, methyl 4-hydroxy phenylacetate, p-cresol, 4-hydroxyphenyl acetic acid, and toluene were more than 10% nonionized in the vehicle (pH = 4). The exposure times used in this study were not provided, but we believe that steady state was obtained (familiarity with authors, appear to be rapidly penetrating chemicals). While the measurement for toluene is based on skin from only one donor, all other compounds were measured on skin from at least two donors (32).

Barber et al., 1992 (66)

The human permeability coefficients of water and urea were taken without modification from Table 1. Urea was more than 10% nonionized in the vehicle. The exposure time was 8 hours, which was sufficient for steady state to be obtained. We learned the following from personal communication with Barber (90): (a) the temperature used in the water permeation studies was 30°C, (b) urea was delivered in a vehicle with pH = 7.1, (c) the membrane was isolated SC, and (d) the authors are not aware of any inaccuracies in the document.

Barry et al., 1985 (74)

Permeability coefficients for the five compounds were calculated directly from the saturated solution flux and solubility measurements provided in Table 2. Anisole, benzaldehyde, benzyl alcohol, and 2-phenyl ethanol were nonionized, and aniline was more than 10% nonionized in the vehicle. The exposure time was 9 hours, which was likely sufficient time for steady state to be obtained. These data were excluded because the skin barrier may have been damaged by contact with 50% aqueous ethanol in the receptor chamber.

Blank, 1964 (58)

The permeability coefficients of ethanol, propanol, pentanol, and octanol were taken from Table I. The exposure time was not provided, but the discussion in the results section indicates that they did examine the data for linearity (which was observed within 24 hours for aqueous vehicles and within 48 hours for nonpolar vehicles), indicating that steady state was achieved.

Blank et al., 1967 (16)

Permeability coefficients for ethyl ether, 2-butanone, 1-butanol, 2-ethoxyethanol, and 2,3-butanediol are taken from Table 1. Except for 2,3-butanediol, Blank et al. report permeability coefficients as a range of values. Values included in the FV database represent the midpoint of the values. Blank et al. report that the permeability coefficient for 2,3-butanediol was $< 0.5 \times 10^{-4}$ cm/h (perhaps this value should be moved to the excluded database).

Blank and McAuliffe, 1985 (49)

The permeability coefficient of benzene was taken from Table I without adjustment. The exposure time was 4 hours. As demonstrated in Fig. 1, this was sufficient time to reach steady state.

Bond and Barry, 1988 (60)

The permeability coefficients of hexanol and water were taken unaltered from the first paragraph of the discussion. The exposure time was 6 hours, which was sufficient for steady state to be attained.

Bond and Barry, 1988 (70)

The permeability coefficient of 5-fluorouracil was taken unaltered from the reported human measurement in Table 1. The exposure time was 60 hours, which was sufficient for steady state to occur, as shown in Fig. 1. From personal communication (69) we learned that the pH of the saturated 5-fluorouracil solution was 4.75, at which the authors believed 5-fluorouracil would be nonionized. However, based on calculations made using SPARC, the dominant species at this pH is a zwitterion. Calculations in SPARC indicate that 5-fluorouracil is ionized at all pH values.

Bronaugh et al., 1981 (64)

The permeability coefficient of N-nitrosodiethanolamine was taken directly from Table I. It is likely that this is the same value included (although incorrectly referenced) in the Flynn database (21). Figure 2 indicates that steady-state penetration occurred between 20 and 40 hours.

Bronaugh and Congdon, 1984 (47)

The permeability coefficient values of hair dyes were taken directly from Table I. Permeability coefficients from pure water were used for 2-amino-4-nitrophenol and 4-amino-2-nitrophenol since these compounds were mostly ionized in the borate buffer (pH = 9.7) used to investigate the other chemicals. At these pH values, all other chemicals were more than 10% nonionized in the vehicle. The exposure time was not specified, but Bronaugh (91) indicated that the exposure times were between 6 and 8 hours and that the vehicle did not evaporate.

Bronaugh et al., 1986 (67)

A permeability coefficient for water was taken as the average of values presented in Table II. This value is not meaningfully different from values reported in other tables in the same paper. The exposure time was 5 hours, which was sufficient for steady state to become established for water.

Chowhan and Pritchard, 1978 (82)

The permeability coefficient of naproxen was calculated by averaging the mean flux from the two experiments reported in Table I and dividing by the concentration, which was assumed to be 5000 μg mL^{-1} (not specified precisely). This procedure was also used by Flynn (21) as described in that paper (but incorrectly referenced to the compound 2-naphthol rather than naproxen). Naproxen was more than 90% ionized in the vehicle (pH = 6.5). The issue of steady state was never addressed. This value is listed in the excluded database because it was more than 90% ionized, and because the compound was applied to skin in an aqueous gel vehicle.

Cornwell and Barry, 1994 (71)

The permeability coefficient of 5-fluorouracil was calculated as the *n*-weighted arithmetic mean (that is, average values were multiplied by the number of samples) of the data in Table 2. This is despite suggestions by the authors that a geometric mean would be more representative of the data distribution. Further work will be necessary to determine whether geometric means are more representative and should be used in developing predictive equations. The drug 5-fluorouracil existed as a zwitterion in the saturated aqueous solution, without a measurable level of nonionized species [see discussion under (60)]. The exposure time was 36 hours, and as Fig. 3 illustrates, steady-state penetration was attained. According to Barry (69), the pH of saturated 5-fluorouracil solutions were 4.75.

Dal Pozzo et al., 1991 (63)

Permeability coefficients for the nicotinic acid derivatives were calculated from flux and concentration values reported in Table II. Two of the compounds (PEG

350 methyl nicotinate and PPG 425 nicotinate) were a polydispersed mixture and so were excluded from the FV database. No flux measurement was given for MEG-methyl nicotinate. Flux of methyl nicotinate and ethyl nicotinate were measured at two concentrations each. The two calculated permeability coefficients for methyl nicotinate were averaged, as were the two permeability coefficients for ethyl nicotinate. All compounds were essentially nonionized in the vehicle. The exposure time was 5 hours, which was deemed to be long enough satisfactorily to approximate steady state for these chemicals.

Dick and Scott, 1992 (61)

Permeability coefficient values for water, mannitol, and paraquat were digitized from Fig. 4. The discrepancy in values shown in Fig. 4 and Table 1 appear to be an error in the typesetting of Table 1. We were unsuccessful at contacting the authors, so we included the values we thought were most likely correct. Water and mannitol were not ionized in the unbuffered vehicle. Paraquat (corresponding anion not specified) exists naturally as a divalent cation and was placed in the provisional database. Paraquat was likely applied as a dichloride salt, since this was the form used by Scott et al. (59). The exposure times were not specified, but it was likely that steady state was reached for these hydrophilic chemicals.

Galey et al., 1976 (55)

Permeability coefficient values for the chemicals investigated by Galey and others were taken directly from values reported in Table II (given in cm s^{-1}). Based on pK$_a$ values calculated in SPARC, water, ouabain, and estradiol were significantly less than 90% ionized in the vehicle (pH $=$ 7), while amphetamine was more than 90% ionized in the vehicle at the same pH. The exposure time was not given, although the authors, who have had extensive experience with membrane transport, report no time dependence in their permeability coefficient measurements.

Hadgraft and Ridout, 1987 (37)

The permeability coefficients for chemicals investigated by Hadgraft and others were calculated from resistances reported in Table 4 ($P_w = 0.36/R_s$, cm s^{-1}). The same conversions were performed by Flynn (21) to incorporate these chemicals into the Flynn database. All the chemicals studied (barbitone, nicotine, phenobarbitone, hydrocortisone, butobarbitone, amylobarbitone, isoquinoline, and salicylic acid) were at least 10% nonionized at the reported measured pH. The authors suggest that steady-state flux values were reported, and a typical plot for phenobarbitone (Fig. 3) shows that data between 70 and 120 hours of exposure were used to determine the permeability coefficient. The permeability coefficient for hydrocortisone was not included in the FV database because the vehicle contained 5% aqueous ethanol, which may have altered the skin.

Harada et al., 1993 (65)

The permeability coefficient of salicylic acid was obtained by averaging the mean flux measurements (made at different pH values), through human breast skin, that are reported in Table 1 and dividing by a concentration of 500 μg mL^{-1}. Flux values for human neck skin were not used. At the most acidic extreme (i.e., pH = 2.0), the skin may have been damaged, so this measurement was not included. The average flux at each of the pH values at which the compound was < 90% ionized (i.e., 3.0, 3.5), were equally weighted to obtain an overall average. (At pH = 3, an average flux was calculated from two fluxes that were reported.) The exposure time, reported as 72 hours, was sufficient (as shown in Fig. 3) to attain steady-state penetration.

Jolicoeur et al., 1992 (33)

The permeability coefficient of etorphine was taken from Table I and also appears in Table II. The experiment was conducted at pH = 7.3 in TRIS buffer, at which the compound was more than 90% ionized based on pK$_a$ calculated in SPARC. Data collected at times between 4 and 24 hours were used to determine the permeability coefficient values, and this probably represents steady-state penetration. It appears that these measurements were made on cadaver skin from only subject.

Kubota and Maibach, 1993 (51)

The permeability values for betamethasone and betamethasone-17-valerate were calculated ($P_{cw} = 1/R_{sc}$) from the stratum corneum resistances (i.e., R_{sc}) reported in Table I of this reference. The authors calculated R_{sc} from permeability (diffusion cell) experiments of split-thickness skin and dermis combined with partition coefficient experiments of whole skin, dermis, and stratum corneum by describing skin as a series of resistances. Both compounds were nonionized. The exposure time of 72 hours was probably sufficient to attain steady-state penetration.

Liu et al., 1994 (56)

The average of the three values of the SC permeability coefficient reported in Table 4 (converted from cm s^{-1} to cm h^{-1}) was used for β-estradiol (designated as E2). The authors calculated stratum corneum permeability coefficients from permeability (diffusion cell) experiments of dermis, split-thickness skin, and split-thickness skin with the stratum corneum stripped, by describing skin as a series of resistances. β-estradiol is nonionized in these experiments. The permeation cell experiments were conducted for only 4 hours, which may not be long enough to achieve steady state. Because time-course data were not reported in the paper, we are unable to definitively assess whether the permeability measurements are steady-state values. Despite this, we included these permeability measurements in the FV database, judging that the measured values were not greatly different from the steady-state values.

Lodén, 1986 (78)

The permeability coefficient of formaldehyde was calculated from the rate of absorption in Table 1 using a 3.7% v/v formaldehyde solution (10% v/v solution of formalin containing 37% formaldehyde and 10–15% methanol). This value was not included in the FV database because formalin may have altered the skin. The compound was more than 10% nonionized in the vehicle (assuming aqueous properties). Exposure times of 20 hours were used in these experiments, and data shown in Fig. 4 indicate that steady-state penetration was achieved.

Megrab et al., 1995 (57)

The permeability coefficient of β-estradiol from saturated solution was calculated from the reported flux (0.015 ± 0.004 μg cm^{-2} h^{-1}) and a solubility ($C_w = 3.48$ μg mL^{-1}) obtained through personal communication (69). Neither this solubility nor the permeability coefficient value were presented in the original publication. β-estradiol does not ionize. The exposure time was at least 25 hours (69).

Michaels et al., 1975 (24)

Permeability coefficients for the chemicals investigated by Michaels and coworkers were taken directly from Table 2. These values are also those included in the Flynn database (21). All chemicals were more than 10% nonionized in the aqueous solution at the specified pH. Information about steady state was never discussed in the experimental portion of the paper. Unless very long exposure times were used, unsteady-state effects are likely to reduce permeability coefficients observed for the largest molecular weight chemicals: ouabain (MW = 584.64) and digitoxin (MW = 764.92).

Morimoto et al., 1992 (25)

The reported permeability coefficients for drugs measured by Morimoto and colleagues were digitized from Fig. 5 in that publication. Buffers were not used, and the chemicals came to a natural state of ionization (pH was measured and reported in Table 3). The ionization of these compounds was calculated in SPARC. Lidocaine, diclofenac sodium, morphine, isoprenaline hydrochloride, and dopamine hydrochloride were more than 90% ionized under experimental conditions. The remaining chemicals (antipyrine, nicorandil, aminopyrine, cyclobarbitone, isosorbide dinitrate, ketoprofen, indomethacin, flurbiprofen, and ibuprofen) were more than 10% nonionized. Additionally, 5-fluorouracil and levodopa existed in a charged but net neutral form (i.e., they were zwitterions). For some of these chemicals, the presence of dermis (750 μm total) might have influenced (i.e., lowered) the observed permeability coefficient value, but no chemicals were excluded for this reason. The exposure time was 10 hours, which closely approximated steady-state penetration for these compounds (see Fig. 2).

Nakai et al., 1995 (unpublished results) (87)

The permeability coefficients of atrazine, benzene, chloroform, tetrachloroethylene, and trichloroethylene were obtained from the averages reported in the tables (the tables were not labeled in these unpublished results). All compounds were nonionized in an unbuffered vehicle, which, according to a personal communication from Nakai (48), had a pH of 7.0 based on pH-sensitive strips. The exposure times for trichloroethylene and tetrachloroethylene was 8 hours, while that for atrazine was 24 hours. These times were sufficient to observe steady-state permeability coefficients for these chemicals.

Parry et al., 1990 (13)

The permeability coefficient of benzoic acid through isolated stratum corneum, determined with simple linear regression, was taken from Table III. This coefficient was determined with nonionized benzoic acid, and no correction was necessary. The exposure time was 4 hours, and steady-state penetration was demonstrated in Fig. 3.

Peck et al., 1995 (43)

The permeability coefficients at 27°C and 39°C were taken from Table 1 (for urea and mannitol), Fig. 4 (for tetraethylammonium bromide), and Figs. 5 and 8 (for corticosterone). At both temperatures the permeability coefficients of corticosterone and tetraethylammonium bromide were digitized. Urea, mannitol, and corticosterone were nonionized, while tetraethylammonium bromide is an ion at the experimental conditions (pH = 7.4). The exposure time was 24 hours, for which steady state was probably reached for these compounds.

Rigg and Barry, 1990 (68)

The permeability of water was taken as the *n*-weighted (that is, average values were multiplied by the number of samples) average of Table 1 values. Two separate permeability coefficient values are listed in the provisional database for 5-fluorouracil, since different experimental protocols were involved in the determination of each. Skin was hydrated for 12 hours before the permeability coefficient of 5-fluorouracil reported in Table II was measured. Skin was not hydrated before the permeability coefficient of 5-fluorouracil reported in Table III was measured. The drug 5-fluorouracil existed as a zwitterion in the saturated aqueous solution, without a measurable level of nonionized species, based on SPARC calculations. The exposure time for water (6 hours) was sufficient for steady-state penetration to be obtained. In both experiments on 5-fluorouracil, exposure time was 24 hours to ensure steady state. From personal communication with Barry (69), we have determined that the pH attained by 5-fluorouracil in saturated solution was typically 4.75.

Ritschel et al., 1989 (77)

The permeability coefficients of coumarin, griseofulvin, and propranolol were taken from Table I. The vehicle volume was small enough that concentration of the penetrating chemical decreased during the experiment. Coumarin and griseofulvin were essentially nonionized in the vehicle, while propranolol was more than 90% ionized in the vehicle based on calculations in SPARC. The exposure time was 48 hours, which is sufficient time for attaining steady state. The hypodermis was removed, but the final skin thickness was not identified.

Roberts et al., 1977 (52)

The permeability coefficients of phenolic compounds were taken from Table 1 with an adjustment of dimensions from cm min^{-1} to cm h^{-1}. Identical values appear in the Flynn database (21). Personal correspondence with Roberts (92) confirmed that these values are more representative than those presented in Table 3, which were measured on skin from a single donor. The pH was unspecified, but all compounds were found to be nonionized in neutral (pH = 7) or slightly acidic (pH = 6) distilled water. The exposure time was not precisely specified although graphical evidence exists (e.g., exposure times shown in Figs. 1 and 4 are 250 and 500 min, respectively).

Roy and Flynn, 1989 (5)

The permeability coefficients of morphine, hydromorphone, codeine, and meperidine are presented in Table II. The permeability coefficients of fentanyl and sufentanil in solution at various pH values are reported in Table IV. The chemicals morphine, hydromorphone, codeine, and meperidine were more than 90% ionized in the vehicle (pH = 7.4), based on SPARC calculations, so these permeability coefficients are listed in the excluded database. Only fentanyl and sufentanil are more than 10% nonionized. The exposure times used for determining the permeability coefficients can be deduced from Figs. 1 and 2: for morphine, 48 hours; for hydromorphone, 50 hours; for codeine, 22 hours; for fentanyl, 8 hours; for sufentanil, 8 hours; and for meperidine, 8 hours. They suggest in this paper that the permeability coefficients for fentanyl and sufentanil were measured on skin that was 3 to 4 times less permeable to fentanyl and sufentanil than typical skins (5). Permeability coefficients for fentanyl and sufentanil are reported again in a subsequent investigation by the same authors (4), who use skin from several subjects. We included these later measurements for fentanyl and sufentanil (4) in the FV database, and did not include the permeability coefficients reported in this reference for fentanyl and sufentanil in any database. It is worth mentioning that Flynn (one of the authors) decided to include the Table II values of fentanyl and sufentanil from the first investigation (5), rather than those of a subsequent investigation (4), in the Flynn database (21).

Roy and Flynn, 1990 (4)

The permeability coefficient values for fentanyl and sufentanil in the FV database were obtained from this reference [see also documentation on Roy and Flynn, 1989 (5)]. Average permeability coefficient values for fentanyl and sufentanil are reported in Table V for what appears to be a large number of measurements from many different skin specimens. The pH of this table is reported to be 7.4, although it appears that several measurements at pH = 8 from Table II and Table III appear to be included. When asked, Flynn was not hopeful that it would be possible to resolve the apparent discrepancy with certainty (96). We assumed that the pH of 7.4 was correct for most of the measurements and divided the average permeability coefficients from Table V by the fraction nonionized at pH = 7.4. Both fentanil and sufentanil are at least 10% nonionized at either pH = 7.4 or pH = 8 based on calculations in SPARC. Roy and Flynn (4,5) reported pK_a values (8.9 for fentanyl and 8.5 for sufentanil at 37°C) that are quite different from the SPARC calculated pK_a values we used (i.e., 7.0 and 6.2 for fentanyl and sufentanil after adjustment to 37°C, respectively), but the source of these values is unknown. The pK_a values that we calculated in SPARC appear to be more consistent with the data than the pK_a values reported by Roy and Flynn (4,5), although neither are precise. Insufficient information is given to know precisely the exposure time, although the authors report that 8 hours was sufficient to reach steady state, for these compounds (5,38).

Permeability coefficients measured at a range of pH for fentanyl and sufen-

Table B1 Calculation of Permeability Coefficients for Nonionized Species of Fentanyl and Sufentanil

pH	Fentanyl			Sufentanil		
	f_{ni} [a]	$P_{cw,obs}$	P_{cw} [b]	f_{ni} [a]	$P_{cw,obs}$	P_{cw} [b]
2.88	8×10^{-5}	0.0003	N/A [c]	5×10^{-4}	0.00046	N/A
5.08	0.013	0.0013	N/A	0.078	0.0025	N/A
6.02	0.103	0.0051	0.0495	0.426	0.0062	0.0146
6.95	0.494	0.0071	0.0144	0.863	0.0101	0.0117
7.43	0.747	0.0127	0.0170	0.950	0.0157	0.0165
7.95	0.907	0.0224	0.0247	0.984	0.0231	0.0235
8.52	0.973	0.0276	0.0284	0.996	0.0298	0.0299
9.04	0.992	0.0349	0.0352	1.0	0.0345	0.0345
9.37	0.996	0.0329	0.0330	1.0	0.0337	0.0337

[a] Calculated using 37°C pK_a values (6.96 for fentanyl and 6.15 for sufentanil), which were calculated at 25°C in SPARC and adjusted to 37°C.
[b] Calculated by dividing $P_{cw,obs}$ (observed permeability coefficient) by f_{ni}.
[c] N/A means not adjusted.

tanil and reported in Table IV are incorporated in Fig. 1 in this chapter. Table B1 documents how permeability coefficient values for the nonionized species (P_{cw}) used in Fig. 1 were determined.

In Fig. 1a the permeability coefficient values of the nonionized species were calculated as the average of all permeability coefficients measurements when more than 99% of the compound was nonionized. For fentanyl the nonionized permeability coefficient is 0.0339 cm h^{-1}, which is the average of measurements at pH = 9.04 and 9.37, and for sufentanil the nonionized permeability coefficient is 0.0327 cm/h, which is the average of permeability coefficient measurements at pH = 8.52, 9.04, and 9.37. Figure 1b compares the observed permeability coefficients with nonionized permeability coefficients that were calculated as the average of adjusted measurements (i.e., $P_{cw} = P_{cw,obs}/f_{ni}$) for which at least 10% of the compound was nonionized. For fentanyl and sufentanil the calculated nonionized permeability coefficients of 0.029 cm h^{-1} and 0.024 cm h^{-1}, respectively, were determined by averaging all of the tabulated P_{cw} that are reported when the pH was between 6.02 and 9.37.

Roy et al., 1994 (38)

The human skin permeability coefficient for morphine was taken directly from Table 1. This chemical was more than 90% ionized at the solution conditions (pH = 7.5) based on calculations in SPARC. Insufficient information is available to determine whether the permeability coefficient of morphine is based on steady- or unsteady-state data. In a prior investigation (5), 48 hours was used as the exposure time.

Roy et al., 1995 (79)

The permeability coefficient value for ketorolac acid is the arithmetic average of the permeability coefficients for the R- and S-enantiomers and the racemic (50: 50) mixture reported in Table 2. The exposure time was at least 30 hours, which was probably sufficient to attain steady state. Ketorolac acid was less than 10% ionized in the vehicle (pH = 2.1) based on calculations in SPARC. The concentration of penetrant in the vehicle decreased during the experiment (by 12% for S-enantiomer, 14% for R-enantiomer, and 19% for racemic mixture). These permeability coefficients are listed in the excluded database because the extremely low pH may have altered the skin.

Sato et al., 1991 (62)

The permeability coefficient of nicorandil was taken from Table I. The experimental procedure for the permeability studies was referenced to an earlier publication (93). Exposure times of approximately 32 hours were used, which was probably sufficient to attain steady state. A saturated aqueous solution of nicorandil was used, but neither the pH nor the concentration of nicorandil were reported. Using a concentration of 0.188 mol L^{-1} (25) for saturated solutions of nicorandil at 37°C and a pK_a value calculated in SPARC, we estimated that pH = 8 for this experiment (see Table A4). At pH = 8, nicorandil is not ionized for the pK_a calculated in SPARC.

Scheuplein et al., 1969 (40)

Permeability coefficients were taken directly from Table 1. All compounds were nonionized. Based on Fig. 2, exposure times were long enough to attain steady state.

Scheuplein and Blank, 1973 (53)

Permeability coefficients were taken directly from Table 1. Exposure times were not given, but based on the authors' discussions of membrane processes in this and their many other papers, we have expect that the measurements were probably at steady state. These chemicals (alcohols) did not ionize.

Scott et al., 1991 (59)

The permeability coefficient values for water, ethanol, mannitol, and paraquat dichloride were taken as the averages reported in Table I. Paraquat was applied as the dichloride salt (paraquat exists as a divalent cation in its natural state). Except for paraquat, $f_{ni} > 0.1$ in the vehicle. Paraquat is a divalent strong electrolyte, which will be fully ionized. The exposure time was 6 hours, which should have been sufficient for water and ethanol to reach steady state. Mannitol and paraquat, which are larger and more polar chemicals, may require this long or longer to reach steady state. The fact that paraquat penetrates as an ion excludes it from the FV database, and it is listed in the provisional database.

Siddiqui et al., 1989 (41)

Permeability coefficients of seven steroids were taken from Table I without modification. The chosen coefficients are based on a fit to the single membrane diffusion model (DM I) rather than the shunt-membrane model proposed (DM II). By personal communication (92), we have confirmed that the vehicle was water, that skin samples from several donors were used to generate the data, and that the temperature at which the measurements were performed was 25°C. The steroids were nonionized. The exposure time of 80 hours was adequate time to reach steady state, as evidenced by Figs. 2 and 3.

Singh and Roberts, 1994 (81)

The lidocaine permeability coefficient is the average of three values reported in the results and discussion section (i.e., 0.0035, 0.0050, and 0.0041 cm/h). Lidocaine was more than 90% ionized in the vehicle (pH = 7.4) based on SPARC calculations. The exposure time was 6 hours, which was probably sufficient to approach or reach steady state.

Singh and Roberts, 1994 (45)

The permeability values for the chemicals investigated by Singh and Roberts were taken from Table 1. As confirmed by Roberts (92), the value for piroxicam was incorrectly represented in that table and the correct value is 0.0034 cm/h, which is consistent with the 50% ionized value and with Fig. 2. According to the authors, diethylamine salicylate ionizes in solution, so that the conjugate base of salicylic acid is actually introduced into solution. Since the penetrating species is nonionized salicylic acid, the log K_{ow} and MW for salicylic acid were used in the regressions. Permeability coefficient values for all chemicals were calculated as 2 times the flux reported when the pH = pK$_a$ (neither pH nor pK$_a$ were specified in the paper). Since $f_{ni} = 0.5$ at pH $= pK_a$, the criterion that $f_{ni} > 0.1$ is met. The authors claim that the data were collected at steady state, although exposure times were not specified.

Southwell et al., 1984 (54)

The permeability coefficient of phenol was calculated as the n-weighted average (that is, average values were multiplied by the number of samples) of the mean flux values reported in Table 1 divided by the concentration (1 g 100 mL^{-1}). Permeability coefficients for methanol, octanol, and caffeine were taken directly from the steady-state diffusion experiment results presented in Table 2. Phenol, methanol, and octanol were not ionized in solution, and based on calculations in SPARC, $f_{ni} > 0.1$ for caffeine, which was buffered at pH $= 7.4$ (69). Phenol was studied with an exposure time of 10 hours, and octanol was studied with an exposure time of 6 hours. As demonstrated in Figs. 3 and 4, penetration of these chemicals was probably at steady state. Although exposure times were not reported for methanol and caffeine, based on the authors' treatment of phenol and octanol, we expect that the measurements for methanol and caffeine were also at steady state.

Swarbrick et al., 1984 (75)

Permeability coefficients were reported by Swarbrick et al. at pH values of 5, 6, and 7 in Table I. At all three pH values, all four carboxylic acids were more than 90% ionized, based on calculations in SPARC, and consequently, these measurements are listed in the excluded database. The exposure time was between 48 and 60 hours.

Williams and Barry, 1991 (72)

The permeability coefficient of 5-fluorouracil was calculated as the average of the control values in Table I (i.e., without penetration enhancer). The drug 5-fluorouracil existed as a zwitterion in the saturated aqueous solution, which had a pH of approximately 4.75 (69), based on calculations in SPARC. In this study the exposure time was 36 hours. The results plotted in Figure 3 indicate that this exposure time is sufficient to ensure steady-state penetration of the drug.

APPENDIX C. LSER PARAMETERS

Quantitative structure–activity relationships (QSAR) are models that relate the structural and electronic features of a molecule to its macroscopic properties. Hansch and Leo (94) have found that many diverse phenomena can be described in terms of a simplified QSAR that uses hydrophobicity alone. A more general approach for relating the hydrophobicity interaction is to analyze free energy based properties in terms of indexes of solute–solvent interactions (26,27). Equations conforming to this formalism are known as linear solvation energy relationships (LSER), and they have been described well by Cramer and colleagues (26), Famini and Peuski (27), and Abraham and coworkers (28). There are many similarities between QSAR and LSER, and indeed they both have roots in more general linear free energy relationships (LFER).

The basic premise of LSER is that a given property can be defined by a linear relationship of two different types of terms that describe the hydrogen bonding and polarizability–dipolarity of chemical interactions. The parameters in LSER analysis, called solvatochromic parameters (because they were historically determined by measuring UV spectral shifts for select dyes), serve as markers of the exoergic solute–solvent interaction phenomena in these separate areas. In general, a molar volume term is also added, and this approach has found particular use in calculation of solubility and partition coefficients (26).

Originally developed to describe solubility of a solute in various solvents (29,95), LSER have found more applicability in the reciprocal sense of correlating a free energy dependent property exhibited by a class of solutes in the same solvent. In this application, the solvatochromic parameters correspond to the solutes rather than the solvent.

The solvatochromic parameters have the following physicochemical interpretation: α is the effective hydrogen-bond acidity, β is the effective hydrogen-bond basicity, π is the solute dipolarity/polarizability, V_x [$\times$ 100 cm^3 mol^{-1}] is the characteristic volume of McGowans (31) calculated from molecular structure alone and is independent of intermolecular forces such as hydrogen bonding. When LSER are applied in this fashion, it is assumed that the difference in the chemical potential that drives the solute through the process, λ, results from van der Waals forces, size effects, polarizability, and the preference of a solute to act as an electron donor and an electron acceptor in hydrogen bonds.

Any site with unshared electrons is a potential hydrogen bond acceptor, although the more strongly basic and the less polarizable the acceptor site, the stronger will be the hydrogen bond. Solvents that are protic are good hydrogen donors, and those that are aprotic may or may not be good hydrogen acceptors. In general, oxygen–hydrogen and nitrogen–hydrogen bonds readily associate in hydrogen bonding, but carbon–hydrogen bonds are too weakly acidic to form hydrogen bonds. As an example, negative ions, which are good H bond acceptors, are strongly solvated by protic solvents but are less soluble in aprotic solvents. By contrast, protic solutes will ordinarily interact by hydrogen bonding with protic solvents.

Polarizability is a measure of the ease with which the electron distribution of a molecule is distorted. It determines the attractive forces (van der Waals forces) that arise between two molecules in close proximity. The electron clouds distort, and an instantaneous dipole arises, which causes attraction. In another context, polarizability measures how well a molecule can stabilize a charge or dipole by means of its dielectric effect. The polarizability term (π) can be divided by the molecular volume to put polarizability on a size-independent basis.

Unlike molar volumes, the volume term of McGowan (V_x) (31) has been calculated so that it is independent of hydrogen bonding interactions. The volume term contributes to estimation of permeability coefficients in two ways. First, as the volume of a chemical increases, its diffusion rate within the stratum corneum will decrease. Second, the solute size contributes to the partitioning of the solute between vehicle and skin due to free energies associated with cavity formation in the skin and in the aqueous vehicle (cavity formation energies in water are larger than in skin). The effects that favor partitioning into the SC are both the solute stratum corneum interactions caused by size-dependent van der Waals dispersion forces and the difference in free energy costs associated with the formation of a cavity necessary to accommodate the solute in either the skin or the vehicle.

The solvatochromic parameters used in this work were from Abraham et al. (28), who determined them by averaging multiple normalized solvent effects on a variety of chemical properties involving many varied types of indicators. These experimentally determined descriptors are now available for well over 1000 solutes (28).

GLOSSARY

a	= Parameter determined by regression of model equation to experimental data
$\hat{a}, \hat{b}$	= Parameters relating K_{cw} to K_{ow}, Eq. (10)
b	= Parameter determined by regression of model equation to experimental data
B	= Parameter measuring the SC/VE permeability ratio
c	= Parameter determined by regression of model equation to experimental data
C_w	= Solute concentration in water
d_1, d_2	= Parameter determined by regression of model equation to experimental data
D_c	= Effective diffusivity of the absorbing chemical in the SC
D_e	= Effective diffusivity of the absorbing chemical in the VE
D_0	= Diffusion constant of hypothetical chemical having zero MV
e_1, e_2	= Parameter determined by regression of model equation to experimental data

E_a	= Activation energy in an Arrhenius type relationship, Eq. (13)
f_{ni}	= Fraction of the total chemical dose that is nonionized in the aqueous solution
F-ratio	= Statistic measuring the predictive power of the regression
ΔH	= Heat of ionization, Table A4
K_a	= Acid dissociation constant
K_{ce}	= Partition coefficient between the SC and the VE for the absorbing chemical
K_{cw}	= Partition coefficient between the SC and water for the absorbing chemical
K_{ew}	= Partition coefficient between the VE and water for the absorbing chemical
K_{ow}	= Octanol–water partition coefficient of the absorbing chemical
$K_{ow,obs}$	= Octanol–water partition coefficient observed for a partially ionized absorbing chemical
K_{ow}^t	= Value of K_{ow} at which the piecewise linear regression changes slope and distinguishes the hydrophilic compounds from the lipophilic compounds
K_T	= Coefficient of thermal expansion
L_c	= Effective thickness of the SC
L_e	= Effective thickness of the VE
m	= Number of different compounds included in the regression equation
MV	= Molecular volume of the absorbing chemical
MW	= Molecular weight of the absorbing chemical
n	= Number of data points included in the regression equation
p	= Number of fitted parameters in the regression equation
P_{cw}	= Steady-state permeability coefficient through the SC from water
P_{cw}^o	= Parameter determined by regression of model equation to experimental data
$P_{cw,calc}$	= Steady-state permeability coefficient through the SC from water calculated using Eq. T1
$P_{cw,obs}$	= Observed experimental steady-state permeability coefficient from water
P_{ew}	= Steady-state permeability coefficient through the VE from water
P_w	= Steady-state permeability coefficient through the SC–VE composite membrane from water
pH	= Negative base 10 logarithm of the hydrogen ion molarity, $-\log_{10}[H^+]$
pK_a	= Negative base 10 logarithm of the acid dissociation constant, $-\log_{10} K_a$
R	= Ideal gas constant

r^2	= Goodness of fit parameter
r^2_{adj}	= Goodness of fit parameter adjusted for the number of fitted parameters and the number of data points
RMSE	= Root mean square error of the regression
SC	= Stratum corneum
SG	= Specific gravity based on density of liquid water at 4°C (i.e., 1.000 g cm^{-3})
T	= Absolute temperature (Kelvin)
T_c	= Critical temperature of the absorbing chemical
T_{ref}	= Absolute temperature (Kelvin) of the liquid density measurement
$U(a - b)$	= Unit step function; $U = 0$ if $a < b$, and $U = 1$ if $a \geq b$
V_x	= Characteristic volume of McGowans (LSER parameter)
VE	= Viable epidermis

Greek

α	= Hydrogen bond acidity of the absorbing chemical (LSER parameter)
β	= Hydrogen bond basicity of the absorbing chemical (LSER parameter)
γ_j^*	= Parameter relating D_c to MV of the absorbing chemical, $j = 1$ or 2
γ_j	= Parameter relating D_c to MW of the absorbing chemical, $j = 1$ or 2
λ	= General linear free energy based property
π	= Dipolarity/polarizability of the absorbing chemical (LSER parameter)
ρ	= Pseudo liquid density of the absorbing chemical
ρ_{ref}	= Experimental liquid density of the absorbing chemical
σ	= Parameter representing MV/MW.

Superscripts

H	= Hydrophilic chemicals
L	= Lipophilic chemicals
*	= Regression constants not statistically different from zero at 95% confidence

REFERENCES

1. C Hansch, A Leo, D Hoekman. Exploring QSAR: Hydrophobic, Electronic, and Steric Constants. Washington, DC: American Chemical Society, 1995.
2. PCModel. Ver. 4.2, Daylight Chemical Information Systems, Inc., Mission Viejo, CA, 1995.
3. M Sznitowska, B Berner, HI Maibach. In vitro permeation of human skin by multipolar ions. Int J Pharm 99:43–49, 1993.

4. SD Roy, GL Flynn. Transdermal delivery of narcotic analgesics: pH, ana-tomical, and subject influences on cutaneous permeability of fentanyl and sufentanil. Pharm Res 7:842–847, 1990.

5. SD Roy, GL Flynn. Transdermal delivery of narcotic analgesics: compara-tive permeabilities of narcotic analgesics through human cadaver skin. Pharm Res 6:825–832, 1989.

6. RP Schwarzenbach, PM Gschwend, DM Imboden. Environmental Organic Chemistry. New York: John Wiley 1993.

7. SPARC (SPARC Performs Automated Reasoning in Chemistry): An Expert System for Estimating Physical and Chemical Reactivity. Ver. Windows Prototype Version 1.1, US EPA (Ecosystem Research Division) and Uni-versity of Georgia, Athens, GA. Athens GA, 1995.

8. SH Hilal, SW Karickhoff, LA Carreira. A rigorous test for SPARC's chemi-cal reactivity models: estimation of more than 4300 ionization pK_a's. QSARDI 14:348–355, 1995.

9. JM Smith, HC Van Ness. Introduction to Chemical Engineering Thermody-namics, 4th ed. New York: McGraw-Hill, 1987.

10. HA Sober, ed. Handbook of Biochemistry. Cleveland, OH; Chemical Rub-ber Company, 1968.

11. F Brescia, J Arents, H Meislich, A Turk. Fundamentals of Chemistry, 3rd ed. New York: Academic Press, 1975.

12. J Crank. The Mathematics of Diffusion. London: Oxford University Press, 1975.

13. GE Parry, AL Bunge, GD Silcox, LK Pershing, DW Pershing. Percutaneous absorption of benzoic acid across human skin. I. In vitro experiments and mathematical modeling. Pharm Res 7:230–236, 1990.

14. RO Potts, RH Guy. Predicting skin permeability. Pharm Res 9:663–669, 1992.

15. M Mulder. Basic Principles of Membrane Technology. Boston: Kluwer Ac-ademic, 1991.

16. IH Blank, RJ Scheuplein, DJ Macfarlane. Mechanism of percutaneous ab-sorption III. The effect of temperature on the transport of non-electrolytes across the skin. J Invest Dermatol 49:582–589, 1967.

17. JR Welty, CE Wicks, RE Wilson, eds. Fundamentals of Momentum, Heat, and Mass Transfer. New York: John Wiley, 1986.

18. RC Reid, JM Prausnitz, BE Poling. The Properties of Gases and Liquids, 4th ed. New York: McGraw-Hill, 1987.

19. PM Elias. Lipids and the epidermal permeability barrier. Arch Dermatolog Res 270:95–117, 1981.

20. BD Anderson, WI Higuchi, PV Raykar. Heterogeneity effects on perme-ability–partition coefficient relationships in human stratum corneum. Pharm Res 5:566–573, 1988.

21. GL Flynn. Physicochemical determinants of skin absorption. In: TR Gerrity, CJ Henry, eds. Principles of Route-to-Route Extrapolation for Risk Assessment. New York: Elsevier, 1990, pp 93–127.
22. GB Kasting, RL Smith, ER Cooper. Effect of lipid solubility and molecular size on percutaneous absorption. In: B Shroot, H Schaefer, eds. Skin Pharmacokinetics. Basel: Karger, 1987, pp 138–153.
23. GB Kasting, RL Smith, BD Anderson. Prodrugs for dermal delivery: solubility, molecular size, and functional group effects. In: KB Sloan, ed. Prodrugs: Topical and Ocular Drug Delivery. New York: Marcel Dekker, 1992, pp 117–161.
24. AS Michaels, SK Chandrasekaran, JE Shaw. Drug permeation through human skin: theory and in vitro experimental measurement. AIChE J 21:985–996, 1975.
25. Y Morimoto, T Hatanaka, K Sugibayashi, H Omiya. Prediction of skin permeability of drugs: comparison of human and hairless rat skin. J Pharm Pharmacol 44:634–639, 1992.
26. CJ Cramer, GR Famini, AH Lowrey. Use of calculated quantum chemical properties as surrogates for solvatochromic parameters in structure–activity relationships. Acc Chem Res 26:599–605, 1993.
27. GR Famini, CA Penski. Using theoretical descriptors in quantitative structure activity relationships: Some physicochemical properties. J Phys Org Chem 5:395–408, 1992.
28. MH Abraham, HS Chadha, GS Whiting, RC Mitchell. Hydrogen bonding part 32. An analysis of water–octanol and water–alkane partitioning and the Δ log P parameter of Seiler. J Pharm Sci 83:1085–1100, 1994.
29. N El Tayar, RS Tsai, B Testa, PA Carrupt, A Leo. Partitioning of solutes in different solvent systems: the contribution of hydrogen-bonding capacity and polarity. J Pharm Sci 80:590–598, 1991.
30. MH Abraham, HS Chadha, RC Mitchell. The factors that influence skin penetration of solutes. J Pharm Pharmacol 47:8–16, 1995.
31. MH Abraham, JC McGowan. The use of characteristic volumes to measure cavity terms in reversed phase liquid chromatography. Chromatographia 23:243–246, 1987.
32. BD Anderson. Department of Pharmaceutics and Pharmaceutical Chemistry, University of Utah, personal communication, 1995.
33. LM Jolicoeur, MR Nassiri, C Shipman, HK Choi, GL Flynn. Etorphine is an opiate analgesic physicochemically suited to transdermal delivery. Pharm Res 9:963–965, 1992.
34. RJ Scheuplein, IH Blank. Permeability of the skin. Physiol Rev 51:702–747, 1971.
35. T Dutkiewicz, H Tyras. A study of the skin absorption of ethylbenzene in man. Br J Ind Med 24:330–332, 1967.

36. T Dutkiewicz, H Tyras. Skin absorption of toluene, styrene, and xylene by man. Br J Ind Med 25:243, 1968.

37. J Hadgraft, G Ridout. Development of model membranes for percutaneous absorption measurements. I. Isopropyl myristate. Int J Pharm 39:149–156, 1987.

38. SD Roy, S-Y Hou, SL Witham, GL Flynn. Transdermal delivery of narcotic analgesics: comparative metabolism and permeability of human cadaver skin and hairless mouse skin. J Pharm Sci 83:1723–1728, 1994.

39. JMP Statistical Discovery Software. Ver. 3.1, SAS Institute, Inc., Cary, North Carolina, 1995.

40. RJ Scheuplein, IH Blank, GJ Brauner, DJ MacFarlane. Percutaneous absorption of steroids. J Invest Dermatol 52:63–70, 1969.

41. O Siddiqui, MS Roberts, AE Polack. Percutaneous absorption of steroids: relative contributions of epidermal penetration and dermal clearance. J Pharmacokinet Biopharm 17:405–424, 1989.

42. BD Anderson, PV Raykar. Solute structure–permeability relationships in human stratum corneum. J Invest Dermatol 93:280–286, 1989.

43. KD Peck, A-H Ghanem, WI Higuchi. The effect of temperature upon the permeation of polar and ionic solutes through human epidermal membranes. J Pharm Sci 84:975–982, 1995.

44. RJ Scheuplein. Mechanism of percutaneous absorption III. The effect of temperature on the transport of nonelectrolytes across the skin. J Invest Dermatol 49:582–589, 1967.

45. P Singh, MS Roberts. Skin permeability and local tissue concentrations of nonsteroidal anti-inflammatory drugs after topical application. J Pharmacol Exp Ther 268:144–151, 1994.

46. DR Lide, ed. CRC Handbook of Chemistry and Physics. Boca Raton, FL: CRC Press, 1996.

47. RL Bronaugh, ER Congdon. Percutaneous absorption of hair dyes: correlation with partition coefficients. J Invest Dermatol 83:124–127, 1984.

48. J Nakai. Health Canada, personal communication, 1995.

49. IH Blank, DJ McAuliffe. Penetration of benzene through human skin. J Invest Dermatol 85:522–526, 1985.

50. MS Roberts. Percutaneous absorption of phenolic compounds. Ph.D. thesis, University of Sydney, Sydney, Australia, 1976.

51. K Kubota, HI Maibach. In vitro percutaneous permeation of betamethasone and betamethasone 17-valerate. J Pharm Sci 82:1039–1045, 1993.

52. MS Roberts, RA Anderson, J Swarbrick. Permeability of human epidermis to phenolic compounds. J Pharm Pharmacol 29:677–683, 1977.

53. RJ Scheuplein, IH Blank. Mechanism of percutaneous absorption. IV. Penetration of nonelectrolytes (alcohols) from aqueous solutions and from pure liquids. J Invest Dermatol 60:286–296, 1973.

54. D Southwell, BW Barry, R Woodford. Variations in permeability of human skin within and between specimens. Int J Pharm 18:299–309, 1984.

55. WR Galey, HK Lonsdale, S Nacht. The in vitro permeability of skin and buccal mucosa to selected drugs and tritiated water. J Invest Dermatol 67:713–717, 1976.

56. P Liu, WI Higuchi, A-H Ghanem, WR Good. Transport of beta-estradiol in freshly excised human skin in vitro: diffusion and metabolism in each skin layer. Pharm Res 11:1777–1784, 1994.

57. NA Megrab, AC Williams, BW Barry. Oestradiol permeation across human skin, silastic and snake skin membranes—the effects of ethanol water co-solvent systems. Int J Pharm 116:101–112, 1995.

58. IH Blank. Penetration of low-molecular-weight alcohols into skin I. Effect of concentration of alcohol and type of vehicle. J Invest Dermatol 43:415–420, 1964.

59. RC Scott, MA Corrigan, F Smith, H Mason. The influence of skin structure on permeability: an intersite and interspecies comparison with hydrophilic penetrants. J Invest Dermatol 96:921–925, 1991.

60. JR Bond, BW Barry. Hairless mouse skin is limited as a model for assessing the effects of penetration enhancers in human skin. J Invest Dermatol 90:810–813, 1988.

61. IP Dick, RC Scott. Pig ear skin as an in-vitro model for human skin permeability. J Pharm Pharmacol 44:640–645, 1992.

62. K Sato, K Sugibayashi, Y Morimoto. Species differences in percutaneous absorption of nicorandil. J Pharm Sci 80:104–107, 1991.

63. A Dal Pozzo, G Donzelli, E Liggeri, L Rodriguez. Percutaneous absorption of nicotinic acid derivatives in vitro. J Pharm Sci 80:54–57, 1991.

64. RL Bronaugh, ER Congdon, RJ Scheuplein. The effect of cosmetic vehicles on the penetration of N-nitrosodiethanolamine through excised human skin. J Invest Dermatol 76:94–96, 1981.

65. K Harada, T Murakami, E Kawasaki, Y Higashi, S Yamamoto, N Yata. In-vitro permeability to salicylic acid of human, rodent, and shed snake skin. J Pharm Pharmacol 45:414–418, 1993.

66. ED Barber, NM Teetsel, KF Kolberg, D Guest. A comparative study of the rates of in vitro percutaneous absorption of eight chemicals using rat and human skin. Fundam Appl Toxicol 19:493–497, 1992.

67. RL Bronaugh, RF Stewart, M Simon. Methods for in vitro percutaneous absorption studies. VII. Use of excised human skin. J Pharm Sci 75:1094–1097, 1986.

68. PC Rigg, BW Barry. Shed snake skin and hairless mouse skin as model membranes for human skin during permeation studies. J Invest Dermatol 94:235–240, 1990.

69. BW Barry. School of Pharmacy, University of Bradford, personal communication, 1996.

70. JR Bond, BW Barry. Limitations of hairless mouse skin as a model for in vitro permeation studies through human skin: hydration damage. J Invest Dermatol 90:486–489, 1988.

71. PA Cornwell, BW Barry. Sesquiterpene components of volatile oils as skin penetration enhancers for the hydrophilic permeant 5-fluorouracil. J Pharm Pharmacol 46:261–269, 1994.

72. AC Williams, BW Barry. Terpenes and the lipid-protein-partitioning theory of skin penetration enhancement. Pharm Res 8:17–24, 1991.

73. D Dick, KME Ng, DN Sauder, I Chu. In vitro and in vivo percutaneous absorption of ^{14}C-chloroform in humans. Hum Exp Toxicol 14:260–265, 1992.

74. BW Barry, SM Harrison, PH Dugard. Vapour and liquid diffusion of model penetrants through human skin; correlation with thermodynamic activity. J Pharm Pharmacol 37:226–236, 1985.

75. J Swarbrick, G Lee, J Brom, NP Gensmantel. Drug permeation through human skin II: Permeability of ionizable compounds. J Pharm Sci 73:1352–1355, 1984.

76. JE Wahlberg. Percutaneous absorption of sodium chromate (51Cr), cobaltous (58Co), and mercuric (203Hg) chlorides through excised human and guinea pig skin. Acta Derm Venereol 45:415–426, 1965.

77. WA Ritschel, A Sabouni, AS Hussain. Percutaneous absorption of coumarin, griseofulvin and propranolol across human scalp and abdominal skin. Methods Find Exp Clin Pharmacol 11:643–646, 1989.

78. M Lodén. The in vitro permeability of human skin to benzene, ethylene, glycol, formaldehyde, and n-hexane. Acta Pharmacol Toxicol 58:382–389, 1986.

79. SD Roy, DJ Chatterjee, E Manoukian, A Divor. Permeability of pure enantiomers of ketorolac through human cadaver skin. J Pharm Sci 84:987–990, 1995.

80. MR Moore, PA Meredith, WS Watson, DJ Sumner, MK Taylor, A Goldberg. The percutaneous absorption of lead-203 in humans from cosmetic preparations containing lead acetate, as assessed by whole-body counting and other techniques. Food Chem Toxicol 18:399–405, 1980.

81. P Singh, MS Roberts. Dermal and underlying tissue pharmacokinetics of lidocaine after topical application. J Pharm Sci 83:774–781, 1994.

82. ZT Chowhan, R Pritchard. Effect of surfactants on percutaneous absorption of naproxen I: comparisons of rabbit, rat, and human excised skin. J Pharm Sci 67:1272–1274, 1978.

83. A Fullerton, JR Andersen, A Hoelgaard. Permeation of nickel through human skin in vitro—effect of vehicles. Br J Dermatol 118:509–516, 1988.

84. MH Samitz, SA Katz. Nickel–epidermal interactions: diffusion and binding. Env Res 11:34–39, 1976.

85. O Norgaard. Investigations with radioactive Ag[111] into the resorption of silver through human skin. Acta Derm Venereol 34:415–419, 1954.

86. B Baranowska-Dutkiewicz. Absorption of hexavalent chromium by skin in man. Arch Toxicol 47:47–50, 1981.

87. J Nakai, I Chu, D Moir, RP Moody. Dermal absorption of chemicals into freshly-prepared and frozen human skin, 1995.

88. RL Bronaugh, TJ Franz. Vehicle effects on percutaneous absorption: in vivo and in vitro comparisons with human skin. Br J Dermatol 115:1–11, 1986.

89. RL Cleek, AL Bunge. A new method for estimating dermal absorption from chemical exposure. 1. General approach. Pharm Res 10:497–506, 1993.

90. ED Barber. Health and Environmental Laboratory, Eastman Kodak Company, personal communication, 1996.

91. RL Bronaugh. Food and Drug Administration, personal communication, 1996.

92. MS Roberts. Department of Medicine, The University of Queensland, personal communication, 1996.

93. K Sato, K Sugibayashi, Y Morimoto, H Omiya, N Enomoto. Prediction of the in-vitro human skin permeability of nicorandil from animal data. J Pharm Pharmacol 41:379–383, 1989.

94. C Hansch, A Leo. Exploring QSAR: Fundamentals and Applications in Chemistry and Biology. Washington, D.C.: American Chemical Society, 1995.

95. P Meyer, G Maurer. Correlation and prediction of partition coefficients of organic solutes between water and an organic solvent with a generalized form of the linear solvation energy relationship. Ind Eng Chem Res 34: 373–381, 1995.

96. GL Flynn. College of Pharmacy, University of Michigan, personal communication, 1996.

4

Partitioning of Chemicals into Skin: Results and Predictions

Brent E. Vecchia* and Annette L. Bunge
Colorado School of Mines, Golden, Colorado, U.S.A.

I. INTRODUCTION

The accurate prediction of dermal absorption using physicochemical properties of penetrating molecules is a long-term goal of the study of percutaneous absorption. One physicochemical property, the stratum corneum–water partition coefficient, represents the capacity of the stratum corneum (SC) for a compound relative to its aqueous concentration and is a required input parameter for some dermal absorption models. Experimental values of SC–water partition coefficient (K_{cw}) do not exist for many compounds with the potential for dermal absorption, and thus there is a need for structure–activity based methods for estimating K_{cw} using more readily available parameters (e.g., octanol–water partition coefficients, K_{ow}). The quality of any predictive equation is limited by the quality of the data used in its development. It is therefore important to review critically each measurement in the database of K_{cw} values prior to the development of predictive equations. In this chapter, we describe a new collection of K_{cw} data and several predictive equations developed from these data, including equations providing information about the mechanisms of SC–water partitioning. We have identified several criteria that are essential for meaningful K_{cw} values, and data that were to be used to develop predictive equations were required to meet these criteria. Finally, this database of K_{cw} values was compared to previously published equations for estimating K_{cw}.

---

* *Current affiliation*: Blakely Sokoloff Taylor & Zafman LLP, Denver, Colorado, U.S.A.

II. DATA VALIDATION CRITERIA

SC–water partition coefficients are determined in relatively simple experiments. Most K_{cw} have been determined by equilibrating an accurately weighed sample of SC (often desiccated) with an aqueous solution of the absorbing compound, which is labeled with ^{14}C or ^{3}H. After sufficient time, the solute concentration in the aqueous solution and in the SC (frequently determined by liquid scintillation counting of radioactivity of the solubilized SC) are determined. The partition coefficient is calculated from the ratio of the SC and water concentrations at equilibrium. All of the partition coefficient data reviewed in this study were measured with only slight variations from this protocol. As part of this study, we did not consider partition coefficients that were calculated by fitting mathematical models to penetration data produced in diffusion cell experiments. Because K_{cw} experiments are relatively easy, there are a large number of measurements reported in the literature that can be used to develop predictive equations.

   All of the K_{cw} data examined in this study were measured in vitro (by the partitioning technique, not the permeation technique) with human skin from aqueous solution. Prior to developing predictive equations, a critical review process was used to validate that measurements met certain quality criteria. Every effort was made to extract data from only original references. Data reported without pertinent details were reserved for future analysis, if the missing information could not be obtained from the original authors. The K_{cw} database is divided into three collectively exhaustive (taken together they contain all measurements we have considered) and mutually exclusive (measurements appearing in one database do not appear in others) divisions: (a) a fully validated database (containing the measurements used to develop predictive equations), (b) a provisional database (reserved for analysis if additional necessary information can be obtained), and (c) an excluded database (containing measurements that do not meet the validation criteria).

   The fully validated (FV) database contains partition coefficient values that meet six criteria: (a) the partition coefficient must be based on the hydrated volume of SC or be converted to that basis, (b) the ionized state of the partitioning compound must be known, and the fraction that is nonionized (f_{ni}) must be greater than 0.9, (c) a valid log K_{ow} (either a recommended value from Hansch and colleagues (1) or else calculated using the Daylight software (2), which was developed from the Hansch database of validated K_{ow} values) must be available, usually for the nonionized compound, (d) reasonable evidence must be provided that equilibrium was established, (e) the temperature must be specified and must be between 20 and 40°C, and (f) the exposure solution must not compromise the SC (more than water itself).

A. Expression of the Stratum Corneum Concentration

A review of the literature reveals that the concentration of the partitioning compound in the SC are represented in different ways producing reported partition coefficient values ($K_{cw,rep}$) with multiple definitions. To be dimensionally consistent with the differential material balances (i.e., mathematical models) describing the penetration of chemicals through a skin membrane, the partition coefficient when multiplied by the aqueous concentration (C_w), given as solute mass per volume of solution, should have units of solute mass per volume of fully hydrated stratum corneum. That is, the dimensionally consistent definition of K_{cw} is

$$K_{cw} = \frac{C_{c,eq}}{C_w} \tag{1}$$

in which $C_{c,eq}$ is the mass of solute per volume of the hydrated SC that is in equilibrium with C_w. However, many researchers have reported a different partition coefficient, K'_{cw}, calculated using the dry mass of the SC as the basis as follows:

$$K'_{cw} = \frac{C'_{c,eq}}{C_w} \tag{2}$$

In Eq. (2) $C'_{c,eq}$ is the mass of solute per mass of the dried SC that is in equilibrium with C_w.

Partition coefficient measurements based on the dry-mass of SC (K'_{cw}) can be related to K_{cw} through the ratio of the aqueous solution density (ρ_w) and ρ'_c defined as the mass of dry SC over hydrated skin volume as follows:

$$K_{cw} = K'_{sc} \frac{\rho'_c}{\rho_w} \tag{3}$$

The value of ρ'_c is not known precisely, although several investigations provide information for making a reasonable estimate. The most decisive measurements come from the study by Raykar and colleagues (3), who measured the water uptake of dry SC over a period of 48 to 72 hours. Raykar and colleagues found that an average of 2.91 g of water was absorbed per g of dry SC over a period of 48 to 72 hours when the temperature was maintained at 37°C. This result is based on 73 measurements of water uptake using skin samples of various anatomical locations from at least 21 different people. To use this information to estimate ρ'_c, we must know the density of dry SC. According to Bronaugh and Congdon (4), Scheuplein (5) reported that the density of dry SC is 1.32 g of dry SC per mL of dry SC. Using 1.3 g mL^{-1} as the density of dry SC and assuming that

2.91 g of water are absorbed per g of dry SC, we calculate that 1 g of dry SC should have a hydrated volume of 3.68 mL (=2.91 g of water/[1 g mL^{-1}] + 1 g of dry SC/[1.3 g mL^{-1}]). Consequently, $\rho_c' \approx 0.27$ g/mL (=1 g dry SC/3.68 mL of hydrated SC). Unfortunately, the Scheuplein report (5) has a limited circulation, and we were not able to verify the 1.32 value for the dry SC density. However, from these calculations it is easy to show that ρ_c' is relatively insensitive to the value for the dry SC density within a reasonably expected range (i.e., for a dry SC density of 1.0 to 1.5 g mL^{-1}, ρ_c' is between 0.26 and 0.28 g mL^{-1}).

Megrab and others (6) found that approximately 3.7 grams of water were absorbed per gram of dry SC over a period of 48 hours, leading to a ρ_c' value of 0.22 g mL^{-1} (assuming that the density of dry SC is 1.3 g mL^{-1}). In their review paper, Scheuplein and Blank (7) report that occluded SC slowly imbibes 5 or 6 times its dry weight in water. If we again assume that the dry density of SC is close to 1.3 g mL^{-1}, then 0.15 g mL$^{-1} < \rho_c' < 0.17$ g mL^{-1}. After 24 hours of hydration, Scheuplein and Blank (8) state that approximately 2 g of water are absorbed per g of dry SC, which gives $\rho_c' = 0.36$ g mL^{-1}.

Although it is not known precisely, it is evident that ρ_c' is significantly different from unity and is probably between 0.2 and 0.3 g mL^{-1}. We estimate that $\rho_c' = 0.25$ g mL^{-1} is representative of SC that has been exposed to excess hydration for periods long enough to ensure complete hydration. Based on this value, $K_{cw} = 0.25 \, K_{cw}' = (0.25 \text{ g mL}^{-1}/1.0 \text{ g mL}^{-1}) \, K_{cw}'$.

A few authors report their partition coefficient data in terms of hydrated mass of SC (referred to as wet mass in Tables A1 and A2). In this case, K_{cw} is calculated from $K_{cw,rep}$ using the adjustment factor of ρ_c/ρ_w, where ρ_c is the mass of hydrated SC per volume of hydrated SC. If the density of dry SC is 1.3 g mL^{-1} and 2.91 g of water absorb in 1 g of dry SC, then $\rho_c = 1.06$ g mL^{-1} = ([1 g dry SC + 2.91 g water]/[1 g dry SC/[1.3 g mL^{-1}] + 2.91 g water/(1 g mL^{-1})]). If 1 g of SC absorbs 5 g of water, then $\rho_c = 1.04$ g mL^{-1}. Consequently, in adjusting partition coefficients based on hydrated mass of SC, we have assumed that $\rho_c/\rho_w \approx 1$ and thus K_{cw} is approximately the same whether based on hydrated mass or hydrated volume of SC.

B. Effect of Ionization

SC–water partition coefficients are available for compounds that are nonionized, charged, and net neutral (zwitterionic). However, the effects of chemical ionization on the partition coefficient are not well documented except for a few isolated studies. Smith and Anderson (9) studied partitioning of ionized (anionic) and nonionized lauric acid (log $K_{ow} = 4.2$) and found that ionized and nonionized species both partitioned into the SC, although the nonionized form partitioned more strongly (9). Specifically, K_{cw} = 5060 for the nonionized species, which was approximately 70 times larger than K_{cw} = 72.2 for the anionic form.

Later we will examine one way to evaluate the effect of ionization, which is to compare K_{cw} data for partially ionized chemicals to an equation regressed to fit the data from only nonionized chemicals. To do so, the fraction of the compound that is nonionized (f_{ni}) was determined from pK_a values calculated in SPARC and adjusted for temperature by the methods described in Chapter 3. When the pH was not reported, it was calculated from the solute concentration and pK_a using the general treatment of simultaneous equilibrium that was also discussed in Chapter 3. In a partition coefficient experiment, the concentration can change until an equilibrium distribution is reached between the SC and the solution. Consequently, the final equilibrium concentration is the appropriate solution concentration for calculating the natural pH of the solution.

C. Selection of log K_{ow}

Later in this chapter we describe equations in which K_{cw} data of the nonionized species were regressed with log K_{ow} values determined for the nonionized form of the absorbing chemical. The preferred log K_{ow} values were the recommended *star* (★) values from Hansch et al. (1), which were "measured as or converted to the neutral form." When these recommended values were not available, Daylight software (2) was used to calculate surrogates that are consistent with the Hansch database of recommended values. Because ionic species are quite water soluble, K_{ow} values for neutral species are frequently larger than those measured for partially ionized chemicals, and consequently K_{ow} values for ionized compounds were not used.

D. Equilibration Time

While a long equilibration time is desirable when measuring solvent–solvent partition coefficients, prolonged contact between skin and a solvent (i.e., the vehicle) may alter the degree to which the skin can absorb a distributing solute. Roskos and Guy (10) suggested that 6 hours of equilibration in water was sufficient to hydrate the SC fully. However, more than 6 hours may be necessary to establish equilibrium with extremely hydrophilic compounds and for molecules that diffuse slowly into the SC (e.g., high molecular weight compounds). Whenever it is known, we have reported the exposure time in our database. For the K_{cw} value to be judged as fully validated, reasonable evidence is required to indicate that equilibrium was attained. This requirement was satisfied if the authors monitored concentration over time until there were no further changes, or if the authors have shown in a previous publication that the exposure time was sufficient to establish equilibrium for similar compounds.

III. MODEL DEVELOPMENT

Equations are developed in this section that will be used for later analysis of the K_{cw} database. First, we develop a simple model that has been used commonly to represent K_{cw} data (i.e., the conventional model). Later we analyze the database with a model that allows for potentially different mechanisms for partitioning into the hydrophilic and lipophilic domains of the SC and with a linear solvation–energy relationship (LSER) model. These last two models are mechanistically based, allowing insight into the mechanisms of chemical partitioning into the SC.

A. Development of the Conventional Model

Solvent–solvent partition coefficients are generally related to other solvent–solvent partition coefficients using linear free-energy relationships (11). SC–water partition coefficients are frequently analyzed in terms of octanol–water partition coefficients (12,13) using a linear free-energy relationship of the form

$$\log K_{cw} = a + b \log K_{ow} \tag{4}$$

In using this equation we assume that the energetics of solvation in octanol and SC are similar for compounds with a wide spectrum of hydrophobicity. Differences in cavity formation energies between the SC (which is an amorphous solid or liquid crystalline) and octanol (which is liquid) may lead to differences in the effect of size. This can be investigated with an equation that also accounts for differences in molecular size. For example,

$$\text{long } K_{cw} = \alpha + b \log K_{ow} + d\,\text{MW} \tag{5}$$

in which molecular weight (MW) is used as a surrogate for molar volume, a more pertinent representation of molecular size. In writing Eq. (5) we have assumed that $\log K_{cw}$ is linearly related to MW as might occur through an activated Arrhenius process.

B. Development of the Lipophilic–Hydrophilic
Domain Model

Morphologically the SC is a heterogeneous membrane composed of protein and lipid domains that are revealed histologically as a mosaic of cornified epidermal cells containing cross-linked keratin filaments and intercellular lipid-containing regions (14). The cellular proteins are not themselves a homogeneous domain, but the differences within this milieu are small when compared with differences between the lipids and this protein phase. Researchers have speculated that compounds partition into the protein phase as well as into the intercellular lipids (3) to different extents that depend on the lipophilic character of the compound. A change in uptake mechanism can occur if the selectivity of the protein and lipid domains is different for solutes with varying lipophilicity. Such a difference is expected, since these domains have significantly different polarities.

At a simple level, a two-phase partitioning model allows for a different partitioning into the hydrophilic and lipophilic domains of the SC. Several organic solvents (e.g., octanol, isopropylmyristate) may reasonably represent the lipid domain. Our starting place is to assume that these or other solvents can also represent partitioning into the protein domain but with a different dependence. That is, we begin with a model containing a different linear dependence on log K_{ow} for hydrophilic and lipophilic compounds (the distinction between hydrophilic and lipophilic compounds will be defined by the regression) as follows:

$$\log K_{cw} = a_H + b_H \log K_{ow} \quad \text{when} \quad K_{ow} < K_{ow}^{\ddagger}$$

$$\log K_{cw} = a_L + b_L \log K_{ow} \quad \text{when} \quad K_{ow} \geq K_{ow}^{\ddagger} \quad (6)$$

in which $K_{ow}^{\ddagger}$ is the value of K_{ow} at which the piecewise linear regression changes slope, distinguishing the hydrophilic compounds from the lipophilic compounds. Eq. (6) can be written generally as

$$\log K_{cw} = a_H + b_H \log K_{ow} + (b_L - b_H)(\log K_{ow} - \log K_{ow}^{\ddagger}) U(K_{ow} - K_{ow}^{\ddagger}) \quad (7)$$

where $U(K_{ow} - K_{ow}^{\ddagger})$ is the unit step function defined as follows:

$$U(K_{ow} - K_{ow}^{\ddagger}) = 0 \quad \text{if} \quad K_{ow} \leq K_{ow}^{\ddagger}$$

$$= 1 \quad \text{if} \quad K_{ow} > K_{ow}^{\ddagger} \quad (8)$$

In Eq. (7) the intercept is a_H and the slope is b_H when $K_{ow} \leq K_{ow}^{\ddagger}$ and the intercept is $a_L = a_H - (b_L - b_H) \log K_{ow}^{\ddagger}$ and the slope is b_L for $K_{ow} > K_{ow}^{\ddagger}$.

In an even simpler model, we assume that hydrophilic compounds partition into the SC at a constant value that is independent of lipophilic character (i.e., $b_H = 0$), which gives

$$\log K_{cw} = a_H + b_L (\log K_{ow} - \log K_{ow}^{\ddagger}) U(K_{ow} - K_{ow}^{\ddagger}) \quad (9)$$

For this expression, a_H is the average log K_{cw} for the hydrophilic compounds; b_L is the slope and $(a_H - b_L \log K_{ow}^{\ddagger})$ the intercept for the regression line representing lipophilic compounds.

This analysis of SC–water partition coefficients in terms of two partitioning domains is simplistic. In using either Eq. (7) or Eq. (9), we assume that the mechanisms for SC–water partitioning of hydrophilic and lipophilic compounds switch abruptly and completely at $K_{ow} = K_{ow}^*$. More sophisticated models can be developed to extend these simple models as needed and justified by the data.

C. Development of the Linear Solvation-Energy Relationship Model

The theoretical foundations of linear solvation energy relationships (LSERs) and have been described in several references (15–17) and also in Appendix C of Chapter 3. In Chapter 3, LSER models were used to interpret and predict SC permeability coefficients. In this chapter we use LSER models to analyze and predict K_{cw} data. A similar analysis has been done before on smaller databases that were not critically evaluated for quality (e.g., the study of Abraham et al. (18)).

Linear solvation energy relationships have been successful at estimating free energy dependent properties exhibited by a class of solutes in the same solvent. Partitioning of organic solutes into the SC from water is a free energy based property that can be analyzed with LSER models. We propose the following general model for analysis of K_{cw} data:

$$\log K_{cw} = \log K_{cw}^0 + a\,\alpha + b\,\beta + c\,\pi + d\,V_x \tag{10}$$

in which the four LSER parameters (α, β, π, and V_x) are measured and tabulated for solutes rather than the solvent. The LSER parameters have the following physicochemical interpretations: α is the effective hydrogen-bond acidity, β is the effective hydrogen-bond basicity, π is the solute dipolarity/polarizability, and V_x is the characteristic volume of McGowans (19), which is calculated from molecular structure alone and is independent of intermolecular forces such as hydrogen bonding. Conventionally, V_x is used in units of $cm^3\ mol^{-1}$ divided by 100 so that V_x values have similar magnitude to the other LSER parameters.

Equation (10) is based on the assumption that solute distribution between the SC and water arises from differences in Van der Waals forces, size effects, polarizability, and the preference of a solute to act as an electron donor and an electron acceptor in hydrogen bonds with the water and with skin. Models for K_{cw} are developed by multilinear regression analysis of K_{cw} data with LSER parameters to determine the coefficients $\log K_{cw}^0$ and a through d.

IV. RESULTS AND DISCUSSION

A. Examination of the Data

The validation criteria were applied to a collection of known K_{cw} values, which are listed in Appendix A. Details on selection and validation of the K_{cw} data from the original publication are given in Appendix B. Table A1 contains the K_{cw} values that satisfied all of the criteria. These compounds span a fair level of diversity as measured by MW and log K_{ow}. LSER parameters were found for the chemicals enclosed within brackets (e.g., [benzene]), and these compounds were included in the LSER analysis.

Table A2 lists the provisional K_{cw} data. Partition coefficients for chemicals with $f_{ni} < 0.9$ or with an unknown f_{ni} were included in the provisional database unless they were otherwise excluded. Several chemicals studied by Surber et al. (20,21) are weak acids or bases. Unfortunately, neither the final equilibrium solution concentration nor the pH were reported or could be determined by communication with the authors (Surber, 1996). In some cases, f_{ni} will be approximately 1 at the natural pH regardless of solute concentration, and these compounds are listed in Table A1. For other compounds, f_{ni} is quite sensitive to the natural pH, which cannot be calculated without knowing the solute concentration. These measurements are listed in Table A2, because these chemicals are probably partially ionized, but to an unknown extent.

Table A3 contains the excluded K_{cw} data. Measurements for chemicals enclosed within brackets (e.g., [alachlor]) were made using sieved, powdered human stratum corneum (PHSC) from plantar calluses. These K_{cw} data were excluded because they may not accurately represent K_{cw} for SC that is not powdered or from callus. This issue will be discussed more fully later. The measurement for nicorandil was excluded because it was also measured using plantar callus, although not powdered. Cyclosporin-A (MW = 1201) was excluded because it is much larger than the other compounds included in the database and may partition by different mechanisms than the smaller compounds. Measurements for PCB were excluded because they were measured for a polydispersed mixture of chlorinated biphenyls rather than for a single compound.

Table A4 lists K_{cw} measurements from the FV database with known LSER parameters along with LSER parameters for the chemical. Table A5 summarizes the calculated f_{ni} and the natural pH attained by aqueous solutions of ionizable compounds at known and unbuffered concentrations.

Figure 1 shows $K_{cw,rep}$ values with no adjustment for basis from the FV and provisional databases, and several measurements from the excluded database (e.g., cyclosporin-A, nicorandil, and PCB), plotted as a function of log K_{ow}. The partition coefficients of cyclosporin-A (an excluded measurement) and hydroxypregnenolone (a provisional measurement) are plotted at an artificial value of log $K_{ow} = -4$, since log K_{ow} values were not available for these compounds. Unless stated otherwise, hydroxypregnenolone and cyclosporin-A are not included in subsequent figures. One value for lauric acid [from Smith and Anderson (9)] is the largest $K_{cw,rep}$ value in the FV database, and it is identified in Fig. 1. The PCB (arochlor 1254, 54% chlorine) measurement is plotted at the average of the *star* (★) values of log K_{ow} for 14 variously substituted pentachlorobiphenyls listed in

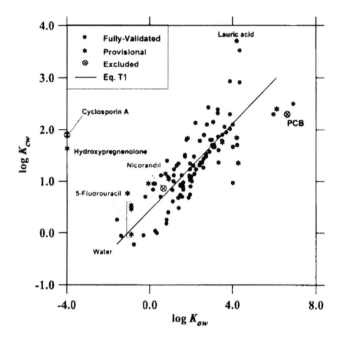

Figure 1 $K_{cw,rep}$ in the FV and provisional databases along with several $K_{cw,rep}$ values from the excluded database plotted as a function of log K_{ow}. Hydroxypregnenolone and cyclosporin-A, which lack appropriate log K_{ow}, are plotted at log $K_{ow} = -4$.

the Hansch database (2). Figure 2 shows the data from Fig. 1 plotted as a function of log K_{ow} after adjustment for basis (i.e., all data are reported in terms of the volume of hydrated SC). The K_{cw} values span four orders of magnitude but are more restricted than the permeability coefficient values listed in Chapter 3, which varied by six orders of magnitude.

B. Data Regression Analysis

Table 1 summarizes the results of linear regressions of the partition coefficient data to the various model equations developed above. These regressions were performed by standard procedures using JMP (22). Uncertainties in the coefficients of the input parameters, given as the standard error of the coefficients, are listed in parentheses. Table 1 also lists regression statistics: r^2, r^2_{adj}, RMSE, and the F-ratio. The r^2_{adj} statistic is analogous to r^2 but allows for more relevant comparisons between models with different numbers of fitted parameters [JMP User's Guide, (22)]. Specifically, $(1 - r^2)$ = error in sum of squares/total sum of squares and $(1 - r^2_{adj}) = (1 - r^2)(n - 1)/(n - p)$, where n is the number of data points and

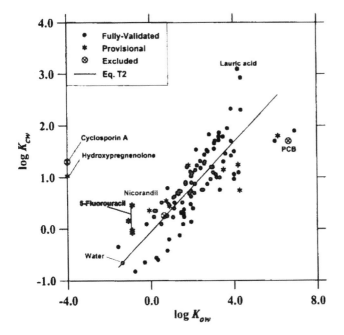

Figure 2 K_{cw} from the FV and provisional databases along with several K_{cw} values from the excluded database plotted as a function of log K_{ow}. Hydroxypregnenolone and cyclosporin-A, which lack appropriate log K_{ow}, are plotted at log $K_{ow} = -4$.

p is the number of parameters. RMSE is the root mean square error of the model, which is zero if the model fits the data perfectly. The F-ratio is defined as the ratio of the sum of squares for the model divided by the degrees of freedom for the model and sum of squares for the error divided by the degrees of freedom for the error. F-ratio $= 1$ when there is zero correlation with the parameters, and it is large for regressions with good predictive power. Because the number of fitted parameters is in the denominator of the F-ratio, changes in the F-ratio with an increase in the number of parameters reflect the effect of the number of fitted parameters on the predictive power. Thus an equation with a larger number of parameters might give a higher r^2 but a lower F-ratio than an equation with fewer parameters. This would indicate that the improvement in predictive power (as indicated by a larger r^2) was not as large per parameter as for the equation with fewer parameters.

C. Analysis with the Conventional Model

Equation T1 in Table 1 was developed by regressing the conventional model, Eq. (4), to the $K_{cw,rep}$ data in the FV database (Table A1). Equation T2 was devel-

Table 1 Comparison of Data Regressions to Various Model Equations

No.	Data set[a]	n/m[b]	Model Eq.	$\log K_{cw}$[c,d]	r^2	r^2_{adj}	RMSE	F-ratio
T1	FV	97/76	4	$\log K_{cw} = 0.439\ (0.079) + 0.418\ (0.031) \log K_{ow}$	0.658	0.654	0.427	183
T2	FV	97/76	4	$\log K_{cw} = -0.059\ (0.078) + 0.434\ (0.031) \log K_{ow}$	0.680	0.677	0.421	202
T3	FV	97/76	5	$\log K_{cw} = -0.146\ (0.090) + 0.395\ (0.030) \log K_{ow} - {}^{*}0.00066\ (0.00030)\ \text{MW}$	0.698	0.691	0.392	107
T4	FV	97/76	7	$\log K_{cw} = -0.104\ (0.207) + 0.301\ (0.303) \log K_{ow} + 0.120\ (0.315)\ (\log K_{ow} + 0.5)\ U\ (K_{ow} - 0.32)$	0.682	0.675	0.40	100
T5	FV	97/76	9	$\log K_{cw} = -0.290\ (0.091) + 0.432\ (0.031)\ (\log K_{ow} + 0.5)\ U\ (K_{ow} - 0.32)$	0.679	0.675	0.402	199
T6	From Ref. 18	22/22	10	$\log K_{cw} = -0.027 - 0.374\pi + 0.334\alpha - 1.674\beta + 1.869\ V_x$	0.943	—	—	—
T7	LSER	38/37	10	$\log K_{cw} = -0.16\ (0.3) - {}^{*}0.27\ (0.2)\pi + {}^{*}0.42\ (0.3)\alpha - 3.15\ (0.5)\beta + 2.22\ (0.2)V_x$	0.830	0.810	0.325	41
T8	LSER	38/37	modified 10	$\log K_{cw} = -0.043\ (0.248) - 2.893\ (0.463)\beta + 1.977\ (0.181)V_x$	0.816	0.805	0.330	77
T9	LSER	38/37	4	$\log K_{cw} = -0.255\ (0.108) + 0.513\ (0.048) \log K_{ow}$	0.758	0.751	0.373	113

[a] FV = Fully validated database; LSER = Subset of FV database for which LSER parameters were available (see Table A4).
[b] n = number of data points, m = number of different compounds.
[c] The uncertainties expressed within parenthesis are reported as standard error in the coefficients.
[d] Coefficients indicated with an asterisk (*) are not meaningfully different from zero at the 95% confidence level.

oped by regressing Eq. (4) to the K_{cw} data, which are based on the volume of hydrated SC. Both Eqs. T1 and T2 are based on a database that is adequately large (97 values for 76 different compounds) and diverse for predictive estimates to be relevant. Equations T1 and T2 are shown in Fig. 1 and 2, respectively. There is little difference in the goodness-of-fit for Eqs. T1 and T2 (i.e., $r^2 = 0.66$ and 0.68, respectively); within statistical uncertainty, the coefficients multiplying log K_{ow} in Eqs. T1 and T2 are equal. This and the limited improvement in the regression statistics for Eq. T2 over Eq. T1 occur because a large number of the measurements in the FV database were adjusted from the basis of dry SC mass to hydrated SC volume (i.e., $K_{cw} = 0.25K_{cw,rep}$). Consistent with this, the chief difference between Eqs. T2 and T1 is that log K_{cw} differs by -0.498 ($= -0.059 - 0.439$), which corresponds to K_{cw} values that are approximately 0.32 ($= 10^{-0.498}$) times $K_{cw,rep}$, which is similar to the adjustment factor $\rho_c'/\rho_w = 0.25$.

Equation T3 in Table. 1 includes differences in MW by regression of Eq. (5) to K_{cw} values in the FV database. The asterisk on the MW term designates that MW does not significantly affect K_{cw} at the 95% confidence level, and MW will no longer be considered in the remaining analyses of K_{cw}. Comparing Eq. T2 to related equations for permeability coefficients listed in Table 1 of Chapter 3, the coefficient multiplying the log K_{ow} term is modestly smaller for K_{cw} than for permeability coefficients (i.e., 0.43 ± 0.03 compared to 0.51 ± 0.04), although the difference is probably not significant.

Importantly, predictive equations like Eq. T2 should not be used for estimating K_{cw} for compounds that are very different from those used to develop the database. That is, appropriate bounds on log K_{ow} and MW need to be set for compounds for which K_{cw} values are estimated. In the FV database, 95% of the chemicals have $-1.1 \leq \log K_{ow} \leq 5.3$ with equal percentages higher and lower than these bounds. Likewise, 80% of the chemicals have $0.3 \leq \log K_{ow} \leq 3.9$. Based on the accuracy of the regression equations at describing the existing data, we recommend the less conservative lower bound (log $K_{ow} = -1.1$) and the more conservative upper bound (log $K_{ow} = 3.9$) for log K_{ow}. A reasonable lower bound for MW is that of water (i.e., the lowest MW chemical in the database, MW $= 18$). In the FV database, 10% of chemicals have MW ≥ 465 and 2.5% have MW ≥ 500. The less conservative bound (i.e., MW ≤ 500) is chosen, since MW was shown to have a small effect on K_{cw}. Generally, the regression equations developed from the entire FV database will provide a reasonable estimate of K_{cw} for aqueous organic compounds with log K_{ow} in the range ($-1 < \log K_{ow} < 4$) and MW in the range $18 < $ MW < 500.

D. Trends in the K_{cw} Databases

In developing Eq. T2, we have made no assumptions regarding the relative extents of partitioning of ionized and nonionized compounds. As such, Eq. T2 can be used as a reference equation for studying the influence of various effects on the ratio of experimental to calculated values ($K_{cw}/K_{cw,calc}$).

Figure 3 shows the effects of temperature and ionization relative to the prediction by Eq. T2 for the K_{cw} values from the FV and provisional databases. Measurements above the upper dashed line are underestimated by Eq. T2 by more than an order of magnitude, while those below the lower dashed line are overestimated by more than an order of magnitude. With the exception of three measurements [lauric acid from Smith and Anderson (9), HC-21-yl-octanoate from Raykar et al. (3), and indomethacin from Surber et al. (21)], all K_{cw} values in the FV and provisional databases can be predicted within one order of magnitude (i.e., within a factor of ten). There appears to be less uncertainty in estimation of K_{cw} compared to estimation of SC permeability coefficients, even though the database of permeability coefficients is larger.

As quantified by averages of the residuals (defined as the log K_{cw} − log $K_{cw,calc}$), measurements taken at temperatures cooler than 30°C are over estimated (on average) by Eq. T2 (which was developed using the entire database), while measurements made at temperatures warmer than 30°C are under estimated (on

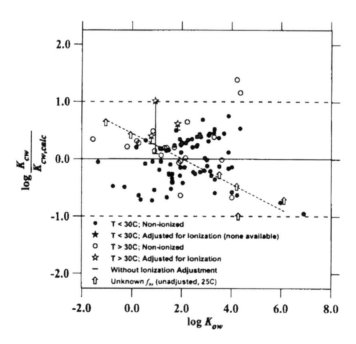

Figure 3 K_{cw} in the FV and provisional databases compared to values calculated using Eq. T2 ($K_{cw,calc}$), with different temperatures and levels of ionization designated.

average). For data measured at $T < 30°C$, the average residual is -0.037 ($n = 76$, standard deviation $= 0.39$), indicating that on average $\log K_{cw,calc} > \log K_{cw}$. For $T \geq 30°C$, the average residual is 0.140 ($n = 20$, standard deviation $= 0.41$), indicating that on average $\log K_{cw,calc} > \log K_{cw}$. The standard deviations are large compared to the mean because there are other significant sources of variability. As a result the effect of temperature on SC–water partitioning is not entirely resolved, although the effect is almost certainly smaller than observed in the permeability coefficient data (see Chapter 3). The solubility in any given solvent usually (but not always) increases with increasing temperature. The direction of the temperature effect is determined by the solute and the solvent and can change magnitude and sign (i.e., solubility increasing or decreasing with temperature) over a range of temperatures (11). There is no simple means of anticipating the effect of temperature on all compounds, but for many compounds the solubility in octanol and water increase similarly as temperature increases. Based on these results, variation in experimental temperature is probably a minor contributor to the total variability in the K_{cw} database.

The dashed line with negative slope represents the regression of $\log (K_{cw}/K_{cw,calc})$ with $\log K_{ow}$ of partially ionized chemicals, including compounds of unknown f_{ni}. This regression [$\log (K_{cw}/K_{cw,calc}) = 0.44 - 0.22 \log K_{ow}$] provided a reasonably good fit to the data with $r^2 = 0.81$. Stars, designating measurements adjusted for ionization (by dividing K_{cw} by f_{ni}), are connected to a horizontal dash marking the unadjusted value. For several measurements from Surber et al. (20,21) it was not possible to calculate f_{ni}, although it is known that $f_{ni} < 1.0$. These chemicals are designated in Fig. 3 by arrows pointing in the direction that an adjustment for ionization would change $K_{cw}/K_{cw,calc}$. For all of these compounds, adjusting for ionization by dividing K_{cw} by f_{ni} increases $\log (K_{cw}/K_{cw,calc})$.

Although the number of data points for a partially ionized compound is quite small ($n = 9$), the trend seems to be that $K_{cw}/K_{cw,calc} > 1$ for the more hydrophilic chemicals (i.e., $\log K_{ow} <$ about 2), and $K_{cw}/K_{cw,calc} < 1$ for the more lipophilic ionizing chemicals (i.e., $\log K_{ow} >$ about 2). Furthermore, adjusting the more lipophilic chemicals by f_{ni} seems to improve the predictability (i.e., $K_{cw}/K_{cw,calc}$ moves closer to 1 after adjusting with f_{ni}), while adjusting measurements for the more hydrophilic chemicals degrades predictability (i.e., $K_{cw}/K_{cw,calc}$ moves away from 1). Three measurements for 5-fluorouracil were not included in this analysis because $\log K_{ow}$ for 5-fluorouracil was measured with the ionized species present, whereas only nonionized species were used in $\log K_{ow}$ measurements for the other nine chemicals.

The observations just described are consistent with the hypothesis that the relative amounts of nonionized and ionized chemical forms partitioning into the SC depends on K_{ow}. The partition coefficient of the nonionized species ($K_{cw,ni}$) of a lipophilic chemical is likely to be significantly larger than its ionized species ($K_{cw,i}$). In this case, increasing the extent of ionization would decrease the quantity of chemical that absorbs. If the nonionized form of a molecule partitions into the SC at least 100 times more than its ionic form, then the partition coefficient of the nonionized species ($K_{cw,ni}$) can be estimated from that observed for a mixture of the nonionized and ionized species (K_{cw}) as $K_{cw,ni} \approx K_{cw}/f_{ni}$.

However, ionized chemicals do absorb into SC to a limited extent. Since nonionized hydrophilic chemicals absorb into the SC to a lesser extent than more lipophilic chemicals, the nonionized hydrophilic species may absorb into skin only slightly more readily than its ionized form. Furthermore, additional chemical interactions may arise between charged molecules, like amino acids, and the protein domain of the SC, which is also composed of amino acids. As a consequence, for more hydrophilic ionizing chemicals, $K_{cw,ni}$ may be of similar magnitude to $K_{cw,i}$, which would cause $K_{cw,ni} \approx K_{cw}$. In this case, K_{cw}/f_{ni} would overestimate the magnitude of $K_{cw,ni}$.

Figure 4 shows $K_{cw}/K_{cw,calc}$ for the FV and provisional databases according to the year that the measurements were collected ($K_{cw,calc}$ was calculated using Eq. T2). There is no systematic trend over the period of time shown. The uncertainty in measuring partition coefficients is nearly the same (perhaps somewhat

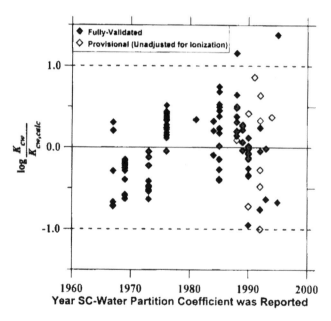

Figure 4 K_{cw} in the FV and provisional databases compared to those calculated using Eq. T2 ($K_{cw,calc}$) plotted as a function of the year in which the data were published.

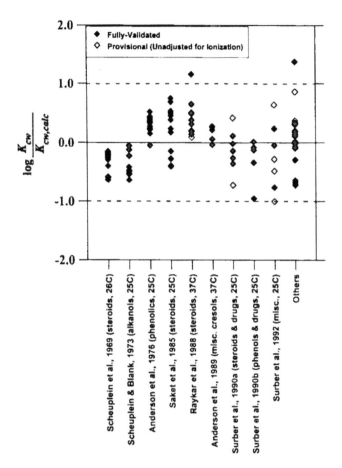

Figure 5 K_{cw} in the FV and provisional databases compared to those calculated using Eq. T2 ($K_{cw,calc}$), as reported in several prominent investigations.

larger) in recent measurements as in those measured as early as 1967. It is difficult to guess the sources of variance, but as is shown in Fig. 5, K_{cw} measured in particular laboratories can be systematically different from measurements made for similar compounds in other laboratories.

Assuming that log K_{ow} effects were dominant, one would expect that partition coefficients measured in all studies would be randomly scattered about the line log $K_{cw}/K_{cw,calc} = 0.0$. Some research groups report K_{cw} measurements that are predominantly higher or lower than expected from analysis of the other validated and nonionized measurements (since the partition coefficients for partially ionized compounds were not used to develop Eq. T2). Whether this is due to the skin source, skin preparation, or an unidentified experimental protocol is not known. For example, partition coefficients for the steroids (23) and alkanols (8) are all overestimated by Eq. T2. Conversely, all partition coefficients for slightly modified hydrocortisone steroids measured by Raykar and colleagues (3) are all systematically underestimated. Using similar procedures as Raykar and colleagues (3), K_{cw} values reported by Anderson et al. (24) for miscellaneous cresols are predominantly underestimated. The partition coefficients labeled as Others show the expected pattern, which is $K_{cw}/K_{cw,calc} \approx 1$ on average. It may be of some relevance that laboratories reporting larger than calculated K_{cw} also report larger than predicted permeability coefficients; laboratories reporting smaller values of K_{cw} than calculated also report smaller permeability coefficients (compare Fig. 5 here to Fig. 7 in Chapter 3). This systematic variation in measurements from different laboratories is responsible for a significant fraction of the total variability in the K_{cw} data.

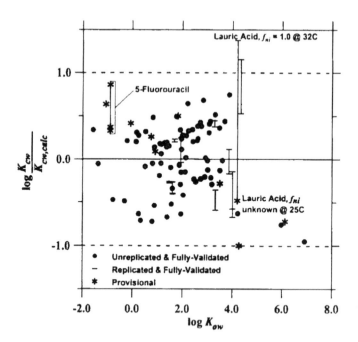

Figure 6 K_{cw} in the FV and provisional databases compared to those calculated using Eq. T2 ($K_{cw,calc}$) with replicated measurements designated.

Figure 6 compares calculations from Eq. T2 with replicated K_{cw} (i.e., multiple K_{cw} measurements for the same compound from different investigations) and unreplicated measurements that appear in the FV and provisional databases. Twenty-six replicate measurements for eleven different compounds are shown. Four of the replicated measurements (three for 5-fluorouracil and one for lauric acid) are from the provisional database. These replicated measurements provide the opportunity to observe differences (which should be small) in K_{cw} values for the same compound. Except for the replicate measurements on lauric acid, all replicate estimates agree within an order of magnitude. The measurement for lauric acid from Smith and Anderson (9) is quite inconsistent with another measurement for lauric acid (21). Additionally, this measurement deviates from the calculated value by more than any other measurement in the database and is one of only three measurements that are misestimated by more than an order of magnitude. While differences this large are alarming, this figure shows that differences of this magnitude are relatively infrequent. Nevertheless, the results

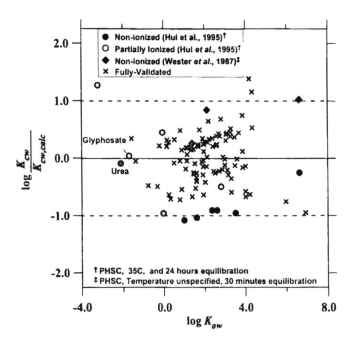

Figure 7 K_{cw} values compared to those calculated using Eq. T2 ($K_{cw,calc}$) with measurements made using sieved, powdered, human plantar callus SC (PHSC) designated from FV measurements.

Table 2 Equations for Estimating K_{cw} for Human Skin

Model No.	Eq. source	Chemical class	n^a	SC–Water partition equation	r^2	Data source	Data range log K_{ow}^b	MW^b
1	Brown & Hattis (31)^c	N/A	N/A	$K_{cw} = 1.277 + 0.1208 K_{ow}$	N/A	N/A	N/A	N/A
2	Cleek & Bunge (29)	not fit to data^d	—	$\log K_{cw} = 0.74 \log K_{ow}$	—	38^d	-3/6	18/765
3	Cleek & Bunge (29)	alcohols, acids, steroids, phenols	42	$\log K_{cw} = -0.006 + 0.57 \log K_{ow}$	0.72	13, 23, 39	-0.3/4.2	46/363
4	Cleek & Bunge (29)	alkanols & acids	12	$\log K_{cw} = -0.26 + 0.72 \log K_{ow}$	0.85	39	-0.3/3.1	46/144
5	Cleek & Bunge (29)	alkanols & acids	12	$\log K_{cw} = 0.60 \log K_{ow}$	0.72	39	-0.3/3.1	46/144
6	El Tayar et al. (12)	alkanols & steroids	22	$\log K_{cw} = 0.10 + 0.51 \log K_{ow}$	0.94	23, 40	-0.8/4.2	32/363
7	McKone & Howd (41)	alkanols	8	$K_{cw} = 0.64 + 0.25 K_{ow}^{0.8}$	0.90	8	-0.8/3.0	32/130
8	McKone (42)	not fit to data	—	$K_{cw} = 0.5 + 0.11 K_{ow}$	—	not fit to data	not fit to data	
9	Raykar et al. (3)	esters of hydrocortisone	11	$^eK_{cw} = 6.29 K_{ow}^{0.24} + 0.0225 K_{ow}^{0.91}$	—	3	1.4/4.3	418/519
10	Roberts et al. (30)	misc.^{f}	45	$\log K_{cw} = -0.024 + 0.590 \log K_{ow}$	0.84	7, 43–46	-0.8/5.5	32/519
11	Roberts et al. (13)	aromatic alcohols	21	$\log K_{cw} = -0.1 + 0.57 \log K_{ow}$	0.98	43, 45	0.8/3.7	94/198
12	Roberts et al. (13)	alkanols & acids	13	$\log K_{cw} = -0.1 + 0.66 \log K_{ow}$	0.96	39	-0.3/3.1	46/144
13	Roberts et al. (13)	steroids	14	$\log K_{cw} = 0.6 + 0.37 \log K_{ow}$	0.76	23	1.5/4.2	273/363

	Reference	Compounds			Ref.	Equation		
14	Saket et al. (37)	hydrocortisone esters & cortisone esters	11	N/A	37	[g] $\log K_{cw} = -0.343 + 0.761 \log K_{ow}$	-0.1/6.1	194/363
				0.97		[h] $\log K_{cw} = -0.280 + 0.739 \log K_{ow}$		
15	Surber et al. (47)	phenols, PCB, DDT, steroids, drugs[i]	13	0.81	20, 47	$\log K_{cw} = 0.87 + 0.26 \log K_{ow}$	-0.1/6.4	151/363
16	Surber et al. (21,47)	phenols, PCB, DDT	6	0.90	47	[j] $\log K_{cw} = 0.69 + 0.28 \log K_{ow}$	0.3/6.4	151/355
17	Surber et al. (20)	steroids, drugs[c]	7	N/A	20	[h] $\log K_{cw} = 0.725 + 0.344 \log K_{ow}$	-0.1/6.1	194/363
				0.74		[l] $\log K_{cw} = 0.987 + 0.254 \log K_{ow}$		
18[m]	Hui et al. (25)	drugs, pesticides, steroids	12	0.90	25	$\log K_{cw} = -2.04 + 0.078(\log K_{ow})^2 + 0.868 \log MW$	-3.2/6.6	60/363

[a] n = the total number of data points and also the number of compounds.

[b] Smallest value/largest value.

[c] Brown and Hattis (31) and several subsequent authors [e.g., Shatkin and Brown (32) and Chinery and Gleason (33)] incorrectly reference this correlation to Roberts and colleagues (34). The actual source is unknown.

[d] Derived from the permeability coefficient correlation of the Flynn permeability database (48) by assuming that the log K_{ow} term represented the SC–water partition coefficient.

[e] Based on separate fits of partitioning into the protein and extracted lipids.

[f] Including steroids, alcohols, acids, and various pharmaceuticals.

[g] This equation is algebraically reorganized from the equation $\log K_{ow} = 0.45 + 1.313 \log K_{cw}$, which fits the data in this reference. The equation published in this reference ($\log K_{cw} = 0.45 + 1.313 \log K_{ow}$) reversed K_{cw} and K_{ow}.

[h] This equation was developed by regression of the data in this reference to an equation in the form of Eq. (4). This equation is plotted in Figs. 9 and 10 as Model 14.

[i] The drugs are acitretin, diazepam, and caffeine.

[j] A regression of the data provided in this reference indicates that this is the correct equation. The published equation ($\log K_{cw} = 0.69 - 0.28 \log K_{ow}$) does not fit the data.

[k] This equation is algebraically reorganized from the published equation ($\log K_{ow} = -2.11 + 2.91 \log K_{cw}$).

[l] This equation was developed by regression of the data in this reference to an equation in the form of Eq. (4). This equation is plotted in Figs. 9 and 10 as Model 17.

[m] This equation was developed with measurements from powdered human stratum corneum (plantar callus).

in Fig. 6 indicate that measurements differing by a factor of two, or even a factor of five, are not necessarily meaningfully different within the other sources of uncertainty.

Until now we have ignored the set of SC–water partition coefficients from Table A3 that were measured using powdered human stratum corneum (PHSC). PHSC was prepared by grinding excised foot callus in the presence of liquid nitrogen and then collecting the fraction retained between 50 and 80-mesh sieves (25). The chemical differences between callus SC and uncallused SC, as well as possible physical alterations introduced by grinding and sieving, raise questions regarding the relevance of partition coefficient measurements made using PHSC. In Fig. 7 we compare PHSC measurements to those in the FV database, which except for nicorandil were measured using excised SC that was not callused (e.g., from abdomen, torso, or breast) or ground.

The majority of the PHSC measurements in Fig. 7 were taken from Table 2 in Hui et al. (25), which contains several puzzling results. Some of the more significant inconsistencies are that (a) the reported natural pH attained by some compounds after addition to distilled water do not make sense (e.g., for urea, which should be very weakly basic, the authors report an unexpectedly low pH of 2.1); (b) the reported pK_a values are very different from those calculated by SPARC (e.g., for atrazine Hui et al. reported $pK_a = 8.15$ compared to 2.0 from SPARC and 1.7 from the SRC PhysProp Database (26), and for urea Hui et al. reported $pK_a = 0.18$, which is similar to 0.1 from the SRC PhysProp Database (26) but quite different from 1.6 from SPARC); and (c) in a few cases the listed pK_a values do not make sense at all (e.g., the nonionizable chemicals, hydrocortisone, and PCB are erroneously reported to have $pK_a = 7.0$). Nevertheless, the characterization of ionic conditions of chemicals in their study is mostly in agreement with our calculations. Also, except for dopamine, values of $\log K_{ow}$, reported to be from Hansch and Leo (27), are mostly consistent with the recommended *star* ($\star$) values (1). The dopamine value listed by Hui et al. ($\log K_{ow} = -3.4$), which, according to Hansch et al. (1), was not corrected for ionization, is quite different from $\log K_{ow} = -0.05$ calculated using Daylight (2).

PHSC from plantar callus probably contains more protein material than uncallused SC. If this is true, partitioning into the protein domains may be more important for PHSC than for uncallused SC. As a result, K_{cw} measurements in PHSC would probably be smaller than measurements in uncallused SC, except for highly hydrophilic chemicals, which might partition more into PHSC, which contains more polar material than uncallused SC. These expectations generally agree with the data from Hui et al. (25), which are consistently overestimated by Eq. T2 except for compounds with $\log K_{ow} <$ about zero. The three PHSC measurements reported by Wester et al. (28) are inconsistent with the data from Hui et al. (25) and with our hypothesis that PHSC contain more polar material than uncallused SC. Based on these limited data, it is impossible to show conclusively whether K_{cw} measurements using PHSC accurately represent uncallused SC.

E. Analysis with the Lipophilic–Hydrophilic Domain Model

Some investigators have theorized that the existence of multiple domains within the stratum corneum should present regions with different polarities. In the simplest theory, the SC consists of polar and nonpolar regions. Hydrophilic chemicals should partition primarily in the polar region, while lipophilic chemicals should partition to the nonpolar regions. Partitioning between the nonpolar regions and water would be expected to increase with a measure of lipophilic character such as log K_{ow}. However, K_{ow} of the absorbed compound should affect partitioning between the polar region and water much less.

Here we examine the database of partition coefficients for trends indicating the presence of more than one partitioning mechanism. Figures 1 and 2 show that K_{cw} values increase approximately linearly with increases in log K_{ow}. There appears to be more uncertainty in the K_{cw} values measured for the most lipophilic and most hydrophilic compounds in the database. Because of this uncertainty, it is difficult to decide whether K_{cw} values for hydrophilic compounds deviate from the linear relationship observed for the lipophilic compounds in a statistically meaningful way.

One way to test the hypothesis of polar and nonpolar regions is to regress a mathematical model based on two-phase partitioning [e.g., Eqs. (7) or (9)] to the FV nonionized database and to compare the regression statistics with those from the conventional model (i.e., Eq. T2). We found that r^2 values for regressions to Eq. (7) were nearly constant over the range $-2 <$ log $K_{ow}^* < 1.5$. There is a local minimum in r^2 at log $K_{ow} = 0.0$, which was essentially the same as the r^2 value obtained with the conventional model, Eq. T2. A local maximum in r^2 occurs at log $K_{ow}^* = -1.5$. An optimum log K_{ow}^* may exist in the region $-1.5 <$ log $K_{ow}^* < 0.0$, but much more data for hydrophilic compounds is required to make decisive conclusions. The r^2 statistic improves as log $K_{ow}^* > 1$ because the analysis customizes the fit to the log $K_{ow} <$ log K_{ow}^* and log $K_{ow} >$ log K_{ow}^* regions. This does not indicate a change in uptake mechanism, because the log K_{ow} dependence is not very different for $K_{ow} < K_{ow}^*$ and $K_{ow} \geq K_{ow}^*$ [i.e., $b_H \cong b_L$ in Eq. (7)].

The simplified two-mechanism model in Eq. (9) was also regressed to the FV database. In this model, log K_{cw} is constant for all compounds with log $K_{ow} <$ log K_{ow}^*. In this case, the r^2 statistic improved monotonically with decreasing log K_{ow}^* in the range $-1 <$ log $K_{ow}^* < 1$, indicating that the regressed data are more consistent with the conventional model, Eq. T2. Thus the K_{cw} values in the FV database do not support the hypothesis that the SC uptake of hydrophilic compounds is mechanistically different from the SC uptake of lipophilic compounds. This observation could be real or an artifact of the small number of hydrophilic compounds with a rather narrow range of log K_{ow}.

From the analysis of the larger ($n = 170$) database of permeability coefficient measurements described in Chapter 3, we concluded that penetration mechanisms for hydrophilic and lipophilic compounds may be different. For the permeability coefficient data, the optimum log K_{ow}^{*} was approximately -0.5. Using log $K_{ow}^{*} = -0.5$ and Eqs. (7) and (9) to analyze the partition coefficients produced Eqs. T4 and T5, respectively, which are listed in Table 1. Consistent with earlier calculations, the coefficient b_{H} in Eq. (7) was not statistically significantly different from zero. As expected, Eq. T5 is quite similar to Eq. T2, since only a few compounds in the database have log $K_{ow} < -0.5$.

F. Analysis with the LSER Model

The LSER parameters, all from Abraham et al. (17), were calculated by averaging multiple normalized solvent effects on a variety of chemical properties involving many varied types of indicators. Abraham and colleagues (18) developed Eq. T6 in Table 1 by regression of Eq. (10) to a small database of SC–water partition coefficients ($n = 22$). We repeated this analysis on a subset of the data in the FV database for which LSER parameters were available. In Table A1, the names of chemicals with LSER parameters available are enclosed in brackets (e.g., [benzene]). Table A4 lists the LSER parameters and K_{cw} for these chemicals, which are mostly the lower MW chemicals from the FV database. In the resulting equation, Eq. T7, the terms labeled with asterisks (π and α) were not statistically significant at the 95% confidence level. In this regression, 83% of the variability in log K_{cw} can be attributed to variation in the LSER parameters (π, α, β, and V_x). For Eq. T8, developed by excluding the statistically insignificant terms and regressing the same database, 82% of the variability in log K_{cw} variations can be attributed to variations in only β and V_x.

Equation T8, which describes K_{cw} values of mostly low MW compounds, provides insight into the mechanism of SC partitioning. The two most important LSER parameters for predicting SC–water partitioning, the hydrogen-bonding basicity, and the molecular volume were also the most important parameters for predicting SC permeability coefficients. There is a particularly high level of variability in the LSER regression coefficients, but generally the values are as good as can be expected taking into account the rather large experimental errors that can be associated with the difficult determination of partition coefficients in biological tissues. In addition, the chemical diversity of the solutes used in the LSER regression is by no means optimal, which makes precise specification of the coefficients difficult. Prediction of SC permeability coefficients and SC–water partition coefficients would be greatly improved if measurements were made on compounds with optimally different combinations of LSER parameters (i.e., different forms of chemical diversity), as determined by statistical model discrimination guidelines.

The partition coefficient values for compounds in the LSER database were also analyzed using the conventional model (i.e., using log K_{ow} alone) to measure the improvement in fit from the more complicated LSER analysis. The result is Eq. T9. Comparing regression statistics, the predictive power of Eq. T8 is modestly greater (i.e., r^2 is larger) than Eq. T9 although the gain is small relative to the additional parameters (i.e., the F-ratio is smaller).

Figure 8 compares K_{cw} values for compounds in the LSER database with estimates made using either Eq. T8, based on LSER parameters, or Eq. T9, based on log K_{ow}. To indicate the variability of the entire database, we have included the K_{cw} values from the FV database compared with estimates from Eq. T2. The range of log K_{ow}, represented in the LSER database is similar to the entire valid database. However, the LSER database is composed of functionally simple,

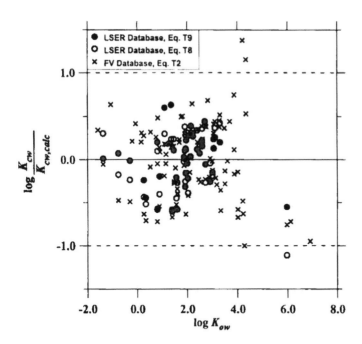

Figure 8 K_{cw} values in the LSER database divided by values $K_{cw,calc}$ calculated using Eq. T8 (developed from the LSER database and based on LSER parameters) compared to those calculated using Eq. T9 (developed from the LSER database and based on log K_{ow}) and to K_{cw} in the FV database divided by $K_{cw,calc}$ calculated using Eq. T2 (developed from the FV database and based on log K_{ow}).

low-MW compounds, primarily phenols. Although Eq. T8 does fit the LSER database moderately better than Eq. T9 (with only log K_{ow}), we believe that K_{ow} is the single most relevant parameter for estimating the SC–water partition coefficient.

G. Estimation of the Stratum Corneum Diffusion Coefficient

The ratio of the SC diffusion coefficient to SC thickness is another useful combination of parameters that appears in dermal absorption models (29). We estimate this ratio using principles that apply to membranes and models developed for predicting K_{cw} and the SC permeability coefficient. As discussed in Chapter 3, penetration through the SC membrane is frequently modeled as a solution-diffusion process (e.g., Ref. 29 and many others). In this case, the steady-state permeability coefficient for crossing the SC from an aqueous vehicle (P_{cw}) into an infinite sink depends upon the diffusivity of the chemical in the SC (D_c), the SC thickness (L_c), and K_{cw} as given by

$$P_{cw} = \frac{K_{cw}D_c}{L_c} \tag{12}$$

Logarithmic transformation and rearrangement of Eq. (12) gives the relationship, $\log (D_c/L_c) = \log P_{cw} - \log K_{cw}$.

Equation T1 from Table 1 of Chapter 3 provides reasonable estimates for $\log P_{cw}$:

$$\log P_{cw} = -2.44(0.12) + 0.514(0.04) \log K_{ow} - 0.0050\,(0.0005)\,MW \tag{13}$$

which can be combined with Eq. T2 for log K_{cw} to produce the following expression for D_c/L_c:

$$\log(D_c/L_c) = -2.408(0.14) + 0.098(0.05) \log K_{ow} \tag{14}$$
$$-0.0050(0.0005)\,MW$$

Eq. (14) represents the difference between Eqs. (13) and T2, and the standard errors, contained within parenthesis, were calculated as the square root of the sum of the squared standard errors for the log K_{ow} or MW terms in Eqs. (13) and T2. According to Eq. (14), D_c/L_c depends weakly on log K_{ow}.

Alternatively, an expression for the SC diffusion coefficient can be developed using LSER based equations for log P_{cw} and log K_{cw}. Here, we combine the LSER model listed in Table 3 of Chapter 3:

$$\log P_{cw} = -2.22(0.20) - 3.28(0.41)\,\beta = 1.73(0.17)\,V_x \tag{15}$$

with Eq. T8 to obtain the following expression for D_c/L_c:

$$\log\left(\frac{D_c}{L_c}\right) = -2.177 \ (0.38) - {}^*0.377 \ (0.62) \ \beta - 2.47 \ (0.25) \ V_x \qquad (16)$$

In this case, hydrogen bond basicity β does not contribute significantly to $\log(D_c/L_c)$ as indicated by the asterisk. Thus, according to Eq. (16), D_c/L_c depends on only molecular size as represented by V_x.

Equations (14) and (16) are two of only a few published equations for estimating D_c. Roberts and colleagues (30) presented an equation analogous to Eq. (16) as well as an equation similar to Eq. (14) but with an additional term for the number of hydrogen bonding groups. These authors also developed equations for estimating lag time to penetrate the SC (from which D_c/L_c can be calculated) that include either the number of hydrogen bonding groups or LSER parameters. Interestingly, molecular size given in terms of MW was not a statistically significant factor in their equations for D_c/L_c (30).

H. Comparing the Database with Other Published Models

Table 2 lists eighteen equations presented in the literature for estimating the K_{cw} for human skin. Also listed are the chemical classes upon which each equation was trained, a reference to the training data, and the range of MW and log K_{ow} of the training data. The equations from Brown and Hattis (Model 1), McKone (Model 7), McKone and Howd (Model 8), Raykar et al. (Model 9), and Hui et al. (Model 18) were not developed in the conventional form. Except for Brown and Hattis (31), these references can be consulted for the theoretical justification of these alternative forms. In Figures 9 and 10, the equations from Table 2 as well as Eq. T2 are plotted as functions of log K_{ow}.

Model 1, from Brown and Hattis (31) and also appearing in several subsequent publications (32,33), is incorrectly attributed to a paper by Roberts and colleagues (34). The equation from Brown and Hattis (31) does not appear in Roberts et al. (34) nor do the data in Roberts et al. (34) fit the Model 1 equation. Since Roberts does not remember developing any such equation (35), it is unlikely that Model 1 comes from a different publication by Roberts. So far we have been unable to find the original source of Model 1.

Model 2 was developed by splitting the semitheoretical equation developed by Potts and Guy (36) for estimating permeability coefficient data into separate solution and diffusion components as suggested by Eq. (12). This division is theoretically possible to the extent that parameters in the Potts and Guy equation can be assigned physicochemical significance.

Two equations are presented for Model 14. The first (log $K_{cw} = -0.343 + 0.761$ log K_{ow}) is an algebraic rearrangement of the equation (log $K_{ow} = 0.45 + 1.313$ log K_{cw}), which fits the data in Saket et al. (37). The equation presented

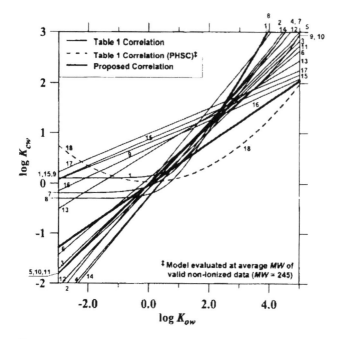

Figure 9 Eighteen published equations for estimating K_{cw} compared with Eq. T2 plotted as a function of log K_{ow}.

by Saket et al., log K_{cw} = 0.45 + 1.313 log K_{ow}, incorrectly reversed the roles of K_{cw} and K_{ow} and did not fit the data. The second equation, log K_{cw} = −0.28 + 0.739 log K_{ow}, was developed by regression of the data in Saket et al. (37) to an equation in the form of Eq. (4). It is this second equation that is designated as Model 14 in Figs. 9 and 10.

Two equations are also presented for Model 17. The first, log K_{cw} = 0.725 + 0.344 log K_{ow}, is an algebraic rearrangement of the published equation log K_{ow} = −2.11 + 2.91 log K_{cw}. The second equation log K_{cw} = 0.987 + 0.254 log K_{ow}, was developed by regression of the data in this Ref. 20 to Eq. (4). This second equation is designated as Model 17 in Figs. 9 and 10.

All eighteen of the equations listed in Table 2 account for the effect of lipophilicity through log K_{ow}, and only Model 18, developed using PHSC, included a term for the molecular size (i.e., MW). Consistent with our analysis, this indicates that MW has little influence on skin partitioning for the range of MW studied in these databases. Most of the equations in Table 2 are of the conventional form, Eq. (4), linearly correlated with log K_{ow} with or without a

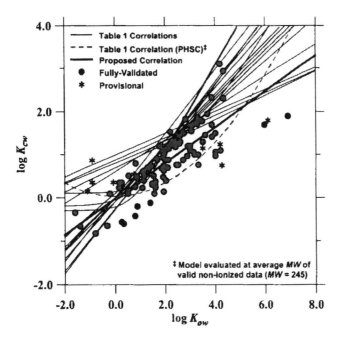

Figure 10   Eighteen published equations for estimating K_{cw} compared to experimental values from the FV and provisional databases.

constant term (i.e., intercept). Among the models with conventional form, values for the slope (b) and intercept (a) depended upon the data used in the development of each model. The slope varies from a low value of 0.25 (for Model 17) to a high value of 0.761 (for Models 2 and 14). Most other models, and the two models developed from the largest databases (i.e., Models 3 and 10) have slope values near 0.6. The intercept varies from a low value of -0.343 (for Model 14) to a high value of 0.99 (for Model 17). For several of the models b was forced to be zero. The two models developed from the largest databases (Models 3 and 10) have intercepts that are almost zero (i.e., -0.006 and -0.024, respectively). Intercept and slope values are not independent. Typically, equations with large values for the slope have smaller intercepts (e.g., Model 14) and equations with small values for the slopes have larger intercepts (e.g., Model 17).

As shown in Fig. 9, except for Model 18, the 18 equations from Table 2 are in relative agreement in the range $1 < \log K_{ow} < 4$ but begin to differ from one another at the high and low extremes of $\log K_{ow}$. The equations in Table 2 can be divided into two groups: those in which $b < 0.4$ and those in which $b >$

0.5. Models 13, 15, 16, and 17 predict a weaker dependence on log K_{ow} (i.e., b < 0.4) than do the other equations. Models 13, 16, and 17 are derived from entirely independent data sets without overlapping compounds, and Model 15 contains all compounds used in developing Models 16 and 17. Thus this apparent weak dependence on log K_{ow} cannot be attributed to one influential set of data. Model 9 predicts a weak K_{ow} dependence (i.e., the coefficient multiplying log K_{ow} is 0.24) for log K_{ow} < about 2.0 but a stronger K_{ow} dependence (i.e., the coefficient multiplying log K_{ow} is 0.91) when log K_{ow} > about 3. The remaining equations including two based on large and relatively diverse databases, Model 3 ($b = 0.57$ from 42 measurements) and 10 ($b = 0.59$ from 45 measurements), predict a stronger dependence on log K_{ow}. In Eq. T2, $b = 0.42$, which is intermediate to these models. Many of the models intersect near log $K_{ow} = 2.5$, but this has unknown significance.

Figure 10 compares the K_{cw} values from the FV and provisional databases (hydroxypregnenolone not shown) to the 18 models in Table 2 and Eq. T2. The measurements for partially ionized compounds, which are distinguished from measurements for nonionized compounds, were not adjusted for ionization. Since Eq. T2 was derived using the FV database, it provides the best overall fit of the nonionized data over the entire range of log K_{ow}. The other equations tend to overestimate the SC–water partition coefficients because these equations were primarily developed using K_{cw} values expressed on the basis of dry SC mass. Thus most of the equations in Table 2 were developed on K_{cw} values that were approximately four times larger than the data shown in Fig. 10. If these equations had been derived for data based on the wet SC volume, then on average the equation estimates would shift down by log(1/4) = −0.60.

It is evident from Fig. 10 that more data are needed at the extreme values of log K_{ow}. Few measurements are available for highly lipophilic compounds, and the existing measurements are more variable. It is not surprising then that model predictions vary widely for log K_{ow} > about 3 to 4. Similarly, the database is not sufficient to judge whether Models 1, 7, and 8 make appropriate predictions for hydrophilic compounds.

I. Final Considerations and Recommendations

There are an ample number of K_{cw} measurements in the literature to make reasonable estimates for moderately lipophilic compounds. However, more measurements are needed at large and small values of K_{ow} (i.e., log K_{ow} > 4 and log K_{ow} < 0, respectively).

The usual procedure in partition coefficient experiments is to measure the mass of absorbed chemical in the SC (M_c). The concentration of chemical in the SC (i.e., the mass of chemical in the SC over the volume of hydrated SC) is then calculated by dividing M_c by the cross-sectional area and the hydrated thickness of the SC [i.e., $C_c = M_c/(L_c A)$]. Unfortunately, the hydrated thickness, L_c, is difficult to measure precisely, introducing uncertainty. Notably, in mathematical models of dermal absorption, L_c and K_{cw} always appear as the product (i.e., $K_{cw}L_c$). This is significant, since experimental determination of $K_{cw}L_c$ involves readily determined quantities: the absorbed mass of chemical in the SC, the cross-sectional area of the skin sample, and the equilibrium concentration of chemical in the aqueous solution [i.e., $K_{cw}L_c = M_c/(AC_w)$]. Thus in model calculations, as well as for improved experimental reliability, it is preferred to report the product ($K_{cw}L_c$) rather than K_{cw} alone.

V. CONCLUSIONS

A set of data validation criteria, based upon physicochemical influences on the SC–water partition coefficient, was used to improve predictive estimates of K_{cw} data. The goal was to develop the most mechanistically relevant predictive model that could be supported by the assembled data. We found that the data are consistent with simple models in terms of log K_{ow} but do not support more complicated models such as those based on LSER parameters or two mechanisms of partitioning. Values of K_{cw} can be estimated within an order of magnitude using simple models involving log K_{ow}.

Measurements of K_{cw} from different laboratories for the same compound can sometimes differ by more than an order of magnitude. Different ways of expressing the concentration of the partitioning chemical in the SC is one source of variation among K_{cw} values. While differences in temperature and different amounts of ionization appear to contribute to this variation, these trends cannot be completely understood using the present set of data. More data for chemicals at the large and small extremes of K_{ow} and for chemicals that ionize is needed the better to understand these effects. Although causes could not be identified, certain laboratories appear to measure K_{cw} values that are systematically higher or lower than predicted by an equation developed using the entire FV database.

Eighteen equations from the literature were compared to the validated K_{cw} data presented here and to an equation developed from that database. Many of these equations provide adequate estimates. The main difference between equations, most of which are in the form of the conventional model, is the dependence on K_{ow}. Differences in the K_{ow} dependence for different data sets require further investigation. Finally, for improved experimental reliability it is preferred to report the quantity $K_{cw}L_c = M_c/(AC_w)$ rather than K_{cw} alone.

APPENDIX A. SC–WATER PARTITION COEFFICIENTS AND INPUT PARAMETERS

Table A1 Fully Validated Database of SC–Water Partition Coefficients

Compound[a]	log K_{ow}[b]	MW	T(°C)[c]	$K_{cw,exp}$[d]	K_{cw}[e]	Basis[f]	f_{nig}	pH[h]	t_{equi}[i]	Ref. No.
Acetaminophen	0.51	151.2	25	5.0	1.3	DM	1	<[7.0]	6–24	47
Aldosterone	1.08	360.4	26	6.8	1.7	DM[k]	1	ND	N/A	23
4-Amino-2-nitrophenol	1.53	154.1	25	13.0	2.5	DV[l]	1	N/A[m]	48	4
Atrazine	2.61	215.7	25	79.4	19.9	DM	1	N/A	24[n]	21
[Benzene]	2.13	78.1	31	30.0	7.5	DM	1	ND	N/A	49
[Benzo[a]pyrene]	5.97	252.3	25	199.5	49.9	DM	1	ND	24[n]	21
[Benzoic acid]	1.87	122.1	35	4.8	4.8	WV	1	2.75	24	50
[Benzyl alcohol]	1.10	108.1	25	4.1	4.1	WM[o]	1	ND	48[o]	45
Betamethasone	1.94	392.5	37[j]	5.6	1.4	DM	1	4.5	72	51
Betamethasone 17-valerate	3.60	476.0	37[j]	113.9	28.5	DM	1	4.5	72	51
[p-Bromophenol]	2.59	173.0	25	27.2[p]	27.2	WM	1	[5.3][q]	48	43
[Butanoic acid]	0.79	88.1	25	1.5	0.4	DM[k]	1	N/A	N/A	39
[Butanol]	0.88	74.1	25	2.5	0.6	DM[r]	1	ND	N/A	8
Chloramphenicol	1.14	323.1	25	10.0	2.5	DM	1	ND	24[n]	21
[Chlorocresol]	3.10	142.6	25	50.4[p]	50.4	WM	1	[5.3][q]	48	43
[p-Chlorophenol]	2.39	128.6	25	20.4[p]	20.4	WM	1	[5.3][q]	48	43
[o-Chlorophenol]	2.15	128.6	25	13.8[p]	13.8	WM	1	[4.6][q]	48	43
Chloroxylenol	[3.48]	115.5	25	60.8[p]	60.8	WM	1	[5.3][q]	48	43
Cortexolone	2.52	346.5	26[j]	23.0	5.8	DM[k]	1	ND	N/A	23
Cortexone	2.88	330.5	26[j]	37.0	9.3	DM[k]	1	ND	N/A	23
Corticosterone	1.94	346.5	26[j]	17.0	4.3	DM[l]	1	ND	N/A	23
Cortisone	1.47	360.5	25	8.1	2.0	DM	1	ND	48	37
Cortisone	1.47	360.5	26[j]	8.5	2.1	DM[k]	1	ND	N/A	23
Cortisone acetate	2.10	402.0	25	20.0	5.0	DM	1	ND	48	37
Cortisone butyrate	[1.76]	430.5	25	63.1	15.8	DM	1	ND	48	37
Cortisone hexanoate	[2.82]	458.5	25	269.2	67.3	DM	1	ND	48	37
Cortisone octanoate	[3.87]	486.5	25	851.1	212.8	DM	1	ND	48	37

Compound		MW	T							
[m-Cresol]	1.96	108.1	25	10.6[p]	10.6	WM	1	[5.5][q]	48	43
[o-Cresol]	1.95	108.1	25	10.6[p]	10.6	WM	1	[5.6][q]	48	43
[p-Cresol]	1.94	108.1	25	10.6[p]	10.6	WM	1	[5.6][q]	48	43
[p-Cresol]	1.94	108.1	37	22.0	5.5	DM[r]	1	4	N/A	24
[4-Cyanophenol]	1.60	129.1	25	7.9	2.0	DM	>0.90	<[6.9]	6–24	47
DDT	6.91	354.5	25	316.2	79.1	DM	1	ND	6–24	47
Diazepam	2.99	284.8	25	63.1	15.8	DM	1	N/A	6	20
2,4-Dichlorophenol	3.06	163.0	25	45.4[p]	45.4	WM	1	[4.4][q]	48	43
β-Estradiol	4.01	272.4	26	46.0	11.5	DM[k]	1	ND	N/A	23
β-Estradiol	4.01	272.4	25	125.9	31.5	DM	1	ND	6	20
β-Estradiol	4.01	272.4	32	9.3	9.3	WM	1	ND	48	6
Estriol	2.45	288.4	26	23.0	5.8	DM[k]	1	ND	N/A	23
Estrone	3.13	270.4	26	46.0	11.5	DM[k]	1	ND	N/A	23
[Ethanol]	−0.31	46.0	25	0.9	0.2	DM[t]	1	ND	N/A	8
[p-Ethylphenol]	2.58	122.2	25	18.3[p]	18.3	WM	1	[5.7][q]	48	43
[Heptanoic acid]	[2.41]	130.2	25	60.3	15.1	DM[k]	1	N/A	N/A	39
[Heptanol]	2.72	116.0	25	30.0	7.5	DM[t]	1	ND	N/A	8
[Hexanoic acid]	1.92	116.2	25	12.0	3.0	DM[t]	1	N/A	N/A	39
[Hexanol]	2.03	102.2	25	10.0	2.5	DM[t]	1	ND	N/A	8
Hydrocortisone (HC)	[1.61]	362.5	25	7.1	1.8	DM	1	ND	48	37
Hydrocortisone (HC)	1.61	362.5	37	6.9	1.7	DM	1	ND	48	37
Hydrocortisone (HC)	1.61	362.5	26	7.0	1.8	DM[k]	1	ND	N/A	23
Hydrocortisone (HC)	1.61	362.5	25	9.5	2.4	DM	1	ND	6	20
HC-21-yl-acetate	[1.16]	404.0	25	17.0	4.2	DM	1	ND	48	37
HC-21-yl-N,N-dimethylsuccinate	[0.88]	489.6	37	12.0	3.0	DM	1	ND	48–72	3
HC-21-yl-hexanoate	[3.28]	460.6	37	207.9	52.0	DM	1	ND	48–72	3
HC-21-yl-hexanoate	[3.28]	460.6	25	245.5	61.4	DM	1	ND	48	37
HC-21-yl-hydroxy hexanoate	[1.29]	476.6	37	20.0	5.0	DM	1	ND	48–72	3
HC-21-yl-octanoate	[4.34]	488.7	37	3423.3	855.8	DM	1	ND	48–72	3
HC-21-yl-octanoate	[4.34]	488.7	25	812.8	203.2	DM	1	ND	48	37
HC-21-yl-pentanoate	[2.75]	446.0	25	125.9	31.5	DM	1	ND	48	37
HC-21-yl-pimelamate	[0.82]	503.6	37	25.0	6.3	DM	1	ND	48–72	3
HC-21-yl-propionate	[1.69]	418.5	37	30.0	7.5	DM	1	ND	48–72	3

Table A1 Continued

Compound[a]	log K_{ow}[b]	MW	$T(°C)$[c]	$K_{cw,rep}$[d]	K_{cw}[e]	Basis[f]	f_{nig}	pH[h]	t_{equil}[i]	Ref. No.
HC-21-yl-propionate	[1.69]	418.5	25	32.4	8.1	DM	1	ND	48	37
HC-21-yl-succinamate	[0.17]	461.6	37	9.0	2.3	DM	1	ND	48-72	3
4-Hydroxybenzyl alcohol	0.25	124.1	37	9.0	2.3	DM[k]	1	4	N/A	24
α-(4-Hydroxyphenyl)acetamide	[−0.21]	151.2	37	5.0	1.3	DM[k]	1	4	N/A	24
17 α-Hydroxyprogesterone	3.17	330.5	26[j]	40.0	10.0	DM[k]	1	ND	N/A	23
[4-Iodophenol]	2.91	220.0	25	63.1	15.8	DM	1	<[7.0]	6-24	47
Lauric acid	4.20	200.3	32	5060	1270	DM	1	3.2	24-48	9
[Methanol]	−0.77	32.0	25	0.6	0.2	DM[r]	1	ND	N/A	8
Methyl 4-hydroxyphenyl acetate	[1.15]	166.0	37	13.0	3.3	DM[k]	1	4	N/A	24
Methyl HC-21-yl-pimelate	[2.20]	518.6	37	136.0	34.0	DM	1	ND	48-72	3
Methyl HC-21-yl-succinate	[1.38]	476.6	37	22.0	5.5	DM	1	ND	48-72	3
Methyl hydroxybenzoate	1.96	152.1	25	7.9	7.9	WM[o]	1	ND	48°	45
[β-Naphthol]	2.70	144.2	25	33.4[p]	33.4	WM	1	[5.2][q]	48	43
2-Nitro-p-phenylenediamine	0.53	153.1	25	13	2.5	DV[l]	1	N/A[m]	48	4
[m-Nitrophenol]	2.00	139.1	25	12.1[p]	12.1	WM	1	[4.8][q]	48	43
[p-Nitrophenol]	1.91	139.1	25	12.8[p]	12.8	WM	1	[3.9][q]	48	43
N-Nitrosodiethanolamine	[−1.58]	134.1	32	1.8	0.5	DM	1	ND	96	52
[Octanoic acid]	3.05	144.2	25	141.3	35.3	DM[k]	1	N/A	N/A	39
[Octanol]	3.00	130.2	25	50.0	12.5	DM[k]	1	ND	N/A	8
[Pentanoic acid]	1.39	102.1	25	3.0	0.8	DM[k]	1	N/A	N/A	8
[Pentanol]	1.56	88.0	25	5.0	1.3	DM[r]	1	ND	N/A	8
4-Pentyloxyphenol	3.50	180.2	25	79.4	19.9	DM	1	<[6.2]	6-24	47
[Phenethyl alcohol]	1.36	122.2	25	4.8	4.8	WM[o]	1	ND	48°	45
[Phenol]	1.46	94.1	25	5.4[p]	5.4	WM	1	[5.4][q]	48	43
o-Phenylenediamine	0.15	108.1	25	6.9	1.3	DV[l]	1	[7.6]	48	4
Pregnenolone	4.22	316.5	26[j]	50.0	12.5	DM[k]	1	ND	N/A	23
Progesterone	3.87	314.5	26[j]	104.0	26.0	DM[k]	1	ND	N/A	23
Progesterone	3.87	314.5	25	199.5	49.9	DM	1	ND	6	20
[Propanoic acid]	0.33	74.1	25	1.0	0.3	DM[k]	1	N/A	N/A	39
[Propanol]	0.25	60.0	25	1.1	0.3	DM[r]	1	ND	N/A	8

[Resorcinol]	0.80	110.1	25	1.8[p]	1.8	WM	—	1	[5.4][q]	48	43
Testosterone	3.32	288.4	26	23.0	5.8	DM[k]	—	1	ND	N/A	25
Testosterone	3.32	288.4	25	39.8	10.0	DM	—	1	ND	6	20
[Thymol]	3.30	150.2	25	72.7[p]	72.7	WM	—	1	[6.0][q]	48	43
2,4,6-Trichlorophenol	3.69	197.5	25	89.0[p]	89.0	WM	—	1	[3.6][q]	48	43
[Water]	−1.38	18.0	25	0.9	0.2	DM[r]	—	1	ND	N/A	8
[3,4-Xylenol]	2.23	122.2	25	16.9[p]	16.9	WM	—	1	[5.8][q]	48	43

[a] The compound investigated. Compounds contained within brackets (e.g. [Benzene]) were used in LSER analysis.

[b] Reported log K_{ow} are taken from the Hansch Starlist (1), unless contained within brackets (e.g., for chloroxylenol [3.48]), in which case they were calculated using Daylight software (2).

[c] Temperature at which the partition coefficient was measured.

[d] Reported SC–water partition coefficient prior to adjustment for basis.

[e] SC–water partition coefficients adjusted to the basis of hydrated-SC volume.

[f] Basis on which the observed SC–water partition coefficient was reported. SC concentration expressed relative to mass of dry SC (DM), to mass of hydrated SC (WM), or to volume of dry SC (DV), or to volume of hydrated SC (WV).

[g] The nonionized fraction determined from pK, values calculated in SPARC (53) at 25°C and adjusted to the experimental temperature as shown in Table A5.

[h] Reported solution pH unless contained within brackets (e.g., for p-bromophenol [5.3]), in which case the pH was calculated from the reported chemical concentration and calculated pK, values (as given in Table A5). Chemicals that do not dissociate are indicated by ND when pH was not reported.

[i] Time allowed for the absorbing chemical to equilibrate with the SC.

[j] The temperature for measuring the SC–water partition coefficients is assumed equal to that used by these authors in experiments to determine P_{cw}.

[k] Scheuplein and coworkers used the same experimental procedure as described in Ref. 40, which used a dry-SC volume basis.

[l] Bronaugh and Congdon adjusted their measured values by a factor of 1.32 to place it on a dry-SC volume basis. We first convert their data to a DM basis by multiplying by 1.32 g dry SC/mL dry SC and adjusting by the 0.25 conversion factor.

[m] For 4-amino-2-nitrophenol and 2-nitro-p-phenylenediamine the pH attained cannot be calculated because the concentration was not provided. Although, if the starting pH was near 7 then the pH will change in a direction that f_{ni} is always 1.

[n] The individual times varied. Most equilibration times reported were 24 hours (21).

[o] There is reasonable evidence to assume that values are consistent with previous measurements (e.g., Ref. 43).

[p] $K_{s,sep}$ values for phenolics were determined by a desorption technique in which the solution and SC concentration were measured after solute-laden SC was allowed to equilibrate with a solute-free aqueous solution (43).

[q] The concentrations were consistently dilute (~1% [w/v]) but not reported. We have used a concentration of 1% (w/v) to calculate the pH and the nonionized fraction. In all cases, $f_{ni}=1$.

[r] Values for n-alcohols and water reported by Scheuplein (54) have a DM basis as explained in the reference.

[s] These authors cite Raykar and colleagues (3) for experimental protocol. We assume that the basis is the same as in Ref. 3.

Table A2 Provisional Database of SC–Water Partition Coefficients

Compound[a]	log K_{cw}[b]	MW	$T(°C)$	$K_{cw,exp}$[c]	K_{cw}[d]	Basis[e]	f_{ni}[f]	pH[g]	t_{equil}[h]	Ref. No.
Acitretin	[6.12]	326.4	25	251.2	62.8	DM	N/A	N/A[i]	24	20
Caffeine	-0.07	194.2	25	9.1	2.3	DM	N/A	N/A[i]	6	20
5-Fluorouracil (+ − + −)	-0.89	130.1	32	0.9	0.9	WM	<0.1	4.75[j]	24	55
5-Fluorouracil (+ − + −)	-0.89	130.1	25	3.4[k]	0.9	DM[k]	<0.1	N/A[i]	24[l]	21
5-Fluorouracil (+ − + −)	-0.89	130.1	20	2.9	2.9	WM	<0.1	4.75[j]	4	56
HC-21-yl-hemipimelate (−)	[1.82]	504.6	37	68.0	17.0	DM	0.78	4	48–72	3
HC-21-yl-hemisuccinate (−)	[0.91]	462.5	37	11.0	2.8	DM	0.12	5.5	48–72	3
4-Hydroxyphenyl acetic acid	0.75	152.1	37	14.0	3.5	DM[m]	0.73	4	N/A	24
Hydroxypregnenolone[n]	N/A	N/A	26°	43.0	10.8	DM[p]	1	ND	N/A	23
Ibuprofen	3.50	206.3	25	56.2[k]	14.1	DM[k]	N/A	N/A[i]	24[l]	21
Indomethacin	4.27	357.8	25	22.4[k]	5.6	DM[k]	N/A	N/A[i]	24[l]	21
Lauric acid	4.20	200.3	25	69.2[k]	17.3	DM[k]	N/A	N/A[i]	24[l]	21
Uracil	-1.07	112.1	25	5.8[k]	1.5	DM[k]	N/A	N/A[i]	24[l]	21

[a] The compound investigated. All positive (+) and negative (−) ionic charges (for the chemical at experimental conditions) are indicated. For example, 5-fluorouracil with two positive and two negative charges is indicated by (+ − + −).

[b] Reported log K_{ow} are taken from the Hansch Starlist (1), unless contained within brackets, in which case they were calculated (2).

[c] Reported SC–water partition coefficient prior to adjustment for basis.

[d] SC–water partition coefficients adjusted to the basis of hydrated-SC volume.

[e] Basis on which the observed SC–water partition coefficient was reported. SC concentration expressed relative to mass of dry SC (DM), to mass of hydrated SC (WM), or to volume of dry SC (DV), or to volume of hydrated SC (WV).

[f] The nonionized fraction determined from pK_a values calculated in SPARC (53) at 25°C and adjusted to the experimental temperature as shown in Table A5.

[g] Reported solution pH unless contained within brackets, in which case the pH was calculated from the reported chemical concentration and calculated pK_a values (as given in Table A5). Chemicals that do not dissociate are indicated by ND when no pH was reported.

[h] Time allowed for the absorbing chemical to equilibrate with the SC.

[i] The natural pH depends on the chemical concentration, which was not reported.

[j] This pH was attained by a saturated 5-fluorouracil solution (57).

[k] This information was provided by personal communication with the corresponding author (58).

[l] The times varied with the chemical, but most of the equilibration times reported were 24 hours (21).

[m] These authors cite Raykar and colleagues (3) for experimental protocol. We assume that the basis is the same as in Ref. 3.

[n] Hydroxypregnenolone is provisional because we could not unambiguously identify the molecular formula for this steroid.

[o] The temperature for measuring the SC–water partition coefficients is assumed to equal that used by these authors in experiments to determine P_{cw}.

[p] Scheuplein and coworkers used the same experimental procedure as Ref. 40, which used a dry-mass SC basis.

Table A3 Excluded Database of SC–Water Partition Coefficients

Compound[a]	log K_{ow}[b]	MW	$T(°C)$	$K_{cw,rep}$[c]	K_{cw}[d]	Basis[c]	f_m[f]	pH[g]	t_{equil}[h]	Ref. No.
[Alachlor]	3.52	270.0	35	[12.0]	3.0	DM	1	6.6	24	25
[Aminopyrine]	1.00	231.3	35	[0.68]	0.2	DM	0.91	4.8	24	25
[Atrazine]	2.61	215.7	35	[5.5]	1.4	DM	1	6.2	24	25
[Benzene]	2.13	78.1	N/A	199.0	49.8	DM	1	ND	0.5	28
Cyclosporin-A[i]	N/A	1201.0	25	79.4	19.9	DM	<0.1	N/A	24[j]	21
[Dopamine hydrochloride] (+)	[−0.05][k]	153.2	35	[10.0]	2.5	DM	<0.1	4	24	25
[2,4-D] (−)	2.81	221.0	35	[17.8]	4.4	DM	<0.1	5.9	24	25
[Glycine] (+−)	−3.21	75.0	35	[3.1]	0.8	DM	<0.1	4.5	24	25
[Glyphosate] (−−−+)	[−1.7][l]	169.0	35	[0.9]	0.2	DM	<0.1	7.3	24	25
[Hydrocortisone]	1.61	362.5	35	[1.5]	0.4	DM	1	8.1	24	25
[Malathion]	2.36	330.0	35	[4.7]	1.1	DM	1	5.2	24	25
Nicorandil[m]	[0.65]	211.2	37	7.25	1.8	DM	1	[7.2]	24	59
[p-Nitroaniline]	1.39	138.1	N/A	25.6	6.4	DM	1	[7.0]	0.5	28
[PCB] (54% chlorine)[n]	6.62°	325.1	35	1175.0	293.7	DM	1	6.5	24	25
PCB (54% chlorine)[n]	6.62°	325.1	25	199.5	49.9	DM	1	ND	6–24	47
[PCB] (54% chlorine)[n]	6.62°	325.1	N/A	22300	5560	DM	1	ND	0.50	28
[Theophylline]	−0.02	180.2	35	[0.4]	0.1	DM	0.33	7.6	24	25
[Urea]	−2.11	60.1	35	[0.5]	0.1	DM	1	2.1[p]	24	26

[a] The compound investigated. Compounds contained within brackets (e.g., [Alachlor]) were measured using sieved, powdered human stratum corneum (PHSC) using plantar calluses. All positive (+) and negative (−) ionic charges (for the chemical at experimental conditions) are indicated. For example, glyphosate with three negative charges and one positive charge is indicated by (−−−+).

[b] Reported log K_{ow} are taken from the Hansch Starlist (1), unless contained within brackets (e.g., for dopamine [−0.05]), in which case they were calculated using Daylight software (2).

[c] Reported SC–water partition coefficient prior to adjustment for basis.

[d] SC–water partition coefficients adjusted to the basis of hydrated-SC volume. Values within brackets (e.g., for alachlor [12.0]) were digitized.

[e] Basis on which the observed SC–water partition coefficient was reported. SC concentration expressed relative to mass of dry SC (DM), to mass of hydrated SC (WM), or to volume of dry SC (DV), or to volume of hydrated SC (WV).

[f] The nonionized fraction determined from pK_a values calculated in SPARC (53) at 25°C and adjusted to the experimental temperature as shown in Table A5.

[g] Reported solution pH unless contained within brackets (e.g., for nicorandil [7.2]), in which case the pH was calculated from the reported chemical concentration and calculated pK_a values (as given in Table A5). Chemicals that do not dissociate are indicated by ND when no pH was reported.

[h] Time allowed for the absorbing chemical to equilibrate with the SC.

[i] Cyclosporin-A is excluded because it is much larger than the other compounds in the database and may partition by different mechanisms from smaller compounds.

[j] The individual times varied. Most equilibration times reported were 24 hours (21).

[k] log K_{ow} was calculated for dopamine, not the hydrochloride salt.

[l] log K_{ow} was reported by the authors as recommended by Hansch and Leo (27). However, this value was not designated as recommended in (1).

[m] Excluded because the stratum corneum used was plantar callus (59).

[n] PCB was excluded because it was a polydispersed mixture and not a single compound.

[o] The log K_{ow} for PCB was calculated as the average of 14 Starlist values for variously substituted pentachlorobiphenyls (1).

[p] The authors state that urea was added to distilled water (presumably near pH 7). If true, the reported pH of 2.1 cannot be attained unless a pH adjustment was made (suggesting that pH = 2.1 may be in error). At any pH value higher than 2.1, urea is essentially nonionized. Since it is likely that the actual pH > 2.1, we have assumed $f_{ni} = 1$.

Table A4 LSER Parameters[a] and Permeability Coefficient Values for Chemicals in the LSER Database

Compound	π	α	β	$V_x \times 10^{-2}$ [cm³ mol⁻¹][b]	T (°C)	K_{cw}[c]	Ref. No.
Benzene	0.52	0.00	0.14	0.716	31	7.5	49
Benzo[a]pyrene	1.98	0.00	0.44	1.954	25	49.9	21
Benzoic Acid	0.90	0.59	0.40	0.932	35	4.8	50
Benzyl alcohol	0.87	0.33	0.56	0.916	25	4.1	45
p-Bromophenol	1.17	0.67	0.20	0.950	25	27.2	43
Butanoic acid	0.62	0.60	0.45	0.747	25	0.4	39
Butanol	0.42	0.37	0.48	0.731	25	0.6	8
Chlorocresol	1.02	0.65	0.22	1.038	25	50.4	43
o-Chlorophenol	0.88	0.32	0.31	0.898	25	13.8	43
p-Chlorophenol	1.08	0.67	0.20	0.898	25	20.4	43
m-Cresol	0.88	0.57	0.34	0.916	25	10.6	43
o-Cresol	0.86	0.52	0.30	0.916	25	10.6	43
p-Cresol	0.87	0.57	0.31	0.916	25	10.6	43
p-Cresol	0.87	0.57	0.31	0.916	37	5.5	24
4-Cyanophenol	1.63	0.79	0.29	0.930	25	2.0	47
Ethanol	0.42	0.37	0.48	0.449	25	0.2	8
p-Ethylphenol	0.90	0.55	0.36	1.057	25	18.3	43
Heptanoic acid	0.60	0.60	0.45	1.169	25	15.1	39

Heptanol	0.42	0.37	0.48	1.154	25	7.5	8
Hexanoic acid	0.60	0.60	0.45	1.028	25	3.0	39
Hexanol	0.42	0.37	0.48	1.013	25	2.5	8
4-Iodophenol	1.22	0.68	0.20	1.033	25	15.8	47
Methanol	0.44	0.43	0.47	0.308	25	0.2	8
β-Naphthol	1.08	0.61	0.40	1.144	25	33.4	43
m-Nitrophenol	1.57	0.79	0.23	0.949	25	12.1	43
p-Nitrophenol	1.72	0.82	0.26	0.949	25	12.8	43
Octanoic acid	0.60	0.60	0.45	1.310	25	35.3	39
Octanol	0.42	0.37	0.48	1.295	25	12.5	8
Pentanoic acid	0.60	0.60	0.45	0.888	25	0.8	39
Pentanol	0.42	0.37	0.48	0.872	25	1.3	8
Phenol	0.89	0.60	0.30	0.775	25	5.4	43
2-Phenylethanol	0.91	0.30	0.64	1.057	25	4.8	45
Propanoic Acid	0.65	0.60	0.45	0.606	25	0.3	39
Propanol	0.42	0.37	0.48	0.590	25	0.3	8
Resorcinol	1.00	1.10	0.58	0.834	25	1.8	43
Thymol	0.60	0.27	0.35	1.340	25	72.7	43
Water	0.45	0.82	0.35	0.167	25	0.2	8
3,4-Xylenol	0.86	0.56	0.39	1.057	25	16.9	43

[a] All LSER parameters were reported by Abraham and colleagues (17).
[b] The reported V_x have been divided by a factor of 100 to make the values comparable in magnitude to the other LSER parameters.
[c] Partition coefficient values are from Table A1.

Table A5 Estimates of the Nonionized Fraction of the Absorbing Compound at Experimental Temperature

Compound[a]	pK_a (25°C)[a]	T (°C)	ΔH (kcal mol^-1)[b]	pK_a (T)[c]	C_w (mol L^-1)[d]	pH[e]	f_{ni}[f]	Ref. No.
Acetaminophen	9.8	25	5	9.8	>3.3E-7	<[7.0][g]	1	47
Aminopyrine	4.06	35	10	3.82	—	4.8	0.91	25
p-Bromophenol	9.4	25	5	9.4	0.058[h]	[5.3]	1	43
Chlorocresol	9.55	25	5	9.55	0.070[h]	[5.3]	1	43
o-Chlorophenol	8.25	25	5	8.25	0.078[h]	[4.6]	1	43
p-Chlorophenol	9.43	25	5	9.43	0.078[h]	[5.3]	1	43
Chloroxylenol	9.68	25	5	9.68	0.086[h]	[5.3]	1	43
m-Cresol	10.13	25	5	10.13	0.092[h]	[5.5]	1	43
o-Cresol	10.33	25	5	10.33	0.092[h]	[5.6]	1	43
p-Cresol	10.31	25	5	10.31	0.092[h]	[5.6]	1	43
4-Cyanophenol	7.77	25	5	7.77	>3.9E-7	<[6.9][g]	>0.88	47
2,4-Dichlorophenol	7.67	25	5	7.67	0.061[h]	[4.4]	1	43
p-Ethylphenol	10.29	25	5	10.29	0.082[h]	[5.7]	1	43
4-Iodophenol	9.33	25	5	9.33	>3.2E-7	<[7.0][g]	1	47
Lauric Acid	4.73	32	0	4.73	—	3.2	1	9
β-Naphthol	9.34	25	5	9.34	0.069[h]	[5.2]	1	43
Nicorandil	2.87	37	5	2.72	2.37E-03	[7.2]	1	59

p-Nitroaniline	1.35	N/A	7.5	1.35[i]	3.55E-05	[7.0]	1	28
m-Nitrophenol	8.39	25	5	8.39	0.072[h]	[4.8]	1	43
p-Nitrophenol	6.83	25	5	6.83	0.072[h]	[3.9]	1	43
4-Pentyloxyphenol	10.19	25	5	10.19	>6.1E-7	<[6.2][g]	1	47
Phenol	10	25	5	10	0.106[h]	[5.4]	1	43
o-Phenylenediamine	4.24	25	7.5	4.24	7.72E-04	[7.6]	1	4
Resorcinol	9.86	25	5	9.86	0.091[h]	[5.4]	1	43
Theophiline	8.14	35	10	7.9	—	7.6	0.33	25
Thymol	10.82	25	5	10.82	0.066[h]	[6.0]	1	43
2,4,6-Trichlorophenol	5.94	25	5	5.94	0.051[h]	[3.6]	1	43
3,4-Xylenol	10.44	25	5	10.44	0.082[h]	[5.8]	1	60

[a] pK$_a$ values calculated in SPARC (53) at 25°C using methods described in Chapter 3.

[b] These heats of ionization are approximate values obtained from the literature (61).

[c] Calculated using an integrated form of the van't Hoff equation, Eq. (3) in Chapter 3.

[d] Solution concentration provided only when it was needed to calculate the pH.

[e] The pH was reported in the original paper, unless contained within brackets, in which case it was calculated from pK$_a$ (T) and the solution concentration assuming that pH was 7.0 prior to chemical addition.

[f] The nonionized fraction calculated using Eq. (1) in Chapter 3 when one pK$_a$ is dominant. Otherwise it was determined using a more rigorous solution of simultaneous equilibrium as discussed in the Chapter 3.

[g] The pH was calculated for the lowest solute concentration used to measure partition coefficient.

[h] The concentrations were consistently dilute (~1% [w/v]) but not reported. We have used a concentration of 1% (w/v) to calculate the pH and the nonionized fraction.

[i] The pK$_a$ value at 25°C was used to calculate the pH, since the temperature was not reported.

APPENDIX B.   DOCUMENTATION OF PARTITION COEFFICIENT DATA

This appendix contains specific information about the K_{cw} values included in the three databases in Appendix A. Details are arranged alphabetically by the last name of the first author of each investigation.

Anderson et al., 1989 (24)

This paper reports $K_{cw,rep}$ values for five chemicals [α-(4-hydroxyphenyl) acetamide, 4-hydroxy-benzyl alcohol, 4-hydroxyphenylacetic acid, methyl 4-hydroxyphenylacetate, and p-cresol]. Because 4-hydroxyphenylacetic acid was more than 10% ionized (pH = 4.0), the measurement for this compound is listed in the provisional database. The other compounds were not ionized. The time for equilibration was not specified. We assumed that partition coefficients were based on dry SC mass, because this was the basis in the paper referenced by the authors as describing their experimental procedure (3).

Anderson et al., 1976 (43)

This paper reports $K_{cw,rep}$ values (based on hydrated SC mass) for various phenolic compounds in Table 3. Mean values were based on measurements in up to four SC samples. Measurements at other temperatures (12.6–34.5°C) were reported but were not incorporated into the database, since variation of temperature (over this range) appeared to have little effect on SC–water partitioning. The concentration used was low (frequently about 1% w/v or 1 g/100 mL) but not reported. We assumed a concentration of 1% w/v in estimating the natural pH of the solutions and f_{ni}. This was consistent with the analysis in Chapter 3 of permeability coefficient data for these compounds. Roberts et al. (34) presented data for several of the compounds in Ref. 43. We included the measurements from this reference rather than those from Roberts et al. (34), because it appears that measurements from Roberts et al. (34) were incorporated into the mean values reported in this reference.

Barry et al., 1985 (62)

This paper reports $K_{cw,rep}$ values (based on dry SC mass) for three chemicals (mannitol, hydrocortisone, and progesterone) in Table 1. The equilibration time was 14 days, and it is possible that the SC was altered by this long contact time with water. These compounds were not ionized.

Blank and McAuliffe, 1985 (49)

This paper reports a $K_{cw,rep}$ value (based on dry SC mass) for benzene in Table II (water vehicle). The authors indicate that enough time was allowed to establish equilibrium, but the exact time was not provided.

Bronaugh and Congdon, 1984 (4)

This paper reports $K_{cw,rep}$ values (based on dry SC volume) for o-phenylenediamine, 2-nitro-p-phenylenediamine, and 4-amino-2-nitrophenol in Table I. According to the authors, values were measured on the basis of SC mass and then multiplied by 1.32 g of dry SC per mL of dry SC. These $K_{cw,rep}$ values were adjusted to the hydrated SC volume basis by multiplying $K_{cw,rep}$ by 0.19 (=0.25 g of dry SC per mL of hydrated SC divided by 1.32 g of dry SC per mL of dry SC). Unlike the permeability coefficient experiments by these authors, pH in the partition coefficient experiments was allowed to go to natural levels. Equilibrium solute concentrations were not reported for 2-nitro-p-phenylenediamine and 4-amino-2-nitrophenol. However, if the starting pH was near 7, f_{ni} should be approximately 1. The concentration used in the measurement for o-phenylenediamine was reported in Table II, and we estimate a pH of 7.6 for this measurement (see Table A5). The equilibration time for these measurements was 48 hours.

Bronaugh et al., 1981 (52)

This paper reports a $K_{cw,rep}$ value (based on dry SC mass) for N-nitrosodiethanolamine in Table I. Four days were allowed for equilibration. N-nitrosodiethanolamine was not ionized in the experimental solution.

Cornwell and Barry, 1994 (55)

This paper reports a $K_{cw,rep}$ value (based on hydrated SC mass) for fluorouracil in Table 4 as the average of five control measurements (i.e., without penetration enhancer). The equilibration time was 24 hours. The estimate that $f_{ni} < 0.1$ was based on pK_a values from SPARC at a concentration of 10 mg mL^{-1}. This measurement is listed in the provisional database because $f_{ni} < 0.9$.

Hui et al., 1995 (25)

The $K_{cw,rep}$ values for sieved powdered human stratum corneum (PHSC) trimmed from plantar calluses were digitized from Fig. 4 in this reference. According to one of the authors, data were based on dry SC mass (63). At the natural pH of the solutions, reported in Table 2 of this reference, several of the compounds in the study (i.e, dopamine, glycine, urea, glyphosate, theophylline, aminopyrine, 2,4-D, and alachlor) were partially ionized. Equilibration was attained over 24 hours. These $K_{cw,rep}$ values were placed in the excluded database because they were measured using PHSC.

Kubota and Maibach, 1993 (51)

The $K_{cw,rep}$ values (based on dry SC mass) for betamethasone (BMS) and BMS-17-valerate were calculated as the linear regression slope (zero intercept forced) of the equilibrium SC (y-axis) and aqueous solution concentrations (x-axis) in Fig. 3a and 3b. The individual data were digitized. Neither compound was ionized in the experimental solutions (pH = 4.5). Equilibrium was attained over 72 hours.

Megrab et al., 1995 (6)

The $K_{cw,rep}$ value (based on dry SC mass) for β-estradiol was digitized from Fig. 2 in this reference. Estradiol did not dissociate at the solution pH, which was allowed to go to natural levels. The equilibration time was 48 hours. Partition coefficients were reported on the basis of hydrated SC mass.

Parry et al., 1990 (50)

This paper reports a $K_{cw,rep}$ value (based on hydrated SC volume) for benzoic acid in the results and discussion section. This value was based on the nonionized fraction of benzoic acid. The time allowed for equilibration was 24 hours.

Raykar et al., 1988 (3)

The $K_{cw,rep}$ values (based on dry SC mass) were taken directly from Table IV (labeled there as intrinsic partition coefficients). Average partition coefficients (calculated as the sample size–weighted average for measurements made with skin with different lipid contents) were reported for compounds identified as li-1k (i.e., methyl-hydrocortisone-21-yl-pimelate, hydrocortisone-21-yl-hexanoate, and hydrocortisone-21-yl-octanoate). The $K_{cw,rep}$ value for hydrocortisone-21-yl-hemisuccinate (1d) was measured at pH = 5.5 and for hydrocortisone-21-yl-hemipimelate (1e) was measured at pH = 4.0. These two compounds were partially ionized (i.e., f_{ni} < 0.9) at these conditions, and consequently, $K_{cw,rep}$ values for these compounds are listed in the provisional database. The remaining nine chemicals were not ionized. The time allowed for equilibration was 48–72 hours.

Roberts, 1976 (45)

The $K_{cw,rep}$ values (based on hydrated SC mass) for three compounds (benzyl alcohol, phenethyl alcohol, and methyl paraben) were reported in Roberts' thesis but were not reported in the paper by Anderson and colleagues (43). Although we do not have the thesis, from personal communication with Roberts (35) we determined that these measurements were made using the procedure reported by Anderson and colleagues (43). Table 4.10 in Roberts' thesis contains data for benzyl alcohol, phenethyl alcohol, and methyl hydroxybenzoate in addition to data published by Anderson and colleagues (43) for 18 other compounds. We assumed that equilibration times were comparable if not identical to those in Anderson et al. (43).

Roberts et al., 1975 (34)

This paper reportes $K_{cw,rep}$ values for seven chemicals (benzyl alcohol, phenethyl alcohol, phenol, p-cresol, m-cresol, o-cresol and p-bromophenol). None of these measurements was included in the databases developed here because we believe that the measurements from this reference are included in the results by Anderson and colleagues (43).

Saket et al., 1985 (37)

This paper reports $K_{cw,rep}$ values for hydrocortisone, five esters of hydrocortisone (hydrocortisone acetate, hydrocortisone propionate, hydrocortisone valerate, hydrocortisone hexanoate, and hydrocortisone octanoate), cortisone and four esters of cortisone (cortisone acetate, cortisone butyrate, cortisone hexanoate, and cortisone octanoate). The $K_{cw,rep}$ value for hydrocortisone at 37°C was taken from Table 1 (abdominal skin) in this reference, while the other $K_{cw,rep}$ values (at 25°C for abdominal skin) were taken from Table 2. According to one of the authors (64), the data were based on the dry SC mass, and equilibrium was established over 48 hours. All chemicals were nonionized in the experimental solutions.

Sato et al., 1991 (65)

The $K_{cw,rep}$ value (based on dry SC mass) for nicorandil into human plantar callus is shown in Fig. 3 of this reference. The digitized value of $K_{cw,rep}$ from this figure is not meaningfully different from the value reported in a previous publication (59) and probably is the same measurement reported again. The authors do not indicate that this is a new measurement. Indeed, the skin source and experimental procedure are identical to those of the previous study (59). For this reason, we have included only the measurement from the previous investigation (59) in the database.

Sato et al., 1989 (59)

The $K_{cw,rep}$ value (based on dry SC mass) is reported in Table 4 of this reference. The skin was plantar callus, which probably has more proteinaceous material than abdominal, thigh, or dorsal stratum corneum, and consequently may have more hydrophilic partitioning properties. Twenty-four hours were allowed for the partitioning process to equilibrate. Nicorandil was not ionized at the basic pH attained by the 0.05% (w/v) solution used in these studies. The partition coefficient is listed in the excluded database since plantar callus was used.

Scheuplein et al., 1969 (23)

The $K_{cw,rep}$ values (based on dry SC mass) for several steroids were taken directly from Table I. None of these compounds ionizes. Equilibration time was not provided. The measurement for hydroxypregnenolone is included in the provisional database because we were unable to determine the structure and properties of this compound.

Scheuplein and Blank, 1973 (8)

The $K_{cw,rep}$ values (based on dry SC mass basis) for water and the normal alcohols (methanol through octanol) were taken directly from Table I in this reference. These compounds do not ionize. The time of equilibration was not specified.

Smith and Anderson, 1995 (9)

The $K_{cw,rep}$ value (based on dry SC mass) for the nonionized form of lauric acid was reported in the text as 5060 ± 880. This result is based on measurements made at pH = 3.2, 5.5, and 7.9. The equilibration time was between 24 and 48 hours. The authors adjusted the measurements by a factor of 1 g mL^{-1} to put them on a volume basis. In the database we report the measurement based on dry mass SC and adjust it with the standard conversion factor to the hydrated SC volume basis (i.e., 0.25). The authors estimated that the $K_{cw,rep}$ value for ionized lauric acid was 72.2, which is approximately 70 times smaller than for the nonionized form.

Surber et al., 1992 (21)

This paper examines $K_{cw,rep}$ values for 22 compounds. Six of these compounds (acetamidophenol, cyanophenol, DTT, iodophenol, PCB, and pentyloxyphenol) were reported originally in Surber et al. (47), and seven others (acitretin, caffeine, diazepam, estradiol, hydrocortisone, progesterone, and testosterone) were originally reported in Surber et al. (20). Data for nine compounds (atrazine, benzo[a]-pyrene, chloramphenicol, cyclosporin-A, 5-fluorouracil, ibuprofen, indomethazine, lauric acid, and uracil) were presented for the first time in this reference. Through personal communication with Surber (58), (a) we obtained exact values for the data plotted in Fig. 3 of this reference, (b) we learned that the $K_{cw,rep}$ values were all based on dry SC mass, (c) we were informed that no buffer was used to control the solution pH, and (d) we learned that compound #18 in Fig. 3 of this reference is PCB. According to the authors, enough time was allowed for equilibrium to be established, although the exact time was not reported (21). The mean values obtained from Surber (58) and those shown in Fig. 3 are averages of measurements made at different concentrations. 5-Fluorouracil and cyclosporin-A were ionized independent of their actual solution concentrations (which were not specified). For several compounds (i.e., ibuprofen, indomethazine, lauric acid, and uracil), the exact value of f_{ni} depends upon the concentration, which was not reported. Through personal communication with Surber (58), we learned that the compounds studied in this investigation were measured using radiolabeled drug to give final concentrations about 5 million cpm mL^{-1} water or isopropyl myristate, since some of the compounds damaged the SC at higher concentrations (58). Except for cyclosporin-A, which was excluded for other reasons, measurements for compounds with f_{ni} that was probably less than 1 were placed in the provisional database. In addition, Surber (58) confirmed that the $K_{cw,rep}$ values for PCB and DDT shown in Fig. 3 were interchanged from the correct values presented in the earlier paper (47).

Surber et al., 1990 (20)

The mean values of $K_{cw,rep}$ (based on dry SC mass) for seven chemicals (acitretin, progesterone, testosterone, diazepam, estradiol, hydrocortisone, and caffeine) were taken from Table IV in this reference. The $K_{cw,rep}$ values for acitretin and caffeine are listed in the provisional database, because solution concentrations of these compounds were not reported and f_{ni} could not be determined. Equilibrium was demonstrated by unchanging vehicle concentrations. The time required ranged from 6 hours (for progesterone, testosterone, estradiol, caffeine, hydrocortisone, and diazepam) to 24 hours (for acitretin).

Surber et al., 1990 (47)

This paper reported $K_{cw,rep}$ values (based on dry SC mass) for six chemicals (4-acetamidophenol, 4-cyanophenol, 4-iodophenol, 4-pentyloxyphenol, PCB, and DDT). Since all six were essentially nonionized even at the most dilute concentrations, measurements made at different concentrations could legitimately be combined, and $K_{cw,rep}$ values were taken directly from Table 6. Six to 24 hours were allowed for equilibration. The measurement for PCB is listed in the excluded database because it was measured with a polydispersed mixture rather than a single compound.

Wester et al., 1987 (28)

The $K_{cw,rep}$ values for sieved powdered human stratum corneum (PHSC) from plantar callus were reported in Table 2 of this reference. One of the authors confirmed that measurements were based on dry SC mass (63). Benzene and PCB (54% chlorine) were nonionized, but 4-nitroaniline, at a concentration of 4.9 μg/mL, was partially ionized at its natural pH. The equilibration time was only 30 min. It may be that PHSC reaches equilibrium more rapidly than SC in a membrane form, but the authors did not experimentally examine this question. These measurements were listed in the excluded database because PHSC was used.

Williams and Barry, 1991 (56)

The $K_{cw,rep}$ value (based on hydrated SC mass) for saturated solutions of 5-fluorouracil was the control value (i.e., without penetration enhancer) listed in Table II of this reference. The equilibration time was 4 hours. At the natural pH for a saturated solution at 20°C, 5-fluorouracil is a zwitterion. As a result, this measurement is listed in the provisional database.

ACKNOWLEDGMENTS

This work was supported in part by the United States Environmental Protection Agency under Assistance Agreement Nos. CR817451 and CR822757, by the U.S. Air Force Office of Scientific Research under agreement F49620-95-1-0021, and by the National Institute of Environmental Health Sciences under grant No. R01-ES06825. We thank Dr. Richard Guy for his helpful comments and suggestions.

GLOSSARY

A	= Surface area of chemical exposure
a	= Intercept of equations in the form of Eq. (4)
a, b, c, d	= Parameters determined by regression of model equation to experimental data
b	= Regression coefficient multiplying log K_{ow} in equations with the form of Eq. (4)
C_c	= Concentration of absorbing chemical in the SC in units of mass of chemical per unit volume of hydrated SC

$C_{c,eq}$	= Concentration of absorbing chemical in the SC in equilibrium with an aqueous solution reported in units of mass of chemical per unit volume of hydrated SC
$C'_{c,eq}$	= Concentration of absorbing chemical in the SC in equilibrium with an aqueous solution reported in units of mass of chemical per unit mass of dried SC
C_w	= Aqueous concentration of the absorbing chemical
D_c	= Effective diffusivity of the absorbing chemical in the SC
f_{ni}	= Fraction of the total chemical dose that is nonionized in the aqueous solution
FV	= Fully validated
F-ratio	= Statistic measuring the predictive power of the regression
ΔH	= Heats of ionization as listed in Table A5
K_a	= Acid dissociation constant for the absorbing chemical
K_{cw}	= Equilibrium SC–water partition coefficient for the absorbing chemical calculated in terms of hydrated volume of SC as defined by Eq. (1); may include nonionized and ionized forms of the absorbing solute
K'_{cw}	= SC–water partition coefficient for the absorbing chemical calculated in terms of dry mass of SC as defined by Eq. (2)
$K_{cw,calc}$	= Calculated SC–water partition coefficient for the absorbing chemical
$K_{cw,i}$	= SC–water partition coefficient for the ionized form of the absorbing chemical
$K_{cw,ni}$	= SC–water partition coefficient for the nonionized form of the absorbing chemical
$K_{cw,rep}$	= Reported SC–water partition coefficient without adjustment for basis
K_{ow}	= Octanol–water partition coefficient of the absorbing chemical
K^0_{cw}	= Parameter in LSER model equation, Eq. (10)
K^t_{ow}	= Value of K_{ow} at which the piecewise linear regression changes slope and distinguishes the hydrophilic chemicals from the lipophilic chemicals
L_c	= Effective thickness of the SC
LSER	= Linear solvation-energy relationships
m	= Number of different compounds included in the regression equation
M_c	= Mass of absorbed chemical in the SC
MW	= Molecular weight of the absorbing chemical
n	= Number of data points included in the regression equation
p	= Number of fitted parameters in the regression equation
P_{cw}	= Steady-state permeability of chemical penetration through the SC from water
pH	= Negative logarithm of the hydrogen ion molarity, $= -\log_{10}[H^+]$
pK_a	= Negative logarithm of the acid dissociation constant, $= -\log_{10}[K_a]$
PHSC	= Powdered human SC from callused tissue
r^2	= Goodness of fit parameter
r^2_{adj}	= Goodness of fit parameter adjusted for the number of fitted parameters and the number of data points
RMSE	= Root mean square error of the regression
SC	= Stratum corneum
t_{equil}	= Time for equilibration of the SC and aqueous solution
T	= Absolute temperature (Kelvin)
$U(a - b)$	= Unit step function. $U = 0$ if $a < b$ and $U = 1$ if $a \geq b$
V_x	= Characteristic volume of McGowans (LSER parameter)

Greek

α = Hydrogen bond acidity of the absorbing chemical (LSER parameter)
β = Hydrogen bond basicity of the absorbing chemical (LSER parameter)
π = Dipolarity/polarizability of the absorbing chemical (LSER parameter)
ρ_c = Mass of hydrated SC divided by hydrated SC volume
ρ_c' = Mass of dry SC divided by hydrated SC volume
ρ_w = Density of aqueous solution of absorbing chemical

Subscripts

H   = Hydrophilic
L   = Lipophilic

REFERENCES

1.   C Hansch, A Leo, D Hoekman. Exploring QSAR: Hydrophobic, Electronic, and Steric Constants. Washington, DC: American Chemical Society, 1995.
2.   PCModel. Ver. 4.2, Daylight Chemical Information Systems, Mission Viejo, CA, 1995.
3.   PV Raykar, M-C Fung, BD Anderson. The role of protein and lipid domains in the uptake of solutes by human stratum corneum. Pharm Res 5:140–150, 1988.
4.   RL Bronaugh, ER Congdon. Percutaneous absorption of hair dyes: correlation with partition coefficients. J Invest Dermatol 83:124–127, 1984.
5.   RJ Scheuplein. Molecular structure and diffusional processes across intact epidermis. Contract No. DA18-108-AMC-148(A), U.S. Army Chemical Research and Development Laboratories, Edgewood Arsenal, MD, 1966.
6.   NA Megrab, AC Williams, BW Barry. Oestradiol permeation across human skin, silastic and snake skin membranes—the effects of ethanol water co-solvent systems. Int J Pharm 116:101–112, 1995.
7.   RJ Scheuplein, IH Blank. Permeability of the skin. Physio Rev 51:702–747, 1971.
8.   RJ Scheuplein, IH Blank. Mechanism of percutaneous absorption. IV. Penetration of nonelectrolytes (alcohols) from aqueous solutions and from pure liquids. J Invest Dermatol 60:286–296, 1973.
9.   SW Smith, BD Anderson. Human skin permeability enhancement by lauric acid under equilibrium aqueous conditions. J Pharm Sci 84:551–556, 1995.
10.   KV Roskos, RH Guy. Assessment of skin barrier function using transepidermal water loss: Effect of age. Pharm Res 6:949–953, 1989.
11.   WJ Lyman, WK Keehl, DH Rosenblatt. Handbook of Chemical Property Estimation Methods. New York: McGraw-Hill, 1982.

12. El Tayar N, RS Tsai, B Testa, PA Carrupt, C Hansch, A Leo. Percutaneous penetration of drugs: a quantitative structure–permeability relationship study. J Pharm Sci 80:744–749, 1991.
13. MS Roberts, RA Anderson, DE Moore, J Swarbrick. The distribution of non-electrolytes between human stratum corneum and water. Aust J Pharm Sci 6:77–82, 1977.
14. PM Elias. Lipids and the epidermal permeability barrier. Arch Derm Res 270:95–117, 1981.
15. CJ Cramer, GR Famini, AH Lowrey. Use of calculated quantum chemical properties as surrogates for solvatochromic parameters in structure–activity relationships. Acc Chem Res 26:599–605, 1993.
16. GR Famini, CA Penski. Using theoretical descriptors in quantitative structure activity relationships: some physicochemical properties. J Phys Org Chem 5:395–408, 1992.
17. MH Abraham, HS Chadha, GS Whiting, RC Mitchell. Hydrogen bonding part 32. An analysis of water–octanol and water–alkane partitioning and the Δ log P parameter of Seiler. J Pharm Sci 83:1085–1100, 1994.
18. MH Abraham, HS Chadha, RC Mitchell. The factors that influence skin penetration of solutes. J Pharm Pharmacol 47:8–16, 1995.
19. MH Abraham, JC McGowan. The use of characteristic volumes to measure cavity terms in reversed phase liquid chromatography. Chromatographia 23:243–246, 1987.
20. C Surber, KP Wilhelm, M Hori, HI Maibach, RH Guy. Optimization of topical therapy: partitioning of drugs into stratum corneum. Pharm Res 7: 1320–1324, 1990.
21. C Surber, K-P Wilhelm, HI Maibach, RH Guy. Can health hazard associated with chemical contamination of the skin be predicted from simple in vitro experiments? In: Marks R, Plewig G, eds. The Environmental Threat to the Skin. London: Martin Dunitz, 1992, pp 269–276.
22. JMP Statistical Discovery Software. Ver. 3.1, SAS Institute, Inc., Cary, North Carolina, 1995.
23. RJ Scheuplein, IH Blank, GJ Brauner, DJ MacFarlane. Percutaneous absorption of steroids. J Invest Dermatol 52:63–70, 1969.
24. BD Anderson, PV Raykar. Solute structure–permeability relationships in human stratum corneum. J Invest Dermatol 93:280–286, 1989.
25. X Hui, RC Wester, PS Magee, HI Maibach. Partitioning of chemicals from water into powdered human stratum corneum (callus): a model study. In Vitro Toxicol 8:159–167, 1995.
26. SRC PhysProp Database, http://esc.syrres.com/interkow/physdemo.htm, 2002.
27. C Hansch, A Leo, eds. Substituent Constants for Correlation Analysis in Chemistry and Biology. New York: John Wiley, 1979.

28. RC Wester, M Mobayen, HI Maibach. In vivo and in vitro absorption and binding to powdered stratum corneum as methods to evaluate skin absorption of environmental chemical contaminants from ground and surface water. J Toxicol Environ Health 21:367–374, 1987.

29. RL Cleek, AL Bunge. A new method for estimating dermal absorption from chemical exposure. 1. General approach. Pharm Res 10:497–506, 1993.

30. MS Roberts, WJ Pugh, J Hadgraft. Epidermal permeability: penetrant structure relationships. 2. The effect of H-bonding groups in penetrants on their diffusion through the stratum corneum. Int J Pharm 132:23–32, 1996.

31. HS Brown, D Hattis. The role of skin absorption as a route of exposure for volatile organic compounds (VOCs) in household tap water: a simulated kinetic approach. J Am Coll of Toxicol 8:839–851, 1989.

32. JA Shatkin, HS Brown. Pharmacokinetics of the dermal route of exposure to volatile organic chemicals in water: a computer simulation model. Env Res 56:90–108, 1991.

33. RL Chinery, AK Gleason. A compartmental model for the prediction of breath concentration and absorbed dose of chloroform after exposure while showering. Risk Anal 13:51–62, 1993.

34. MS Roberts, EJ Triggs, RA Anderson. Permeability of solutes through biological membranes measured by a desorption technique. Nature 257:225–227, 1975.

35. MS Roberts. Department of Medicine, The University of Queensland, personal communication, 1996.

36. RO Potts, RH Guy. Predicting skin permeability. Pharm Res 9:663–669, 1992.

37. MM Saket, KC James, IW Kellaway. The partitioning of some 21-alkyl steroid esters between human stratum corneum and water. Int J Pharm 27:287–298, 1985.

38. GL Flynn. Physicochemical determinants of skin absorption. In: TR Gerrity, CJ Henry, eds. Principles of Route-to-Route Extrapolation for Risk Assessment. New York: Elsevier, 1990, pp 93–127.

39. RJ Scheuplein. Mechanism of percutaneous absorption III. The effect of temperature on the transport of non-electrolytes across the skin. J Invest Dermatol 49:582–589, 1967.

40. RJ Scheuplein. Mechanism of percutaneous absorption I. Routes of penetration and the influence of solubility. J Invest Dermatol 45:334–346, 1965.

41. TE McKone, RA Howd. Estimating dermal uptake of nonionic organic chemicals from water and soil: I. Unified fugacity-based models for risk assessments. Risk Anal 12:543–557, 1992.

42. TE McKone. Dermal uptake of organic chemicals from a soil matrix. Risk Anal 10:407–419, 1990.

43. RA Anderson, EJ Triggs, MS Roberts. The percutaneous absorption of phe-

nolic compounds 3. Evaluation of permeability through human stratum corneum using a desorption technique. Aust J Pharm Sci NS5:107–110, 1976.

44. BD Anderson, WI Higuchi, PV Raykar. Heterogeneity effects on permeability-partition coefficient relationships in human stratum corneum. Pharm Res 5:566–573, 1988.

45. MS Roberts. Percutaneous absorption of phenolic compounds. Ph.D. thesis, University of Sydney, Sydney, Australia, 1976.

46. EJ Lien, GL Tong. Physicochemical properties and percutaneous absorption of drugs. J Soc Cosmet Chem 24:371–384, 1973.

47. C Surber, KP Wilhelm, HI Maibach, LL Hall, RH Guy. Partitioning of chemicals into human stratum corneum: implications for risk assessment following dermal exposure. Fund Appl Toxicol 15:99–107, 1990.

48. AL Bunge, GL Flynn, RH Guy. A predictive model for dermal exposure assessment. In: Wang RGM, ed. Drinking Water Contamination and Health: Integration of Exposure Assessment, Toxicology, and Risk Assessment. New York: Marcel Dekker, 1994, pp 347–374.

49. IH Blank, DJ McAuliffe. Penetration of benzene through human skin. J Invest Dermatol 85:522–526, 1985.

50. GE Parry, AL Bunge, GD Silcox, LK Pershing, DW Pershing. Percutaneous absorption of benzoic acid across human skin. I. In vitro experiments and mathematical modeling. Pharm Res 7:230–236, 1990.

51. K Kubota, HI Maibach. In vitro percutaneous permeation of betamethasone and betamethasone 17-valerate. J Pharm Sci 82:1039–1045, 1993.

52. RL Bronaugh, ER Congdon, RJ Scheuplein. The effect of cosmetic vehicles on the penetration of N-nitrosodiethanolamine through excised human skin. J Invest Dermatol 76:94–96, 1981.

53. SPARC (SPARC Performs Automated Reasoning in Chemistry): An Expert System for Estimating Physical and Chemical Reactivity. Ver. Windows Prototype Version 1.1, US EPA (Ecosystem Research Division) and University of Georgia, Athens, GA, 1995.

54. RJ Scheuplein. Molecular structure and diffusional processes across intact epidermis. Contract No. DA18-108-AMC-148(A), U.S. Army Chemical Research and Development Laboratories, Edgewood Arsenal, MD, 1964.

55. PA Cornwell, BW Barry. Sesquiterpene components of volatile oils as skin penetration enhancers for the hydrophilic permeant 5-fluorouracil. J Pharm Pharmacol 46:261–269, 1994.

56. AC Williams, BW Barry. Terpenes and the lipid-protein-partitioning theory of skin penetration enhancement. Pharm Res 8:17–24, 1991.

57. BW Barry. School of Pharmacy, University of Bradford, personal communication, 1996.

58. C Surber. Kantonsspital Basel, Universitätskliniken, Basel, Switzerland, personal communication, 1996.

59. K Sato, K Sugibayashi, Y Morimoto, H Omiya, N Enomoto. Prediction of the in-vitro human skin permeability of nicorandil from animal data. J Pharm Pharmacol 41:379–383, 1989.

60. MS Roberts, RA Anderson, J Swarbrick. Permeability of human epidermis to phenolic compounds. J Pharm Pharmacol 29:677–683, 1977.

61. HA Sober, ed. Handbook of Biochemistry. Cleveland, OH: Chemical Rubber Company, 1968.

62. BW Barry, SM Harrison, PH Dugard. Vapour and liquid diffusion of model penetrants through human skin; correlation with thermodynamic activity. J Pharm Pharmacol 37:226–236, 1985.

63. RC Wester. Department of Dermatology, University of California, San Francisco, personal communication, 1995.

64. IW Kellaway. Welsh School of Pharmacy, University of Cardiff, Wales, personal communication, 1996.

65. K Sato, K Sugibayashi, Y Morimoto. Species differences in percutaneous absorption of nicorandil. J Pharm Sci 80:104–107, 1991.

5

Iontophoresis: Applications in Drug Delivery and Noninvasive Monitoring*

M. Begoña Delgado-Charro and Richard H. Guy
*Centre Interuniversitaire de Recherche et d'Enseignement, Universities
of Geneva and Lyon, Archamps, France, and University of Geneva,
Geneva, Switzerland*

INTRODUCTION

The term iontophoresis typically refers to the transfer of charged substances through a biological membrane under the influence of an electric field (1). This technique is far from new, Leduc having shown nearly 100 years ago that the technique could be used to deliver active drugs across mammalian skin in vivo (2,3). Since then, iontophoresis has been variously used to administer pilocarpine in the diagnosis of cystic fibrosis, to treat hyperhidrosis of palms and soles (tap water iontophoresis), to induce local anaesthesia in the skin and in the external ear canal, to aid penetration of fluoride ions in dentistry, and so on (2–4). Yet it was only 20 years ago, following the initial success of passive transdermal drug delivery, that iontophoresis received attention as a way to expand the range of drugs that could be administered via the skin (5). At the same time, it was fully appreciated (6) that the permeation of water, neutral and zwitterionic compounds could also be enhanced by iontophoresis, thereby expanding its potential applications. This observation, and the recognition of the skin as a permselective

* This article was published in original form in S.T.P. Pharma Sciences 11:403–414, 2001.

membrane, focused attention on convective solvent flow (electroosmosis) as a second mechanism of electrotransport (7). The advantages of iontophoresis as a controlled and versatile drug administration technique were soon identified for peptides and drugs for the treatment of Parkinson's disease, migraine, pain, etc. Within the last 10 years, the symmetrical nature of iontophoresis (i.e., that the passage of current across the skin causes ions to move into and out of the membrane at the same time) has led to its application as a noninvasive method of extracting endogenous substances (8). This so-called reverse iontophoresis procedure is exemplified by the Glucowatch Biographer® recently approved by the FDA for glucose monitoring.

Some excellent reviews on iontophoresis have already been published (1,3,7,9–11); here we attempt to summarize what is known about the mechanisms of transport, the key parameters governing flux, and the pathways of penetration during iontophoresis. The possible fields of application for drug administration and noninvasive sampling are also reviewed.

I. TRANSDERMAL IONTOPHORESIS

A. A Basic Circuit of Iontophoresis

Iontophoresis is the application of a low electrical potential gradient across the skin so as to enhance the penetration of molecules. Its appeal resides on the direct relation that can be established, via the transport number, between the flux of any ion present in the circuit, on either side of the skin, and the current passed (1,9–12):

$$J_i = \frac{I \cdot t_i}{F \cdot z_i} \tag{1}$$

where J_i, t_i, and z_i are, respectively, the flux, transport number, and valence of the ion i. It follows that the sensible way of performing iontophoresis in drug delivery is to apply a constant current, such that the flow of electrons is directly translated into an ion flux.

A basic iontophoretic device consists of a power supply (a battery) connected to an anode and a cathode, which separately contact the skin via an appropriate conducting medium (Fig. 1). A preferred material for the electrodes is Ag/AgCl; such electrodes are reversible and do not induce pH changes in the media in contact with the skin (9,11,13). The main disadvantage of Ag/AgCl electrodes is the consumption and accumulation of chloride ions at the anode and cathode, respectively. For example, the flux of anionic drugs may be progressively reduced due to the increasing chloride at the cathode (14–16). The reaction of some drugs at the electrodes has also been observed (17–21). The use of inert electrodes

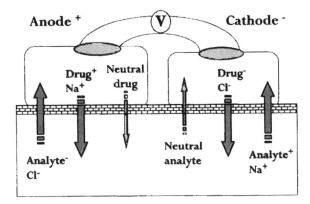

Figure 1 Schema of molecular transport during iontophoresis. Thick black arrows represent electromigration fluxes. Cations present in the anode chamber are delivered through the skin into the body. The anode attracts negatively charged endogenous ions and analytes. Anions present in the cathode migrate through the skin towards the body while endogenous cations and analytes are sampled at this negative chamber. The electroosmotic flow, represented by gray arrows, allows the delivery of neutral molecules into the body from the anode, and their sampling at the cathode.

such as platinum is inadvisable as they result in the hydrolysis of water and progressive and significant pH alterations. This results in risks of skin irritation, and changes in skin permselectivity and/or drug ionization.[1] The alternative use of high ionic strength buffers to keep the pH constant inevitably decreases the transport efficiency of the ion of interest (22). The importance of selecting electrodes and electrode solutions intelligently has been nicely illustrated by Phipps et al. (12).

In basic research, the electrodes are generally held in plastic or glass chambers filled with electrolyte solution. Obviously, for practical applications, electrode formulations based on conductive gels are preferred. Either the electrolyte solution or the gel makes contact with the skin, thus avoiding direct contact with the electrode. More sophisticated designs divide one or another electrode chamber into two compartments communicating via ionic exchange membranes or salt bridges for reasons of delivery efficiency or drug stability (11,17–21).

The electrochemistry of an iontophoretic circuit is such that there is oxidation at the anode and reduction at the cathode. The flux of electrons through the "outside" circuit is exactly balanced by the flux of ions through the "inside"

---

[1] There is only one application of iontophoresis (4) for which Pt electrodes are superior, which is sweat inhibition by tap water iontophoresis and which seems to be dependent on the acidification of the sweat glands.

circuit (Fig. 1). Specifically, oxidation at the Ag/AgCl anode results in the loss of a Cl⁻ from the solution, and this is balanced either by pushing a cation (Na⁺, Drug⁺) through the skin or by the arrival of an anion from beneath the skin (Cl⁻). Reciprocally, reduction at the cathode releases a chloride ion, an extra negative charge, which is neutralized either by the electromigration of an ion (Cl⁻ or Drug⁻ for example) through the skin or by the arrival of cation from beneath. In other words, to deliver n monovalent ions through the skin, n electrons must be generated at the anode (oxidation) and transferred to the cathode, where the reduction consumes these n electrons (9,12). The number of electrons flowing through the external circuit is a direct reflection of the amount of ionic charge flowing through the skin. Herein resides the principle of controlled drug delivery/extraction in iontophoresis.

When an electric potential gradient is established across a membrane, ions on either side will migrate in the direction dictated by their charge. The speed of migration of an ion is determined by its physicochemical characteristics and the properties of the media through which the ion is moving (9,12). The sum of the individual ion fluxes must equal the current supplied by the power source; in other words, there is a "competition" among all the ions present to carry the charge. Obviously, the chances of being a major carrier, and in consequence being efficiently transported through the skin, increase with the electrical mobility and concentration of the ion concerned.

From this derives the concept of transport number (efficiency of transport), which is the fraction of the total charge transported by the ion i (see Equation 1):

$$t_d = \frac{c_d \cdot z_d \cdot \mu_d}{\sum c_i \cdot z_i \cdot \mu_i} \qquad (2)$$

where c, z, and μ refer to concentration, valence, and mobility of either the drug d or any ion i in the system (10). As stated in Eq. (1), the transport number and the intensity of current are the two main parameters controlling the iontophoretic flux. Intensity of current is directly and easily controlled by the power supply. On the other hand, transport numbers depend on the physicochemical properties and concentration of the ion of interest and also on the other ions present. Thus the transport number is a "formulation dependent" parameter. It follows that an important component of iontophoretic development is directed towards increasing the transport number of the drug of interest. Transport numbers are usually constant for a given set of conditions and are usually obtained experimentally (2,23,24). They remain difficult to estimate theoretically, as they are complex functions, not always linear, of the mobility and concentration of all the ions present (2,10,25). Equation (2) has been developed for specific experimental conditions, such as the so-called binary cation situation (10).

Unfortunately, the terms concentration and mobility refer to those inside the skin and not in the donor solution, a precision that makes them tricky to estimate. This introduces some limitations in the use of this equation as a predictive tool.

An important consequence of the symmetry of ionic transport (and the fact that ions move in both directions under both electrodes) is that the efficiency of transport never reaches 100%. That is, one can always anticipate competition with endogenous ions (Na^+, Cl^- . . .) in the in vivo situation. In other words, a drug will never exhibit a transport number of 1 even in the most optimized device in which it is the only ion available for transport at the drug electrode.

B. Advantages and Limitations

As a starting point, transdermal iontophoresis retains all the advantages of the transdermal route: avoidance of first-pass effect and/or GI metabolism/degradation, improved patient acceptance and compliance, easy access, etc. On the contrary, due to the efficient degree of enhancement achieved, iontophoresis enlarges the range of drug candidates for transdermal administration, in particular including polar and/or charged drugs whose passive skin permeation is severely restricted (5).

Further advantages result from the physical nature of the mechanism of transport. Other techniques of enhancement (chemical enhancers, sonophoresis, electroporation) can produce equivalent or even better delivery than iontophoresis, but this is achieved via direct effects on the skin itself (disruption or extraction of stratum corneum lipids, for example). Iontophoresis acts on the molecules themselves and consequently perturbs the membrane barrier to a much smaller extent [as evidenced by the fast skin recovery and only minor irritation usually observed after iontophoresis (26)]. Another advantage of iontophoresis is that drug delivery is less sensitive to the condition of the skin at the application site (unlike passive transdermal administration).

Iontophoretic fluxes depend on the current applied and on the transport number. The first is an independent parameter, while the second depends more on the relative ionic composition on either side of the skin, rather than on the properties of the stratum corneum per se. For example, the large interspecies differences in the passive fluxes of lithium, pyridostigmine, and hydromorphone disappear when these permeants are delivered iontophoretically (12,27). Furthermore, Sekkat et al. (28) have illustrated the potentially improved safety of iontophoresis for drug delivery through a compromised barrier: while the passive delivery of lidocaine through a deliberately compromised skin barrier may be increased by more than two orders of magnitude, the iontophoretic delivery is invariant because the current controls the transport. Iontophoresis may be useful, therefore, for safe transdermal delivery in populations of uncertain barrier maturity (e.g., prematures, neonates). Of course, it must be noted that iontophoresis does not "cure" passive variability; the safe delivery of drugs like lidocaine can be assumed because their passive transport is very small compared to that possible with iontophoresis. For other compounds, such as nicotine, the benefit is less because iontophoretic and passive contributions are of the same order of magnitude (29).

Further benefits of iontophoresis derive from the fast-responding and versatile administration kinetics that may be achieved. Among noninvasive techniques of drug administration, iontophoresis can provide in vivo plasma profiles similar to those obtained following an IV infusion (30).

An externally controlled iontophoretic device offers a unique and simple way to manipulate drug input as required. Thus pulsatile or constant delivery profiles are achievable, an interesting feature for peptides such as LHRH analogs, which display different pharmacological actions depending on their input kinetics. Equally, dosing can be easily adjusted for individual patients by controlling for the total current delivered. This is quite useful in the case of small therapeutic index drugs or those exhibiting a need for careful and individual dose adjustment.

Iontophoresis can therefore make possible the transdermal delivery of drugs previously considered not possible for this route. It can also be used to improve the performance of drugs that are already administered by passive patches. For example, it can be exploited to achieve a faster drug input and action (fentanyl) (31) or to mimic the extremely rapid kinetics of nicotine delivered by a cigarette (29).

The main limitation for iontophoresis resides nevertheless in the range of molecules for which an efficient enhancement is achievable. Equation (2) predicts low transport numbers for large drugs, because of their poor performance as charge carriers. As will be discussed below, for these drugs the main mechanism of electrotransport is convective flow (electroosmosis), which may be considered (to a first approximation) to be independent of the molecular size (7). Unfortunately, this mechanism of transport is much less efficient than electromigration. There have therefore been attempts to combine iontophoresis with other enhancement techniques, such as chemical promoters and electroporation (14,32). While such combinations have induced greater deliveries, they have equally caused more significant damage to the barrier and a consequent loss of the control inherent in the iontophoretic approach.

II. MECHANISMS OF IONTOPHORETIC TRANSPORT

A. Electromigration

The electromigration (electrorepulsion) contribution to iontophoretic transport is a direct result of current application. Electron fluxes are transformed into ionic fluxes via the electrode reactions. Ionic transport proceeds through the skin so as to maintain electroneutrality (1,9,10). Faraday's law applies to steady-state transport and relates the number of ions crossing the membrane to the electric current, the time of current passage, and the charge per ion (1,9):

$$M_i = \frac{T \cdot i_i}{F \cdot z_i} \tag{3}$$

where M_i is the number of moles of the ith ion, T is the time, z is the valence, F is Faraday's constant, and i_i the current carried by the ith species. Given that there is usually more than one ion moving across the barrier, the total number of moles transported (M) by the total current flowing (I) is given by

$$M = \sum_i M_i = \frac{T}{F} \sum_i \frac{i_i}{z_i} \tag{4}$$

where

$$I = \sum_i i_i \tag{5}$$

If the transport number of the drug of interest is t_D, it follows that the number of moles of drug transported is

$$M_D = \frac{t_D \cdot I \cdot T}{f \cdot Z_D} \tag{6}$$

If the transport number is known, the intensity required to deliver a given dose in a fixed time T can be estimated. With this calculated intensity, the area of skin to be contacted by the driving electrode formulation can be deduced, remembering that the upper limit of current density is around 0.5 mA/cm^2 (26). Above this value, discomfort and pain are typically felt. It is important to note, though, that it is the total charge delivered, and not the current density, that determines the dose. That is, the same intensity should always result in the delivery of the same dose, independent of the area used.

The other parameter determining delivery is the transport number. The transport number is a complex function of the distribution and relative mobility of all the mobile ionic species in the membrane, and in the media on either side of the membrane; transport numbers cannot be estimated theoretically (2). On the other hand, a transport number is usually constant for a given set of conditions and can therefore be determined experimentally; for example, by dividing the total amount transported of the ion of interest by the total charge delivered (using Faraday's equation) or from the gradient of the plot of flux versus current intensity [Eq. (1) (2,23,24)].

However, at least qualitatively, the impact of certain physicochemical properties on the transport number can be anticipated. For example, ionic mobility and hence the corresponding transport number decrease with increasing molecular weight or size, as has been observed for oligonucleotides (33) and other miscellaneous compounds (34,35). The presence of other ions in the vehicle decreases the transport number of the drug of interest (22,29,36). Obviously, that effect is more dramatic when the mobility and concentration of the competing coions are high. For this reason, the use of polymeric buffers, which given their size cannot act as competing species (11,37,38), has been proposed.

The concentration of the molecule of interest also plays a role. It must first be stated that "concentration" refers to the ionized species, so the pH and the drug's pKa must be taken into consideration to formulate the vehicle optimally. It should also be noted that there is a pH gradient across the skin (39) and that the degree of ionization of the molecule may change as it makes its way through the membrane. In the simplest case, the drug is formulated so that the vehicle contains no competing coions. In drug administration, minimizing the presence of competing coions avoids competition to carry charge across the skin into the body; for example, when a cationic drug (such as a hydrochloride form) is the only species present in the anode chamber (Figs. 2 and 3). In these conditions, the iontophoretic flux of the drug reaches a limiting value which is ultimately independent of concentration (10,23,24,27). According to Kasting and Keister (25), this maximal value depends on the ratio of diffusivities (or mobilities) of the cation and the main counterion (usually chloride) arriving from beneath the skin. This so-called single-ion situation is advantageous for drug administration in that (a) it allows maximal fluxes to be obtained without necessarily using the

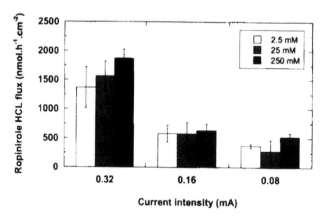

Figure 2 Ropinirole hydrochloride (RHCl) fluxes through neonatal porcine skin measured after 8 hours of in vitro iontophoresis. This figure illustrates the weak dependence on RHCl donor concentration. The donor solution consisted of RHCl dissolved in water. On the contrary, a clear dependence on current intensity is observed. (Data taken from Ref. 23.)

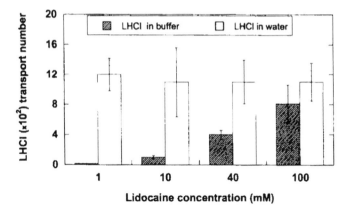

Figure 3 Effect of lidocaine (LHCl) donor concentration on the drug's transport number. LHCl transport number increases with drug concentration in the presence of accompanying buffer. The transport numbers are independent of drug concentration in the so-called "single cation" situation (no competing coions present). (Data taken from Ref. 24.)

highest concentration of drug and (b) it introduces a safety mechanism in that drug flux becomes dependent only on the relative mobilities of the drug and the principal endogenous counterion. For most small drugs, maximal transport numbers on the order of 0.08–0.15 (23,24,27). Nevertheless, there remain insufficient data with which to identify the key physicochemical properties that determine a drug's inherent transport number. It would appear, however, that the available evidence points to a sharp decrease in maximal transport number with increasing molecular weight (34).

In cases where the drug has to be formulated (for example, for stability or solubility reasons) in the presence of a buffer and/or a certain amount of additional electrolyte, then its iontophoretic flux is usually proportional to concentration. This is always the situation, of course, in reverse iontophoresis, for which the "donor" formulation is physiological fluid, and the analytes of interest (fortunately) constitute only a small fraction of the total number of charge carriers available.

Finally, the current profile must be considered. It has been proposed that alternating current might be useful as it would avoid skin polarization (40), but there is simply no significant experimental evidence to support such a claim. On the contrary, discrete pulses of direct current can be used to achieve a pulsatile drug input, a situation that is preferred, pharmacodynamically, for certain drugs. In the end, though, the dose delivered is proportional to the total charge administered and is independent of the profile (time and intensity) used (41).

B. The Convective Contribution: Electroosmosis

In 1980, Gangarosa et al. (6) used the term iontohydrokinesis to describe the transport of water into a tissue as a result of iontophoresis, irrespective of the mechanism of transport. These early experiments demonstrated quite nicely, in vivo, the transport of solvent induced by iontophoresis. Briefly, after applying iontophoresis to mice (in which anode and cathode formulations were spiked with tritiated water) a punch biopsy was taken from the skin beneath the electrodes and analyzed for radioactivity. Both anodal and cathodal iontophoresis increased the transport of water into the tissue, but the accumulation of water was maximal from the anode. These observations opened the field of iontophoresis, up to then considered limited to charged molecules, to uncharged substances as well.

Several mechanisms that could induce the iontohydrokinetic effect were discussed. The two main hypotheses were electroosmosis and hydrated ion shell transport. Electroosmosis results when an electric field is imposed across a membrane that has fixed charge and would be later identified as the main cause of convective solvent flow during transdermal iontophoresis (see below). The second hypothesis, the transfer of the water associated with an ion (sodium ions having a larger hydration shell would carry more water than chloride ions) was shown later on to be unable to explain the experimental data.

Subsequently, Burnette and Marrero (42) measured the transport of the tripeptide TRH through excised nude mouse skin. Surprisingly, the peptide was much more efficiently delivered at pH 8, under which circumstances it is 98% unprotonated, than at pH 4, where it is 99% ionized. These results were explained by a net convective flow of solvent in the anode-to-cathode direction at pH 8, but not at pH 4, and by the fact that TRH was being delivered primarily by electroosmosis under the conditions of this experiment. This effect was therefore related to a pH-dependent cation permselectivity of the skin and the corresponding Donnan exclusion of coions. Given that counterions (cations such as Na^+) were in excess, more momentum was imparted to the solvent in their direction of movement, i.e. anode-to-cathode.

It seems, then, that the skin behaves as an ion-exchange membrane, which is negatively charged at physiological pH, and that its permselectivity induces a net volume flow during iontophoresis, which proceeds from the anode to the cathode (43). In fact, it is well known that application of an electrical potential across a "porous" (ion-exchange) membrane containing fixed charges results in bulk fluid movement in the direction of counterion migration (7,43,44). Burnette and Ongpipattanakul (45) demonstrated the skin cation permselectivity by measuring the flux of the neutral polar marker mannitol during iontophoresis through excised human skin and determining the transport numbers of the main anionic and cationic charge carriers (i.e., Na^+ and Cl^-). At physiological pH, the transport numbers for Na^+ and Cl^- were, respectively, 0.6 and 0.3, values that were constant over the range of current densities studied (78 to 230 $\mu A/cm^2$). Correspondingly, mannitol flux was 10-fold from the anode than from the cathode. The transport of the three species was proportional to the current applied. The partial replacement of sodium by calcium ions resulted in an enhanced chloride flux and a decrease in mannitol transport, an effect perhaps due to the possible binding of calcium to the fixed negative charges on the tissue. Mannitol fluxes were one order of magnitude smaller than those of the ionic charge carriers (even when differences in concentration were considered), which suggested that convective transport was a less efficient transport mechanism than electromigration.

Burnette proposed two contributions to the convective flow term. First the electroosmosis effect: electroosmosis is the net water flow induced by the momentum transfer between ions as they are transported across the membrane and the surrounding solvent molecules. For a membrane such as the skin, which is cation permselective, the momentum transfer from the positive ions is greater than that from negative ions, resulting in net flow from anode to cathode. The second mechanism is the so-called transport number effect (or induced osmotic pressure effect), which has been described and demonstrated by Barry and Hope (46) for plant cells. This results from the difference between ion transport numbers in the membrane and their corresponding values in free solution. If this difference in the transport numbers is high, and if there are substantial stagnant diffusion layers at the membrane–solution interface, a current-induced concentration gradient is created that results in an osmotic flow of water in the same direction as the electroosmotic effect.

A series of papers by Pikal et al. (44,47,48) examined further the fundamentals of the convective contribution, and asserted (44) that the volume flow, J_v (volume·time^{-1}·area^{-1}) is proportional to the potential gradient established by the electric field:

$$J_V = L_{VE} \left(\frac{-d\Phi}{dx} \right) \tag{7}$$

where L_{VE} is the (phenomenological) electroosmotic flow coefficient describing the direction and magnitude of the volume flow resulting from the driving force, $-d\Phi/dx$. By convention, the volume flow is positive when the membrane is negative. Pikal explains the electroosmotic effect via the electrical volume force on the ion atmosphere (electrical double layer). When a membrane pore is lined with fixed or immobile charges, there is a region, the so-called ion atmosphere or electrical double layer, in which the pore fluid contains a nonzero charge density ρ arising from mobile ions. This density, when integrated over the total volume of the pore, must be equal in charge but opposite in sign to the total fixed charge on the pore wall. The ion atmosphere dissipates continuously with increasing distance from the wall. The interaction of the electric field with the ion atmosphere produces an electrical volume force which is the driving force for electroosmosis. On the other hand, and according to Pikal, the induced osmotic pressure or transport number effect only contributes ~5% to the total convective flow.

The main features of this mechanism of transport could now be identified or predicted. For example, it was suggested that the flow of solute with the solvent stream was independent of the solute size. Thus the relative importance of electroosmosis to iontophoretic transport increases with increasing molecular size (7,44). In Pikal's model, the electroosmotic flow is proportional to the product of the pore charge concentration and the square of the pore radius. Thus either a bigger area of negative large pores or a greater negative charge would result in an increased electroosmotic flow. At low pH, the skin loses its cation-permselectivity, and the residual negative charge results in a very reduced (if not reversed in direction (see below) convective flow. Increasing the background electrolyte (NaCl) concentration decreases solvent flow via an effect on the ion atmosphere. In an elegant series of experiments, Pikal actually measured water flow during iontophoresis (47). Values on the order of 4–11 $\mu L \cdot h^{-1} \cdot cm^{-2}$ were found. This volume flow will enhance or retard the flux of the solute of interest, ionic or neutral, depending on whether the flux of the compound interest is with or against the volume "stream" (47).

Pikal (48) also described the relationship between the molar flux of the solute, i, its molar concentration in the donor solution, c_i, and the volume flux, J_{vs}.

$$J_i = J_{vs} \cdot c_i \tag{8}$$

The iontophoretic flux of three small uncharged or zwitterionic solutes (glycine, glucose, and tyrosine), and of two high-molecular-weight anionic (at pH 8.5) species (carboxyinulin and albumin) was considered. Interestingly, the volume fluxes deduced for all these permeants were similar: 2–6 $\mu L \cdot h^{-1} \cdot cm^{-2}$. Surprisingly, carboxyinulin was better delivered from the anode despite its negative charge. This led to the hypothesis of what may be called "wrong-way iontophoresis": large anions, unable to compete efficiently to carry charge across the skin, are better delivered from the anode by electroosmosis. However, significant evidence to support this hypothesis remains elusive.

Electroosmotic flow is a nonequilibrium process that may be analyzed by nonequilibrium thermodynamics (7). Briefly, electroosmosis is an electrokinetic phenomenon, the reciprocal of a "streaming current" (43), which corresponds to the flow of charge and volume that results from the application of a pressure gradient across a permselective membrane. In an iontophoretic experiment, a voltage is applied at a constant pressure that results in flows of charge and volume.

The permselective properties of different skin models have been characterized. The transport numbers of ions have been measured via EMF methods and the Hittorf method (45,47,49,50) Typically these values have indicated a weak cation permselectivity: 0.5–0.6 for the monovalent cations (Na^+, K^+, Cs^+) at pH 7.4 in human skin; 0.25 for Na^+ at pH 4.0 and 0.3 for Ca^{+2}. For Cl^- in human skin, the transport number varies from 0.7 at pH 3 to 0.3 at pH 7.4.

It has also been important to examine possible differences between human skin and the animal models frequently used in iontophoresis. In general, these differences are minor. For example, using hairless mouse skin in vitro, lowering the pH of the bathing solutions eventually reversed the direction of the electroosmotic flow (47,49,51). This is due to the progressive neutralization of the negative charges on the skin and ultimately the apparition of positive ones. At this low pH, the skin therefore becomes anion-permselective (net positive charge), and the convective flow occurs preferentially in the cathode-to-anode direction. The skin is a heterogeneous, complex biological membrane through which iontophoretic transport may proceed via both intercellular and follicular pathways. Thus it is difficult to predict exactly the source of the membrane's charge. However, amino acids, fatty acids, etc. can be anticipated to contribute. In addition, an apparent skin isoelectric point (pI) has been defined as the pH at which the skin has no permselectivity, i.e., at which its net (not the absolute) charge is zero (Fig. 4). The value of the pI has been determined by measuring the flux of mannitol (52,53) or of 5-FU (54) as a function of pH or by the use of scanning electrochemical microscopy (SCEM) (55). The pI values reported are 3.5 (55) and 4.5–4.6

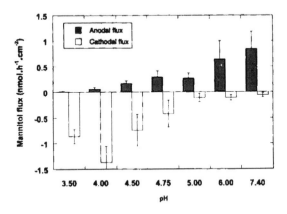

Figure 4   Anodal and cathodal fluxes of the neutral and polar markers of electroosmosis mannitol during iontophoresis at 0.4 mA. The measurements were made in side-by-side cells using dermatomed human skin. The pH of both the epidermal and the dermal bathing solutions were modified. (Data taken from Ref. 52.)

(53) for hairless mouse skin, 3.5–3.75 for neonatal porcine skin (53), 4.4 for adult pig-ear skin (54,52), and 4.8 for human skin (52). It should be noted that, while the complete reversal of the skin's net charge is easily accomplished in vitro, where the pH of both the epidermal and the dermal solutions can be easily altered, it is only possible in vivo to establish a pH gradient from the surface to the physiological value. As a result, while it is possible to optimize electroosmotic transport via pH manipulation of the drug formulation, the degree of change possible is clearly limited (56,57).

The pH is not the only tool with which to modify the skin permselectivity characteristics. Certain cations, for example, adsorb tenaciously inside the skin. The flux of such compounds typically remains constant or even decreases with increasing concentration when delivered in the presence of competing ions, and an important skin reservoir has often been identified. This behavior has been observed for classic drugs such as beta-blockers (24,58), quinine (24), and ropinirole (23), and for peptides such as nafarelin and leuprolide (18,59–62).

It has been suggested that these cationic species are effectively trapped in the skin and that their positive charge neutralizes partially, or even completely, the membrane's negative charge. This obviously results in decreased anode-to-cathode solvent flow and sometimes the reverse, i.e., an increased cathode-to-anode stream of solvent (18). The consequences of this phenomenon on a specific drug's flux depend on the electroosmotic contribution to its iontophoretic transport. That is, small drugs that have appreciable transport numbers are not significantly affected, while for larger drugs (such as nafarelin or leuprolide), which are mainly transported by electroosmosis, the effect can be dramatic.

Electroosmosis exponentially decays with increasing ionic strength of the donor solution (7,56); this is related to an effect on the ion atmosphere. Decreased electroosmosis has also been observed when divalent cations such as calcium are present in significant amount (45,63).

It has been hypothesized that electroosmotic transport is independent of molecular size, provided that the latter does not approach to the dimensions of the transport pathway (7,64). Available experimental data, in general, supports this conclusion, although the transport of water appears to be somewhat higher than expected, perhaps because of its significant passive permeability (6,51). Interestingly, the anode-to-cathode direction of convective solvent flow is maintained, even when chloride ion becomes the major charge carrier across the skin (65). In other words, it is the net charge on the membrane which dictates the direction of electroosmosis, not the nature of the current-carrying species.

In summary, transdermal iontophoresis results in a bulk volume flow that is directly linked to the charge of the skin and that proceeds in the direction of counterion movement (i.e., cations moving in the anode-to-cathode direction). This stream of solvent carries along with it dissolved solutes, thereby permitting the enhanced transport of neutral, and especially polar, molecules. Electroosmosis assists the transport of cations and retards that of anions. The convective flux of a solute is the product of the solvent flow multiplied by the concentration of the solute. The solvent flow is principally electroosmotic and depends on the net charge of the skin; the pH and the ionic strength are the main formulation parameters that can be used to modulate electroosmosis.

C. Combined Electroosmosis and Electromigration Transport

In the preceding sections, the two principal mechanisms of electrotransport have been rather artificially separated. However, iontophoretic fluxes of charged molecules are the sum of these two contributions which are not always easy to separate (66). In a good analogy, Pikal (7) describes a cation as a boat powered by an engine moving downstream along a river; its speed is the vector addition of the engine's power and the river's flow. An anion behaves as the same boat traveling upstream; the engine moves it up the river but now against the current, which reduces its speed. A boat without an engine can obviously only travel downstream. However, an observer on the river bank records only the direction and net speed of transport, and the functioning or not of an engine. It must be kept in mind that formulation parameters such as drug concentration, pH, ionic strength, and accompanying buffers may have an effect on one or both of the contributions. Thus the interpretation of experimental data must be carefully performed on an individual basis. The relative contributions of electroosmosis and electromigration are clearly molecule dependent but may also change for a single molecule with changes in the formulation.

The effect of increasing current intensity is the simplest to understand. Both electromigration and electroosmosis contributions increase with the intensity of current applied (2,12,14,23,27,29) and constitutes the basis for the controlled drug delivery or reverse iontophoretic sampling. The increase in electroosmosis with current is, however, less marked (45,60). For example, the transport of TRH (42) is proportional to current whether the principal mechanism of transport is electroosmosis or electromigration.

The effects of pH are more complex (Fig. 5). First of all, pH will determine the degree of ionization of a weakly acidic or basic permeant and hence the relative contributions to electrotransport. The pH can also have an impact on the convective flow by altering the charge of the skin (42,51–53,66).

Iontophoresis of acyclovir also illustrates the complexity of electrotransport (67). At pH 7.4, this antiviral drug is mainly unionized and effectively transported by electroosmosis from the anode. Cathodal iontophoresis at pH 7.4 is undistinguishable from passive delivery. At pH 3, the drug is 20% in the cationic form

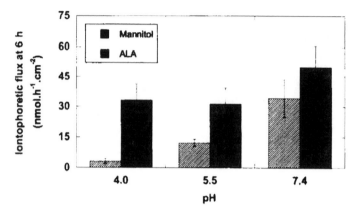

Figure 5 Anodal fluxes ALA (5-aminolevulinic acid) and mannitol through derma-tomed porcine ear skin as a function of the donor pH. Mannitol fluxes are exclusively due to electroosmosis. ALA (pKa values are 4.0 and 8.4) fluxes are due to both electromigration and electroosmosis. The driving concentration of ALA and mannitol were both 15 mM. (Data taken from Ref. 88.)

and delivered most effectively from the anode by electromigration. However the reversal of skin permselectivity at this pH means that the convective mechanism from the cathodal site can also be used to enhance the drug delivery.

Donor drug concentration effects can be subtle. Convective transport of neutral solutes is directly proportional to the concentration. This is obviously the basis for the iontophoretic sampling of glucose (68). Ions are more complex. As discussed above, the flux can either be dependent on donor drug concentration when there are competing coions in the donor formulations (10,25) or independent of the level of drug when it is the only available charge carrier. One may also expect that, if increasing drug concentration leads to a significantly higher ionic strength, then the electroosmotic contribution will be reduced.

The presence of accompanying buffers and electrolyte usually results in decreased drug flux (2,9,10,22). As the number of ions available as charge carriers increases, the transport number of the drug of interest necessarily decreases. The level of effect depends on the relative concentrations and mobilities of the species considered. Thus the use of polymeric buffers has been proposed (11,37,38).

As stated above, both the formulation characteristics and the physicochemical properties of the permeant determine the preponderant mechanism of transport. It is anticipated that, with increasing molecular weight, the principal mechanism of transport will progressively switch from electromigration to electroosmosis.

III. PATHWAYS OF IONTOPHORETIC TRANSPORT

Identification of the pathways of transport during iontophoresis has received some attention (69), and an important role of the follicular or appendegeal pathway has been suggested. Scanning electrochemical microscopy, coupled with video microscopy (70), indicates that the iontophoretic flux of $Fe(CN)_6^{4-}$ is highly localized, with 40 to 60% of the total transport being associated with skin appendages (71). Transport of the uncharged solute, hydroquinone (again monitored by scanning electrochemical microscopy) also followed the appendegeal pathway, implying that the mechanism of electrotransport does not dictate a particular route. The volume flow in one hair follicle was estimated (72) to be 2.5 nL/h, which, taking into account the hair follicle density in hairless mouse skin, suggests a value of 0.25 μL/cm^2·h for a current of 0.1 mA/cm^2. It appears, therefore, that a significant fraction of electroosmotic flow across hairless mouse skin occurs via hair follicles.

Other work, however, suggests the participation of the intercellular route. The in vivo iontophoresis of mercuric chloride in pigs and subsequent examination of the skin tissue by transmission electron microscopy showed that transport across the stratum corneum occurred via an intercellular pathway (73).

The different epidermal distribution of nile-red and calcein observed by confocal microscopy following iontophoresis (74) has indicated that the dominant iontophoretic pathway may be dictated by the physicochemical properties of the permeant. The iontophoretic transport of calcein via nonfollicular and follicular pathways has been quantitatively analyzed using this same technique (75). The data confirms the importance of hair follicles in hairless mouse skin iontophoresis, despite the small fractional area that they occupy; in effect, the follicular route was approximately four times more efficient than the nonfollicular. Nevertheless, in terms of the total transport of calcein, it was clear that a significant amount of nonfollicular transport remains to be explained.

IV. DRUG DELIVERY BY IONTOPHORESIS: EXAMPLE APPLICATIONS

A. Peptides

The use of iontophoresis for peptide transdermal application has been extensively examined (15,16,18,19,36,40–42,59–62,76,77). Iontophoresis facilitates the delivery of these usually charged and/or polar molecules through an otherwise quite impermeable skin barrier. It was soon shown that iontophoresis could provide, noninvasively, plasma levels of certain peptides comparable to those obtained after IV infusion or subcutaneous administration (11,76). Furthermore, iontophoresis reduced the lag time and offered input kinetics ranging from zero-order to pulsatile. The reader is referred to appropriate reviews for more details (59,76).

B. Antimigraine, Antiemetic, and Anti-Parkinson Drugs

The gastrointestinal absorption of the antimigraine drug alnitidan can be difficult during an acute attack due to nausea, vomiting, and gastric stasis (22). The in vitro (through hairless rats) and in vivo (in humans) iontophoretic administration of this drug has been studied (22). The in vitro data established the time, pH, and current dependencies of alnitidan transport. In vivo, the plasma levels obtained after 30 minutes of iontophoresis were equivalent to those observed after a subcutaneous administration, with small intra- and interindividual variability observed.

The coiontophoresis of metoclopramide and hydrocortisone has been proposed to decrease the local irritation caused by the antiemetic as it passes through the skin (78). The pharmacokinetics of metoclopramide after iontophoretic delivery in vivo in man were not modified by the coadministration of hydrocortisone. However, the codelivery of hydrocortisone was sufficient to reduce the erythema and edema observed.

The in vitro and in vivo (in man) iontophoresis of the anti-Parkinson's drug apomorphine has been studied (79,80). In vivo, after one-hour iontophoresis, plasma profiles comparable to a zero-order infusion study were obtained, although the concentrations were subtherapeutic. Deconvolution of the data demonstrated that a steady iontophoretic flux was attained following only 10 to 20 minutes of iontophoresis. Plasma levels decreased rapidly following current interruption. Minor skin irritation and tingling sensations were reported. The feasibility of iontophoretically administering therapeutic doses of ropinirole has also been shown in vitro (23).

C. Analgesics

The iontophoresis of morphine, hydromorphone, fentanyl, and sufentanil has been studied (27,31,81). Fentanyl, of course, has already been formulated as a passive transdermal patch (Duragesic®) for the management of chronic pain. However this method of administration is not appropriate for treating acute postoperative pain as it cannot be patient controlled. The iontophoretic device, E-TRANS®, on the other hand, allows an effective treatment that is patient-actuable (31). In vivo human studies have shown that both a 24-hour continuous and an on-demand pulsed delivery of fentanyl are possible with the E-TRANS®. The amount of fentanyl absorbed, the serum concentration, and the AUC increased with the applied current in either administration mode. Interestingly, the variability observed after iontophoresis was similar to that after IV administration of fentanyl.

D.   Local Delivery: Photodynamic Therapy, Psoriasis, Local Anesthesia, and Anti-Inflammatory Effect

One of the first suggested applications of iontophoresis was the treatment of herpes labialis and keratitis in surface tissues (82,83); in fact, iontophoresis of idoxuridine has been shown to be effective against herpes simplex viral lesions on the lip. More recently, iontophoresis has also effectively increased acyclovir skin permeation in vitro (67,84). The distribution of this antiviral agent after iontophoresis in human skin in vitro is more homogeneous than that obtained after passive application.

The skin distribution of khellin, an agent used for the photochemotherapy of vitiligo and psoriasis, was not significantly improved in vitro by iontophoresis due to its lipophilic and uncharged nature (85). Similarly, iontophoresis of methotrexate (86), a charged but nevertheless large (MW = 454) and poorly aqueous soluble drug, appears unlikely to result in useful clinical benefit.

Administration of anesthesics and anti-inflammatories was another obvious application of iontophoresis. Iontocaine® was the first drug–device (lidocaine) combination approved by the FDA (5,11). An integrated lidocaine device is currently in Phase III clinical trials. In this system, as reported elsewhere, a vasoactive drug is administered with lidocaine to optimize its iontophoretic delivery and to prolong the local anesthesia produced thereby (87).

Finally, iontophoresis has been shown to improve the transport of ALA (5-aminolevulinic acid) and ALA esters in vivo and in vitro (88,89). ALA is a precursor of protoporphyrin IX (PpIX), which can be used as an endogenous photosensitizer in photodynamic therapy of a range of cancers.

E.   Oligonucleotides

The transdermal route has been considered for antisense therapy (90). In an early study, the iontophoresis of bases, nucleosides, and nucleotides was studied in vitro through hairless mouse and human skin. The transport of charged and uncharged species was significantly enhanced in a concentration-dependent manner. ATP and GTP were significantly metabolized, while imidoGTP was more resistant to enzymatic attack (20).

There have been several attempts to deliver oligonucleotides iontophoretically and to establish structure–activity relationships. For example, Oldenburg et al. found that the flux of random oligonucleotides increased with current and donor concentration and decreased with the length of the oligonucleotide (33). The iontophoretic delivery of telomeric oligonucleotides through hairless mouse skin in an intact form was also shown to be dependent on pH and ionic strength (17). Phosphorothioate oligonucleotides are considered to be better candidates for antisense applications in that they are more resistant to nuclease attack. Brand et al. (21) have examined the influence of size and base composition on the iontophoretic flux of well-defined phosphorothioate oligonucleotides. The results suggest that size is not the only determinant of flux, and that sequence and base composition are important. The potential for systemic versus topical therapy would also be dependent on oligonucleotide internalization by keratinocytes, these being the first cells to be encountered (91). Finally, it has been shown that iontophoresis of the oligonucleotide complementary to the p450-3A2 AUG start site results in delivery sufficient to induce metabolic changes in rats in vivo (92). The midazolam-induced sleep times, which are in direct relation to the levels of cytochrome p450-3A2, were significantly longer for rats treated with iontophoresis of the phosphorotioate oligonucleotide. Intact oligonucleotide was found in liver, kidney, and, in smaller concentrations, in spleen and small intestine.

F. Reverse Iontophoresis: A Noninvasive Tool for Clinical Sampling?

As stated previously, iontophoresis is a symmetric process that transports ions across the skin in both directions of the membrane (Fig. 1). That is, endogenous ions will move toward the electrode of opposite sign in accord with their concentration and electrical mobility. Thus the anode will extract and become enriched in the ubiquitous chloride and also in any other anion whose physicochemical properties and concentration result in a reasonable transport number. The same applies at the cathode for sodium and other cations. Equally, uncharged solutes will be transported into the cathodal chamber from inside the body by the anode-to-cathode direction of flow. The first description of concentration-dependent noninvasive "reverse iontophoretic" extraction was published in 1989 for theophylline and clonidine (8). The potential for noninvasive glucose monitoring was rapidly perceived and soon demonstrated first in vitro and then in vivo (68). Reverse iontophoresis demands that two criteria be met: [a] that the extracted amounts respond quickly to changes in the systemic levels of the analyte of interest, and [b] given that the concentration of analyte in the collection medium is necessarily smaller than its endogenous level, and that the extraction process is not selective, a specific and sensitive assay is required. Despite these considerable challenges, the Glucowatch®, a device for the automatic, frequent, and noninvasive monitoring of glucose levels (68,93) has been developed. The impressive technology is sufficiently sensitive to permit measurements based on iontophoretic extraction to be made every 10 minutes (3 minutes of iontophoresis followed by 7 minutes for sensing). The specific biosensor assay is based on the amperometric detection of glucose using the classic Pt-glucose oxidase couple (93). It is true that the device requires a single point calibration made with a conventional fingerstick, but then it offers a continuous 12 hours of noninvasive monitoring with measurement of glycemia values every 20 minutes and alarms set at levels indicating too-high or too-low values to warn of hyper- and hypoglycemie incidents (93).

The approval of the Glucowatch Biographer® by the FDA at the beginning of 2001 has recognized the utility of reverse iontophoresis as a sampling technique as well as reopened the question of what's next. Obviously, few molecules offer the enormous interest and commercial interest of glucose measurements in diabetics. Reverse iontophoresis has been suggested for the noninvasive monitoring of phenylalanine levels in the management of phenylketonuria (57) and as a potential tool for therapeutic monitoring (94,95). Obviously, the feasibility and practicality of this approach will have to be evaluated in individual terms and according to the physicochemical, pharmacokinetic, and clinical properties of each candidate.

ACKNOWLEDGMENTS

We want to thank the "Programme commun en Génie Biomédical" of the EPFL and the Universities of Geneva and Lausanne and the Swiss National Science Foundation (3200-059042.99/1) for financial support.

REFERENCES

1. RR Burnette. Iontophoresis. In: J Hadgraft, RH Guy, eds. Transdermal drug delivery. Developmental issues and research initiatives. New York: Marcel Dekker, 1989, pp 247–291.
2. N Harper-Bellantone, S Rim, ML Francoeur, B Rasadi. Enhanced percutaneous absorption via iontophoresis I. Evaluation of an in vitro system and transport model compounds. Int J Pharm 30:63–72, 1986.
3. P Tyle. Iontophoretic devices for drug delivery. Pharm Res 3:318–326, 1986.
4. K Sato, DE Timm, F Sato, EA Templeton, DS Meletiou, T Toyomoto, G Soos, SK Sato. Generation and transit pathway of H⁺ is critical for inhibition of palmar sweating by iontophoresis in water. J Appl Physiol, 75:2258–2264, 1993.
5. RH Guy. Current status and future prospects of transdermal delivery. Pharm Res 13:1765–1769, 1996.

6. LP Gangarosa, N Park, CA Wiggins, JM Hill. Increased penetration of non-electrolytes into mouse skin during iontophoretic water transport (iontohydrokinesis). J Pharmacol Exp Ther 212:377–381, 1980.
7. MJ Pikal. The role of electroosmotic flow in transdermal iontophoresis. Adv Drug Deliv Rev 9:201–237, 1992.
8. P Glikfeld, RS Hinz, RH Guy. Noninvasive sampling of biological fluids by iontophoresis. Pharm Res 6:988–990, 1989.
9. BH Sage, JE Riviere. Model systems in iontophoresis transport efficacy. Adv Drug Deliv Rev, 9:265–287, 1992.
10. JB Phipps, JR Gyory. Transdermal ion migration. Adv Drug Deliv Rev 9: 137–176, 1992.
11. ER Scott, B Phipps, R Gyory, RV Padmanabhan. Electrotransport systems for transdermal delivery: a practical implementation of iontophoresis. In: DL Wise, ed. Handbook of Pharmaceutical Controlled Release Technology. New York: Marcel Dekker, 2000, pp 617–659.
12. JB Phipps, RV Padmanabhan, GA Lattin. Iontophoretic delivery of model inorganic and drug ions. J Pharm Sci 78:365–369, 1989.
13. C Cullander, G Rao, RH Guy. Why silver/silver chloride? Criteria for iontophoresis electrodes. In: KR Brain, VJ James, KA Walters, eds. Prediction of Percutaneous Penetration. Cardiff: STS, 1993, pp 381–390.
14. CL Gay, PG Green, RH Guy, ML Francoeur. Iontophoretic delivery of piroxicam across the skin. J Control Rel 22:57–68, 1992.
15. PG Green, RS Hinz, C Cullander, G Yamane, RH Guy. Iontophoretic delivery of a series of amino acids and amino acid derivatives across the skin "in vitro." Pharm Res 8:1113–1120, 1991.
16. PG Green, RS Hinz, A Kim, FC Szoka, RH Guy. Iontophoretic delivery of a series of tripeptides across the skin in vitro. Pharm Res 8:1121–1127, 1991.
17. RM Brand, PL Iversen. Iontophoretic delivery of a telomeric oligonucleotide. Pharm Res 13:851–854, 1996.
18. MB Delgado-Charro, RH Guy. Iontophoretic delivery of nafarelin across the skin. Int J Pharm 117:165–172, 1995.
19. LL Miller, CJ Kolaskie, GA Smith, J Riviere. Transdermal iontophoresis of gonadotropin releasing hormone (LHRH) and two analogues. J Pharm Sci 79:490–493, 1990.
20. R Van der Geest, F Hueber, FC Szoka Jr., RH Guy. Iontophoresis of bases, nucleosides and nucleotides. Pharm Res 13:553–558, 1996.
21. RM Brand, A Wahl, PI Iversen. Effects of size and sequence on the iontophoretic delivery of oligonucleotides. J Pharm Sci 87:49–52, 1998.
22. A Jadoul, J Mesens, W Caers, F de Beukelaar, R Crabbe, V Préat. Transdermal permeation of alnitidan by iontophoresis: in vitro optimization and human pharmacokinetic data. Pharm Res 13:1348–1353, 1996.

23. A Luzardo-Alvarez, MB Delgado-Charro, J Blanco-Méndez. Iontophoretic delivery of ropinirole hydrochloride: effect of current density and vehicle formulation. Pharm Res 18:1714–1720, 2001.

24. D Marro, YN Kalia, MB Delgado-Charro, RH Guy. Contributions of electromigration and electroosmosis to iontophoretic drug delivery. Pharm Res 18:1701–1708, 2001.

25. GB Kasting, JC Keister. Application of electrodiffusion theory for an homogeneous membrane to iontophoretic transport through skin. J Control Rel 8:195–210, 1989.

26. PW Ledger. Skin biological issues in electrically enhanced transdermal delivery. Adv Drug Deliv Rev 9:289–307, 1992.

27. RH Padmanabhan, JB Phipps, GA Lattin, RJ Sawchuk. In vitro and in vivo evaluation of transdermal iontophoretic delivery of hydromorphone. J Control Rel 11:123–135, 1990.

28. N Sekkat, YN Kalia, RH Guy. Control of drug delivery across an impaired skin barrier by iontophoresis. Proceedings of 28th International Symposium on Controlled Release of Bioactive Materials, San Diego, 2001, pp 184–185.

29. RM Brand, RH Guy. Iontophoresis of nicotine in vitro: pulsatile drug delivery across the skin? J Control Rel 33:285–292, 1995.

30. P Singh, MS Roberts, HI Maibach. Modelling of plasma levels of drugs following transdermal iontophoresis. J Control Rel 33:293–298, 1995.

31. SK Gupta, M Southam, G Sathyan, M Klausner. Effect of current density on pharmacokinetics following continuous or intermittent input from a fentanyl electrotransport system. J Pharm Sci 87:976–981, 1998.

32. K Sugibayashi, M Kagino, S Numajiri, N Inoue, D Kobayashi, M Kimura, M Yamaguchi, Y Morimoto. Synergistic effects of iontophoresis and jet injector pretreatment on the in vitro skin permeation of diclofenac and angiotensine II. J Pharm Pharmacol 52:1179–1186, 2000.

33. KR Oldenburg, KT Vo, GA Smith, HE Selick. Iontophoretic delivery of oligonucleotides across full thickness hairless mouse skin. J Pharm Sci 84:915–921, 1995.

34. RH Guy, MB Delgado-Charro, YK Kalia. Iontophoretic transport across the skin. Skin Pharmacol Appl Skin Physiol 14(suppl):35–40, 2001.

35. NH Yoshida, MS Roberts. Solute molecular size and transdermal iontophoresis across excised human skin. J Control Rel 25:177–195, 1993.

36. M Clemessy, G Couarraze, B Bevan, F Puisieux. Mechanisms involved in iontophoretic transport of angiotensine. Pharm Res 12:998–1002, 1995.

37. PP Sarpotdar, CR Daniels. Use of polymeric buffers to facilitate iontophoretic transport of drugs. Pharm Res 7:S185, 1990.

38. TS Sahota, RJ Latham, RG Lindford, PM Taylor. Polymer electrolyte materials for use in iontophoretic drug delivery. In: PI Harris, D Chapman, eds.

New Biomedical Materials. Basic and Applied Studies. Amsterdam: IOS Press, 1998, pp 143–148.

39. NN Turner, C Cullander, RH Guy. Determination of the pH gradient across stratum corneum. J Invest Dermatol 3:110–113, 1998.

40. YW Chien, O Siddiqui, Y Sun, WV Shi, JC Liu. Transdermal iontophoretic delivery of therapeutic peptides/proteins I: Insulin. Annals New York Acad Sci 507:32–51, 1987.

41. J Hirvonen, F Hueber, RH Guy. Current profiles regulates iontophoretic delivery of aminoacids across the skin. J Control Rel 37:239–249, 1995.

42. RR Burnette, D Marrero. Comparison between the iontophoretic and passive transport of thyrotropin releasing hormone across excised nude mouse skin. J Pharm Sci 75:738–743, 1986.

43. F Helfferich. Ion Exchange. New York: Dover, 1995, pp 323–338.

44. MJ Pikal. Transport mechanisms in iontophoresis. I. A theoretical model for the effect of electroosmotic flow on flux enhancement in transdermal iontophoresis. Pharm Res 7:118–126, 1990.

45. RR Burnette, B Ongpipattanakul. Characterization of the permselective properties of excised human skin during iontophoresis. J Pharm Sci 76:765–773, 1987.

46. PH Barry, AB Hope. Electroosmosis in membranes: effects of unstirred layers and transport numbers. II. Experimental. Biophysical Journal 9:729–757, 1969.

47. MJ Pikal, S Shah. Transport mechanisms in iontophoresis. II. Electroosmotic flow and transference number measurements for hairless mouse skin. Pharm Res 7:213–221, 1990.

48. MJ Pikal, S Shah. Transport mechanisms in iontophoresis. III. An experimental study of the contributions of electroosmotic flow and permeability change in transport of low and high molecular weight solutes. Pharm Res 7:222–231, 1990.

49. GB Kasting, LA Bowman. DC electrical properties of frozen, excised human skin. Pharm Res 7:134–143, 1990.

50. JD DeNuzzio, B Berner. Electrochemical and iontophoretic studies of human skin. J Control Rel 11:105–112, 1990.

51. A Kim, PG Green, G Rao, RH Guy. Convective solvent flow across the skin during iontophoresis. Pharm Res 10:1315–1320, 1993.

52. D Marro, R Guy, MB Delgado-Charro. Characterization of the iontophoretic permselectivity properties of human and pig skin. J Control Rel 70:213–217, 2001.

53. A Luzardo-Alvarez, M Rodríguez-Fernández, J Blanco-Méndez, R Guy, MB Delgado-Charro. Iontophoretic permselectivity of mammalian skin: characterization of hairless mouse and porcine membrane models. Pharm Res 15:984–987, 1998.

54. V Merino, A López, YN Kalia, RH Guy. Electrorepulsion versus elec-
     troosmosis: effect of pH on the iontophoretic flux of 5fluorouracil. Pharm
     Res 16:758–761, 1999.
55. BD Bath, HS White, ER Scott. Visualization and analysis of electroosmotic
     flow in hairless mouse skin. Pharm Res 17:471–475, 2000.
56. P Santi, RH Guy. Reverse iontophoresis. Parameters determining electroos-
     motic flow: I. pH and ionic strength. J Control Rel 38:159–165, 1996.
57. V Merino, A López, D. Hochstrasser, RH Guy. Noninvasive sampling of
     phenylalanine by reverse iontophoresis. J Control Rel 61:65–69, 1999.
58. J Hirvonen, RH Guy. Iontophoretic delivery across the skin: electroosmosis
     and its modulation by drug substances. Pharm Res 14:1258–1263, 1997.
59. MB Delgado-Charro, RH Guy. Iontophoresis of peptides. In: B Berner, SM
     Dinh, eds. Electronically Controlled Drug Delivery. Boca Raton: CRC
     Press, 1989.
60. MB Delgado-Charro, R Guy. Characterization of convective solvent flow
     during iontophoresis. Pharm Res 11:929–935, 1994.
61. AJ Hoogstrate, V Srinivasan, SM Sims, WI Higuchi. Iontophoretic en-
     hancement of peptides: behaviour of leuprolide versus model permeants. J
     Control Rel 31:41–47, 1994.
62. J Hirvonen, YN Kalia, RH Guy. Transdermal delivery of peptides by ionto-
     phoresis. Nature Biotech. 14:1710–1713, 1996.
63. P Santi, RH Guy. Reverse iontophoresis parameters determining electroos-
     motic flow: II. Electrode chamber formulation. J Control Rel 42:29–36,
     1996.
64. P Singh, M Anliker, GA Smith, D Zavortink, HI Maibach. Transdermal
     iontophoresis and solute penetration across excised human skin. J Pharm
     Sci 84:1342–1346, 1995.
65. D Marro, YN Kalia, MB Delgado-Charro, RH Guy. Optimizing iontopho-
     retic drug delivery: identification and distribution of the charge carrying
     species. Pharm Res, in press.
66. RH Guy, YN Kalia, MB Delgado Charro, V Merino, A López, D Marro.
     Iontophoresis: electrorepulsion and electroosmosis. J Control Rel 64:129–
     132, 2000.
67. NM Volpato, P Santi, P Colombo. Iontophoresis enhances the transport of
     acyclovir through nude mouse skin by electrorepulsion and electroosmosis.
     Pharm Res 12:1623–1627, 1995.
68. G Rao, RH Guy, P Glikfeld, WR LaCourse, L Leung, J Tamada, R Potts,
     N Azimi. Reverse iontophoresis: noninvasive glucose monitoring in vivo
     in humans. Pharm Res 12:1869–1873, 1995.
69. C Cullander. What are the pathways of iontophoretic current flow through
     mammalian skin? Adv Drug Deliv Rev 9:119–135, 1992.
70. ER Scott, AI Laplaza, HS White, JB Phipps. Transport of ionic species in

skin: contribution of pores to overall skin conductance. Pharm Res 10: 1699–1709, 1993.

71. ER Scott, JB Phipps, HS White. Direct imaging of molecular transport through skin. J Invest Dermatol 104:142–145, 1995.

72. BD Bath, ER Scott, JB Phipps, HS White. Scanning electrochemical microscopy of iontophoretic transport in hairless mouse skin. Analysis of the relative contributions of diffusion, migration, and electroosmosis to transport in hair follicles. J Pharm Sci 89:1537–1549, 2000.

73. NA Monteiro-Riviere, AO Inman, JE Riviere. Identification of the pathway of iontophoretic drug delivery: light and ultrastructural studies using mercuric chloride in pigs. Pharm Res 11:251–256, 1994.

74. NG Turner, RH Guy. Iontophoretic transport pathways: dependence on penetrant physicochemical properties. J Pharm Sci 86:1385–1389, 1997.

75. NG Turner, RH Guy. Visualization and quantitation of iontophoretic pathways using confocal microscopy. J Invest Dermatol Symp Proceed 3:136–142, 1998.

76. PG Green. Iontophoretic delivery of peptide drugs. J Control Rel 41:33–48, 1996.

77. L Langkjaer, J Brange, GM Grodsky, RH Guy. Iontophoresis of monomeric insulin analogues in vitro: effects of insuline charge and skin pretreatment. J Control Rel 51:47–56, 1998.

78. M Cormier, ST Chao, SK Gupta, R Haak. Effect of transdermal iontophoresis codelivery of hydrocortisone on metoclopramide pharmacokinetics and skin induced reactions on human subjects. J Pharm Sci 88:1030–1035, 1999.

79. R Van der Geest, M Danhof, HE Bodde. Iontophoretic delivery of apomorphine I: in vitro optimization and validation. Pharm Res 14:1798–1803, 1997.

80. R Van der Geest, T van Laar, JM GubbensStibbe, HE Bodde, M Danhof. Iontophoretic delivery of apomorphine II: an in vivo study in patients with Parkinson's disease. Pharm Res 14:1804–1810, 1997.

81. S Thysman, V Préat. In vivo iontophoresis of fentanyl and sufentanil in rats: pharmacokinetics and acute antinoceptive effects. Anesth Analg 77:61–66, 1993.

82. LP Gangarosa, JM Hill. Modern iontophoresis for local delivery. Int J Pharm 123:159–171, 1995.

83. JM Hill, LP Gangarosa, NH Park. Iontophoretic application of antiviral chemotherapeutic agents. Ann NY Acad Sci, 284:604–612, 1977.

84. NM Volpato, S Nicoli, C Laureri, P Colombo, P Santi. In vitro acyclovir distribution in human skin layers after transdermal iontophoresis. J Control Rel 50:291–296, 1998.

85. B Marconi, F Mancini, P Colombo, F Allegra, F Giordano, A Gazzaniga, G

Orecchia, P Santi. Distribution of khellin in excised human skin following iontophoresis and passive dermal transport. J Control Rel 60:261–268, 1999.

86. MJ Alvarez-Figueroa, MB Delgado-Charro, J Blanco-Méndez. Passive and iontophoretic transdermal penetration of methotrexate. Int J Pharm 212: 101–107, 2001.

87. JE Riviere, NA Monteiro-Riviere, AO Inman. Determination of lidocaine concentrations in skin after transdermal iontophoresis: effects of vasoactive drugs. Pharm Res 9:211–214, 1992.

88. RFV Lopez, MVLB Bentley, MB Delgado-Charro, R Guy. Iontophoretic delivery of 5aminolevulinic acid (ALA): effect of pH. Pharm Res 18:311–315, 2001.

89. S Gerscher, JP Connelly, J Griffiths, SB Brown, AJ MacRobert, G Wong, LE Rhodes. Comparison of the pharmacokinetics and phototoxicity of protoporphyrin IX metabolized from 5-aminolevulinic acid and two derivatives in human skin in vivo. Photochemistry and Photobiology 72:569–574, 2000.

90. RM Brand, PL Iversen. Transdermal delivery of antisense compounds. Adv Drug Del Rev 44:51–57, 2000.

91. RM Brand, K Haase, TL Hannah, PL Iversen. An experimental model for interpreting percutaneous penetration of oligonucleotides that incorporates the role of keratinocytes. J Invest Dermatol, 111:1166–1171, 1998.

92. RM Brand, TL Hannah, J Norris, PL Iversen. Transdermal delivery of antisense oligonucleotides can induce changes in gene expression in vivo. Antisense and Nucleic Acid Drug Dev 11:1–6, 2001.

93. JA Tamada, S Garg, L Jovanovic, KR Pitzer, RO Potts. Noninvasive glucose monitoring. Comprehensive clinical results. JAMA, 282:1839–1844, 1999.

94. B Leboulanger, RH Guy, MB Delgado-Charro. Reverse Iontophoresis of phenytoin: a novel noninvasive monitoring approach? Proceedings of the 7th International Congress of Therapeutic Drug Monitoring and Clinical Toxicology. Ther Drug Monit 23:477, 2001.

95. MB Delgado-Charro, R Guy. Reverse iontophoresis: a tool for valproate monitoring? Proceedings of the 7th International Congress of Therapeutic Drug Monitoring and Clinical Toxicology. Ther Drug Monit 23:476, 2001.

6

Skin Electroporation for Transdermal and Topical Drug Delivery

Véronique Préat and Rita Vanbever
Université Catholique de Louvain, Brussels, Belgium

I. INTRODUCTION

Transdermal delivery offers a number of potential advantages over conventional methods, such as the injectable and oral routes (Hadgraft and Guy, 1989; Smith and Maibach, 1995; Guy, 1996). Potential degradation in the gastrointestinal tract and first pass through the liver are avoided. Transdermal delivery has the potential for sustained and controlled release. Patient compliance may be improved by this user-friendly method.

However, because molecular passage is impeded by the barrier properties of skin, transdermal drug delivery has found limited clinical application (Smith and Maibach, 1995). The barrier properties of skin are attributed primarily to the intercellular lipid bilayers of the stratum corneum. Candidates for passive transdermal delivery therefore share three common traits: effectiveness at relatively low doses, molecular mass less than 400 Da, and lipophilicity. Drugs delivered from conventional passive transdermal patches usually reach therapeutic plasma levels with lag times of hours.

A number of approaches to enhancing and better controlling transport across skin and/or expanding the range of drugs delivered have been investigated (Hadgraft and Guy, 1989; Walters and Hadgraft, 1993; Smith and Maibach, 1995). These involve chemical methods, e.g., chemical enhancers and liposomes, and physical methods, e.g., iontophoresis, electroporation, and sonophoresis, that are based on two strategies: enhancing skin permeability and/or providing a driving force acting on the drug. Recently, the application of short high-voltage pulses has been shown to increase transport across the skin by many orders of magnitude, probably by a mechanism involving electroporation (Prausnitz et al., 1993a; Vanbever et al., 1994, Prausnitz, 1999; Pliquett, 1999; Weaver et al., 1999).

Electroporation (or electropermeabilization) involves the creation of transient aqueous pathways across lipid bilayer membranes by the application of short high-voltage pulses. It is a universal physical phenomenon applying to lipid bilayers of nonliving systems, such as liposomes and red blood cell ghosts, as well as the plasma membranes of living cells, either isolated or part of a tissue. Electrical exposures typically involve electric field pulses that generate transmembrane potentials of 0.5–1.0 V and last for 10 μs to 10 ms. Both dramatic electrical behavior (reversible electrical breakdown) and significant molecular transport occur because of structural rearrangement of the cell membrane (Weaver, 1995; Weaver and Chizmadzhev, 1996). The biological composition and structure of the stratum corneum lipid bilayers make it particularly attractive for electroporation. The stratum corneum contains approximately 100 bilayer membranes in series, and electrical breakdown associated with dramatic increase in transport has been observed for transdermal voltages reaching 30–100 V, which well corresponds to the voltages used for electroporation in cells (Prausnitz et al., 1993; Pliquett et al., 1995).

The aim of this chapter is to give an overall picture of the main features of skin electroporation used for transdermal and topical drug delivery. The enhancement in transport provided, the mechanisms involved, and the control achieved on molecular transport are depicted. The issue of tolerance is discussed.

II. INCREASE IN MOLECULAR TRANSPORT

A. Range of Compounds

Application of high-voltage electric field pulses has been shown to increase transport across and/or into skin for compounds ranging in size from small ions (Pliquett et al., 1996b) to moderate-sized molecules, e.g., calcein (Prausnitz et al., 1993a) (Fig. 1) to macromolecules, e.g., LHRH, heparin, oligonucleotides, and FITC-dextran (up to 40 kDa) (Bommannan et al., 1994; Prausnitz et al., 1995; Zewert et al., 1995; Regnier et al., 1998a; Lombry et al., 2000) (Table 1).

Electroporation of skin can improve transdermal transport of drugs prototypical of transdermal application, e.g., fentanyl (Vanbever et al., 1996a). However, orders-of-magnitude increases in transport have also been demonstrated for (a) hydrophilic molecules, e.g., metoprolol, mannitol (Vanbever et al., 1994, 1998c), (b) highly charged compounds, e.g., calcein, heparin (Prausnitz et al.,

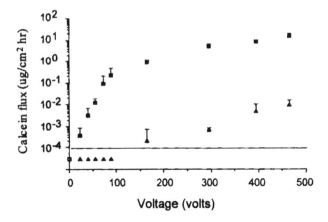

Figure 1   Average transdermal fluxes of calcein during electroporation pulses and iontophoresis. Upper axis indicates pulsing voltages electrically equivalent to continuous dc
voltages on lower axis, assuming no skin structural changes for both methods. The elevated
fluxes caused by electroporation as compared to iontophoresis suggest however important
changes in skin properties induced by pulsing. (From Prausnitz et al., 1993a.)

1993a, 1995) (c) macromolecules with a molecular weight cutoff larger than 40
kDa (Lombry et al., 2000).

   While the transport of 10 nm particles into the skin has been shown using
confocal microscopy (Prausnitz et al., 1996a), transport of microparticles across
skin remains controversial. Zhang et al., (1997) and Hoffman et al., (1995) claim
that electroincorporation, i.e., the delivery of drugs encapsulated in microparticles
using skin electroporation, increased macromolecule transport, whereas Chen et
al., (1999) did not detect any transport of 14 nm to 2.1 μm particles across the
stratum corneum.

   Topical delivery of small dyes (Pliquett and Weaver, 1996a; Johnson et
al., 1998), drugs (Wang et al., 1998a; Badkar et al., 1999), oligonucleotides (Regnier et al., 1998a), DNA plasmid (Dujardin and Préat, 2001), and antigen (Misra
et al., 2001; Dujardin et al., 2000) can also be enhanced by skin electroporation.

   These data suggest that skin electroporation could be used to expand the
range of compounds delivered transdermally or topically to hydrophilic, charged,
and even macromolecular compounds.

B.   Magnitude and Onset Time of Enhancement

Skin electroporation can increase transdermal transport by up to four orders of
magnitude with lag times of only seconds to minutes in vitro showing that mole-

Table 1   Summary of the In Vitro Studies on Transdermal and Topical Drug
Delivery Using Skin Electroporation

Compound	Molecular mass	Charge	Log enhancement ratio	References
Water	18	0	1	Vanbever et al., 1998c
Mannitol	182	0	2	Vanbever et al., 1998c
Vitamin C	176			Zhang et al., 1999
Metoprolol	267	+1	3	Vanbever et al., 1994, Vanbever and Préat, 1995
Alnitidan	302	+1/+2	2	Jadoul et al., 1998b
Methylene blue	320	+1/+2		Johnson et al., 1998
Fentanyl	336	+1	2	Vanbever et al., 1996a,b
FITC	390	−1	?	Lombry et al., 2000
Domperidone	426	+1	2	Jadoul and Préat, 1997
Lucifer yellow	457	−2	4	Chen et al., 1998a
Sulforhodamine	607	−1	3	Pliquett and Weaver, 1996a; Chen et al., 1998a; Vanbever et al., 1999
Calcein	623	−4	4	Prausnitz et al., 1993a, 1994; Pliquett et al., 1996a; Chen et al., 1998a
Erythrosin derivative	1025	−1	4	Praunitz et al., 1993a
Cyclosporine A	1201	0	1	Wang et al., 1998a,b
Oligonucleotide	4000–7000	−15/−24	1	Zewert et al., 1995; Regnier et al., 1998, 1999; Regnier and Préat, 1998, 1999
Heparin	12,000	~ 76	2	Prausnitz et al., 1995
FITC-dextran	4–38,000	—	?	Lombry et al., 2000
Ovalbumin	48,000		—	Dujardin et al., 2000
Nanomicrospheres	10 nm to 45 μm	Highly negative	—	Prausnitz et al., 1996a Chen et al., 1999 Hoffmann et al., 1995

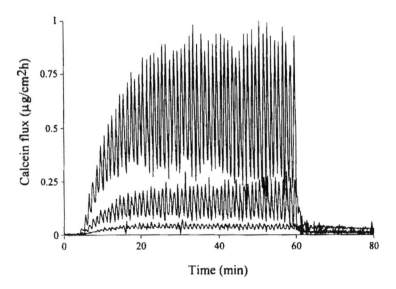

Figure 2 Rapid temporal control of transdermal transport using skin electroporation. The time profiles of calcein transdermal flux due to electroporation at different transdermal voltages are shown. 1 ms pulses were applied to human epidermis in vitro at 1 pulse per minute for 1 h, at 270 V (solid line), 135 V (dashed line), or 115 V (dotted line). (From Prausnitz et al., 1994.)

cules rapidly respond to electric pulses (Fig. 2) (Prausnitz et al., 1994; Pliquett and Weaver, 1996a). The enhancement magnitude and the onset times for transport have been shown to depend on the pulsing protocols and the physicochemical properties of the molecule being transported as well as the skin model (isolated epidermis versus full thickness skin) (Table 1) (Prausnitz et al., 1994; Vanbever and Préat, 1999; Préat et al., 1997).

C. In Vivo Studies

Most of the studies on transdermal drug delivery using skin electroporation have been performed in vitro using standard diffusion cells. However, in vivo studies have confirmed that skin electroporation enhances and expedites transdermal drug delivery (Vanbever and Préat, 1999).

A first limited study assessed transport due to skin electroporation in vivo by measuring serum concentrations of calcein (Prausnitz et al., 1993b). Fluxes at least two orders of magnitude greater than controls were observed. More recently, electroporation of skin was demonstrated to provide in vivo rapid transdermal

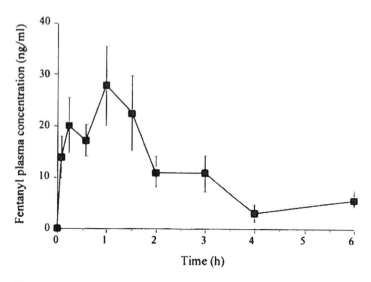

Figure 3   Fentanyl plasma concentrations as a function of time after transdermal delivery using electroporation. Electroporation was carried out using 15 pulses of 250 V (voltage applied at the electrodes) and 200 ms, applied from time 0 to 5 min in hairless rat skin in vivo. Foams at the cathode and anode were soaked with a solution of fentanyl (400 μg/mL in citrate buffer 0.01 M at pH 5). (From Vanbever et al., 1998a, with permission.)

delivery of drug at therapeutic levels. Therapeutic plasma levels of fentanyl in hairless rats were reached within 15 min after the application of high-voltage pulses, with deep analgesia lasting for about an hour (Fig. 3). For comparison, fentanyl delivered by transdermal passive diffusion reached therapeutic plasma levels after several hours, and following iontophoresis or subcutaneous injection after a few minutes as well (Vanbever et al., 1998a).

Electroporation has been investigated as a needle-free immunization method. An immune response was elicited in mice when antigens were delivered by skin electroporation (Misra et al., 1999; Dujardin et al., 2000).

III. MECHANISMS

Experimental and theoretical evidence suggests that the flux increases are caused by transient changes in skin microstructure by a mechanism involving electroporation of the stratum corneum lipid bilayers (Chang et al., 1992; Prausnitz, 1996b; Chizmadzhev et al., 1995; Weaver, 1993, 1995; Weaver and Chizmadzhev, 1996; Weaver et al., 1999). Application of high-voltage pulses to skin dramatically and reversibly increases skin permeability (Prausnitz et al., 1993b; Pliquett et al., 1995a; Vanbever et al., 1999b). New aqueous pathways would be created within stratum corneum by a mechanism involving electroporation of the lipid bilayers. Thermal effects may be involved. Molecular transport through transiently permeabilized skin then occurs due to different mechanisms (Vanbever et al., 1994, 1996b; Chizmadzhev et al., 1995). Electrophoresis during pulses and enhanced diffusion between and after pulses emerged as the two main mechanisms of transport of charged compounds. The contribution of electroosmosis is insignificant. Moreover, the contribution of electrophoresis and diffusion are dependent on the physicochemical properties of the molecule transported as well as the formulation of the drug reservoir (see below).

A. New Aqueous Pathways

Models involving transient creation of aqueous pathways across lipid bilayers have been used to explain the dramatic and often-reversible changes in membrane properties of cells and nonliving systems (e.g., liposomes, red blood cell ghosts) associated with high-voltage pulse application, hence the term "electroporation" (Chang et al., 1992; Weaver, 1995; Weaver and Chizmadzhev, 1996a, b). Although "electropores" are plausible, there is no direct evidence for their existence. It is unlikely that transient pores can be visualized by any present form of microscopy: pores are believed to be small (< 10 nm), sparse ($= 0.1\%$ of surface area), and generally short-lived (microseconds to seconds) (Weaver, 1995). Instead, information regarding pores is entirely indirect, mainly through their involvement in ionic and molecular transport (Weaver, 1993).

The three characteristic features of electroporation in cell membranes are (a) dramatic increases in transmembrane transport, (b) reversibility, and (c) evidence for structural changes in the membrane barrier (Prausnitz et al., 1993a).

Experimental and theoretical data support the hypothesis that high-voltage pulses induce also the creation of new, and/or the enlargement of existing, aqueous pathways in the stratum corneum. Flux measurements in vitro demonstrate increases of up to four orders of magnitude in transdermal transport following high-voltage pulses, the flux increase being achieved in a matter of seconds (Prausnitz et al., 1994). Enhancement in transport across skin occurred for compounds as large as macromolecules (Prausnitz et al., 1995; Zewert et al., 1995; Lombry et al., 2000), suggesting that transport pathways are larger than preexisting pathways.

Skin resistance, which is largely attributed to stratum corneum lipids, has been shown to drop by up to three orders of magnitude on a time scale of microseconds or faster during high-voltage pulses (50 to 100 V across the skin) (Pliquett et al., 1995b). This drop in skin resistance was partially reversible.

Fluorescence microscopy indicates that molecular transport during high-voltage pulsing occurs through the bulk of the stratum corneum away from appendages at localized sites through intercellular and transcellular pathways (Pliquett et al., 1996b; Prausnitz et al., 1996a, Regnier and Préat, 1998; Lombry et al., 2000). The effective fractional aqueous area for small ion transport was approximately 0.1% (Prausnitz et al., 1996b; Pliquett and Weaver, 1996b; Vanbever et al., 1999). At the same amount of charge transferred, drug transport is usually higher following skin electroporation than iontophoresis, suggesting that electrophoresis alone cannot explain drug transport by skin electroporation.

Not all these data can be explained without alteration in skin structure. They are consistent with the electroporation features of single bilayer membranes and therefore with the hypothesis of the creation of new aqueous pathways (Chang et al., 1992; Weaver, 1995; Weaver and Chizmadzhev, 1996; Prausnitz, 1996b; Weaver et al., 1999).

B. Thermal Effects

In addition to the extensive and instantaneous creation of new aqueous pathways by the electric field, local heating of the stratum corneum due to passage of current also contributes to the phenomenon of skin electroporation (Pliquett, 1999). Assuming uniform passage of current through the entire skin surface area, temperature rise during a pulse was estimated insignificant (Prausnitz, 1996). However, during electroporation, the permeability of the stratum corneum becomes spatially heterogeneous (see D), and localized heating might occur at sites of high conductance (Pliquett et al., 1996a, 1999; Prausnitz et al., 1996).

Theoretical modeling taking convection, heat conduction, and latent heat of the skin lipids into account, as well as modeling the involved surface as constant, i.e., the local dissipation regions of 50 µm in radius, resulted in insignificant heat for pulses as long as 10 ms. More accurate modeling considering a propagating heat front, i.e., initial spots of 10 nm in radius with additional 1 nm steps in the immediate vicinity, resulted in contrast in significant heat production (Pliquett, 1999).

Theoretical modeling of the temperature rise has been accompanied by experimental measurements. The use of temperature-sensitive liquid crystals allowed the visualization of the spatial temperature distribution, which did not rise over the entire skin surface but started at small locations and spread thereafter. This method combined with infrared measurements at the skin surface resulted in an estimate of a locally maximal temperature of 100°C (Pliquett, 1999).

One consequence of skin heating is the generation of a sphingolipid phase transition around 70°C. Under these conditions, electroporation of the lipid layers is very probable, because the lipids are in fluid phase and the electric field still present. This phenomenon may explain the propagation of increased conductivity from initial spots, as well as the heat front observed experimentally.

C. Mechanisms of Molecular Transport

During and after the physical disruption of the lipid bilayers of the stratum corneum, molecular transport occurs by electrophoresis, electroosmosis and/or diffusion.

1. Electrophoresis

The main driving force for transport of charged compounds during high-voltage pulses is electrophoresis (Prausnitz et al., 1993a; Vanbever et al., 1994, 1996b). Evidence is the drop in transdermal transport with the "wrong" polarity of the electrodes in the donor solution (Prausnitz et al., 1993a; Vanbever et al., 1994, 1996b; Regnier and Préat, 1999) (Fig. 1). Further evidence is that the transport of neutral compounds is insignificant compared with the transport of charged molecules during pulses (Prausnitz et al., 1993b; Vanbever et al., 1998c).

2. Electroosmosis

The contribution of electroosmosis in transdermal transport during high-voltage pulses is low (Vanbever et al., 1996b, 1998c). The evidence comes from the comparison of anodic and cathodic transport of neutral molecules. Whatever the electrode polarity, transport during pulsing was identical. Moreover, the short time of current application (typically under 1 min) limits the role of electroosmosis in drug transport by skin electroporation.

3. Enhanced Diffusion

Molecular transport by passive diffusion through skin highly permeabilized by electric pulses can be significant in transdermal delivery using skin electroporation (Vanbever et al., 1996b, 1998c). Increases in transport have been seen during pulsing with the polarity such that electrophoresis opposes transdermal transport (Prausnitz et al., 1993a; Vanbever et al., 1996b). Electroporation-enhanced transport of a neutral molecule, mannitol, occurred predominantly after high-voltage pulses by diffusion through permeabilized skin (Fig. 4) (Vanbever et al., 1998c). Although much higher skin permeability is achieved during the pulse, prolonged permeabilization and thereby transport lasts for hours after pulsing in in vitro models (Vanbever et al., 1996b, 1998c; Regnier and Préat, 1999).

D. Routes of Transport

Transdermal transport by high-voltage pulses has been shown to occur through highly localized transport regions (LTRs), distributed over human stratum corneum (Pliquett et al., 1996b; Prausnitz et al., 1996a). These LTRs are not located

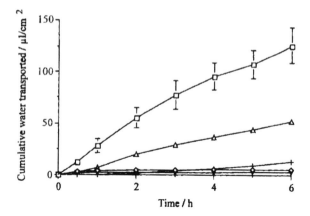

Figure 4 Cumulative mannitol transported across hairless rat skin in vitro following high-voltage pulse application at time 0. Mannitol was present in the donor compartment only during pulsing (○), only after pulsing (◇), or both during and after pulsing (□). Passive mannitol diffusion is also shown (△). (From Vanbever et al., 1998c, with permission.)

at the appendages of the skin for short (1 ms) high-voltage (100 V across the skin) pulses, but when decreasing the voltage and lengthening the pulse, the LTRs often include an appendage (Vanbever et al., 1999; Pliquett et al., 1998). Theoretical modeling suggests that appendageal macropores are possible pathways for electrical current at moderate voltage (Chizmadzhev et al., 1998a,b). The current density and hence drug transport is maximal at the center of the LTR.

The size of the transport regions is strongly dependent on pulse duration. It ranged between 0.1 mm and 0.2–2.5 mm in diameter for short and long pulses, respectively (Pliquett et al., 1996b, 1998; Vanbever et al., 1999). However, LTR size also increased with additional pulses. In contrast, the number of LTRs mainly depended on the voltage of the pulses. While LTR number ranged between 2 and 10 per 0.1 cm² for long medium-voltage pulses, it increased by an order of magnitude when short higher voltages were applied. Transport regions covered between 0.02 and 25% of the skin surface (Pliquett et al., 1996b, 1998; Vanbever at al, 1999).

It has been reported that LTRs are surrounded by localized dissipation regions (LDRs). The additional diffuse ring around the LTR presents a low resistivity where transport of only small ions occurs (Pliquett et al., 1998).

Within an LTR, molecular transport appears to occur through intercellular and/or transcellular pathways (Prausnitz et al., 1996a; Pliquett et al., 1996b, 1998). The contribution of each transport depended on the voltage applied to the skin: the higher the voltage, the higher the corneocyte permeabilization, the more the transcellular transport. As shown in Fig. 5, the molecular weight also influences the route of transport: whereas a small molecular weight compound, e.g., FITC penetrates in the keratinocytes, larger molecules, e.g., FITC-dextran, 38 kDa, remain mainly around the keratinocytes with only a small fraction entering the cell (Lombry et al., 2000; Regnier and Préat, 1998a; Dujardin et al., 2001).

IV. PARAMETERS AFFECTING ENHANCEMENT

Controlled delivery of drugs is of fundamental interest in human therapy. The parameters allowing control of transdermal drug delivery using high-voltage pulses have been investigated (Prausnitz et al., 1993a, 1994; Vanbever et al., 1994, 1996a; Gallo et al., 1997; Regnier et al., 1998). The electrical features of the pulsing protocol, the physicochemical parameters of the drug, and the formulation of the drug reservoir affect delivery magnitude and rate (Préat et al., 1997; Banga 1998, Banga et al., 1999) (Table 2).

A. Electrical Parameters

The influence on molecular transport of the electrical parameters of the pulses, i.e., the pulse waveform and the pulse voltage, time constant, number, and rate, have been extensively examined (Prausnitz et al., 1993a, 1994; Vanbever et al., 1994, 1996a; Gallo et al., 1997; Jadoul et al., 1998; Regnier and Préat, 1998).

1. Pulse Waveform

Skin electroporation experiments reported in the literature generally use generators delivering exponential-decay pulses (Prausnitz et al., 1993b; Vanbever et al., 1994; Bommannan et al., 1994; Pliquett and Weaver, 1996a). However, generators delivering square-wave pulses have been used for electrochemotherapy (Mir and Orlowski, 1999; Heller et al., 1999) and transdermal drug delivery (Gallo et al., 1997). The potential advantage of exponential-decay pulses is to maintain or expand the electroporation-induced high-permeability state and efficiently drift the molecules across the skin by electrophoresis. In contrast to exponentially decaying pulses, square pulses and their parameters remain constant whatever the skin or drug reservoir resistance may be (Vanbever et al., 1995; Vanbever et al., 1996a; Regnier and Préat, 1998; Lombry et al., 2000).

Two different types of exponential-decay pulse protocols are reported in the literature. They can mainly be distinguished by the pulse duration: (a) numerous (> 100) "short" duration (1–2 ms), high-voltage (100 V across the skin) pulses (Prausnitz et al., 1993a, 1994, 1995; Pliquett and Weaver, 1996a), (b) a

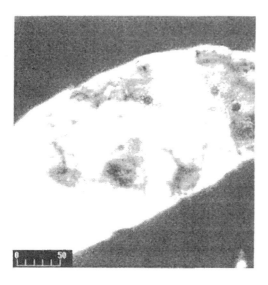

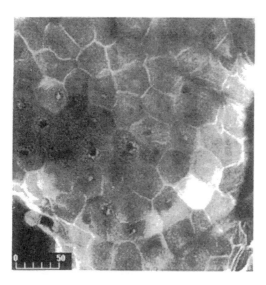

Figure 5   XY-planar LSCM section showing the distribution of FITC (top) and FITC-Dextran (bottom) 38 kDa in intact skin immediately after electroporation. Z = 27 and 24 μm, respectively. Scale bar = 50 μm. (From Lombry et al., 2000, with permission.)

Table 2 Summary of the Parameters Affecting Transdermal
Drug Transport by Skin Electroporation

Parameters	Increase in	Effect[a]
Electrical parameters	Pulse voltage	+
	Pulse number	+
	Pulse length	+
Physicochemistry of drug	Charge	+
	Molecular weight	−
	Conformation	?
	Lipophilicity	−
Formulation of drug reservoir	Competitive ions	−
	Ionization (pH)	+
	Viscosity	−

[a] + positive effect. − negative effect

low number (< 20) of "long" duration (70–1000 ms) medium-voltage pulses (40–70 V across skin) (Vanbever et al., 1994, 1996a, b; Regnier and Préat, 1998). At the same total electrical charge transferred through the skin, a few long pulses allowed generally higher molecular transport than many short pulses (Vanbever et al., 1999, Pliquett et al., 1998).

2. Pulse Voltage, Duration, Number and Rate

The effect of varying pulse voltage, duration, number, and rate on transdermal transport has been extensively studied in vitro, the pulse voltage being the most studied parameter (Prausnitz et al., 1993a, 1994; Vanbever et al., 1994, 1996a, Chen et al., 1998a). Because significant voltage drop occurs within donor and receiver solutions, the voltage applied across the skin is only a fraction of the voltage applied across the electrodes, this fraction depending on the relative resistance of the solutions and the skin. Skin resistance dramatically decreases during a pulse. The greater the voltage, the higher the resistance drop (Pliquett et al., 1995b; Vanbever et al., 1999). Hence transdermal voltages varied between approximately 50 and 10% of voltages applied across the electrodes.

With increased voltage of the pulses, the transdermal flux increases, but it often increases less steeply at large voltage (Prausnitz et al., 1993a; Vanbever et al., 1996a; Pliquett and Weaver, 1996a). As with electroporation experiments with cell suspensions, a trend toward a flux plateau at large voltage was generally observed (Prausnitz et al., 1993a; Gift and Weaver, 1995; Pliquett and Weaver, 1996a; Vanbever et al., 1996a).

When increasing the pulse time constant and the number of pulses, total drug transport increases linearly or not depending on the molecule being transported (Vanbever et al., 1994, 1996a; Zewert et al., 1995). Increasing the pulse rate increased transdermal flux as well (Prausnitz et al., 1994; Vanbever et al., 1994, 1999a; Pliquett and Weaver, 1996a). An example of the effect of pulse voltage and duration on fentanyl transport is shown in Fig. 6. The electrical parameters can be combined in an electrical exposure dose, i.e., the product of voltage by cumulative pulsing time (Gallo et al., 1997; Johnson et al., 1998).

The electrical parameters influence transdermal flux but also onset times for transport. Onset times of transport have been shown to decrease with increasing pulse time constant and rate, but to be independent of voltage (Prausnitz et al., 1994; Vanbever et al., 1999).

These studies demonstrate that control of drug transdermal transport can be attained by controlling the pulse voltage, time constant, number, and rate.

B. Physicochemical Parameters of the Drug

Response of transdermal transport to high-voltage pulses strongly depends on the physicochemical properties of the molecule being transported (Vanbever et

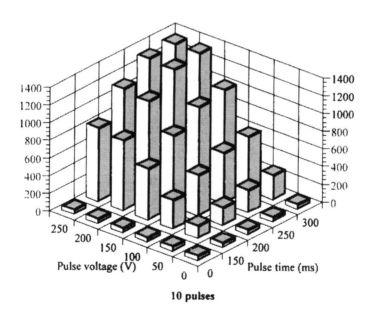

10 pulses

Figure 6 Effect of pulse voltage and time on fentanyl transdermal transport by electroporation. The plot presents values calculated from a response surface equation obtained by factorial design. (From Vanbever et al., 1996a, with permission.)

al., 1996b; Pliquett and Weaver, 1996a). The dependence appeared both in magnitude and in time scale of enhancement.

A high molecular electrical charge seems the most favorable physicochemical property for promoting transdermal transport by skin electroporation. Electrophoresis rapidly drives charged molecules across skin (Prausnitz et al., 1993b; Vanbever et al., 1994). A good example is given by fentanyl (Vanbever et al., 1996b). The enhancement in fentanyl transport due to skin electroporation was approximately of a factor 300 when the drug was in ionized form; in contrast, transport of nonionized fentanyl during pulsing increased only twofold over passive diffusion. Heparin, a highly charged macromolecule, exhibited transdermal transport at therapeutic rates using high-voltage pulses (Prausnitz et al., 1995). Transport of small neutral molecules might however be considered as well, but transport is slow because occurring by passive diffusion after pulsing (Vanbever et al., 1998c).

Another physicochemical parameter influencing transdermal transport using skin electroporation is the molecular volume: keeping the other parameters constant, the greater the molecular size, the lower the transdermal transport. Using FITC-dextran of increasing molecular weight, Lombry et al. (2000) showed that a significant transport and an intracellular penetration were detected after high-voltage pulse application. The absence of cutoff and the delivery of macromolecules of up to 40 kDa suggest that electroporation could be useful for macromolecule delivery. A pulse applied before iontophoresis increases the iontophoretic permeation of peptides; however, increasing their molecular weight decreased the benefit of the pulse (Potts et al., 1997).

C. Formulation of the Drug Reservoir

Increasing the drug concentration of the donor solution enhances drug transport proportionally (Vanbever et al., 1995; Regnier et al., 1998a). Hence the choice of drug concentration allows control of drug delivery. Increasing the co-ion concentration decreased drug transport because of competition in electrophoretic transport (Vanbever et al., 1994, 1996b). Drug transport decreased with increasing viscosity of the solution as well (Vanbever and Préat, 1995). These data point out that careful consideration of formulation variables of the drug reservoir is essential for the optimization of the delivery using skin electroporation.

D. Modifying Agents

The addition of substances into the donor solution can cause an altered rate of transport by skin electroporation. This has been exemplified in vitro with the presence of charged macromolecular compounds or reducing agents in the donor solution (Weaver et al., 1997; Vanbever et al., 1997; Zewert et al., 1999; Ilic et al., 1999).

Heparin, a linear, highly charged macromolecule, significantly altered the transport of two fluorescent dyes during pulsing, and it prolonged the lifetime of aqueous pathways after pulsing, as demonstrated by diminished resistance recovery (Weaver et al., 1997). Mannitol transport due to high-voltage pulses was shown to increase by up to fivefold with the addition of dextran sulfate to the donor solution (Vanbever et al., 1997). These experiments support the hypothesis that macromolecules are introduced into skin aqueous pathways created at high voltage and thereby increase transdermal transport by promoting and/or prolonging the effects of electroporation (Sukharev et al., 1992; Prausnitz et al., 1995).

In contrast to cell membrane electroporation, which can be used to load cells with proteins, DNAs as well as micrometer-sized particles, transdermal transport achieved using skin electroporation diminishes with increasing molecular size. The cross-linked keratin matrix within corneocytes has been hypothesized to cause steric hindrance for the movement of macromolecules across skin. The reduction of its disulfide linkages has been proposed for rendering the stratum corneum more permeable. The addition of sodium thiosulfate to the donor solution induced a dramatic increase in the transcutaneous transport of proteins (up to 150 kDa in molecular size) via the creation of microconduits spanning the stratum corneum, i.e., the weakening of keratin by thiosulfate led to dislodgement of corneocytes and thereby microholes, which supports the hypothesis (Zewert et al., 1999). The combination of thiosulfate and urea allowed the dose of thiosulfate necessary for microconduit creation to be reduced (Ilic et al., 1999). But, before we give further consideration to this method, skin tolerance needs to be evaluated.

V. TOLERANCE

A major factor in the clinical acceptability of electrically enhanced transdermal delivery is its effect on the skin. A few studies have addressed the issue of skin tolerance to electric pulses (see Table 3, Fig. 7). Visual examination, biophysical studies of the stratum corneum, noninvasive bioengineering methods, and investigation of skin ultrastructure have been performed. Overall, the alterations following high-voltage pulses were mild and reversible.

A. Visual Evaluation

Visual evaluation of skin after high-voltage pulse exposure has been carried out in in vivo related studies, i.e., transdermal drug delivery, electrochemotherapy, and gene therapy (Prausnitz et al., 1993b; Vanbever et al., 1998b; Okino et al., 1992; Titomirov et al., 1991). These studies generally reported mild, transient erythema and/or edema over the area of electrical contact with the skin. No burn

Table 3 Summary of the Effects of Electroporation on the Skin

Study	Observations	Reference
Visual aspect	Mild reversible erythema	Rivière et al., 1995
Impedance	Resistance (up to three orders of magnitude on a time scale of microseconds)	Prausnitz et al., 1993a, Pliquett et al., 1995b, 1996c, 1999; Vanbever et al., 1999
FTIR	= Fluidity of the lipid alkyl chain Hydration	Jadoul et al., 1997
DTA	= T° (T_2 and T_{3+4}) in enthalpy → disordering	Jadoul et al., 1998a
X-ray scattering	Lamellar ordering Intralamellar packing	Jadoul et al., 1997
FFEM	Spherical deformations, rough surfaces, disorganization of the lamellae, appearance in a network-like structure and vesicles	Jadoul et al., 1998a; Gallo et al., 1999
TEWL	Mild reversible increase (impairment of barrier function)	Vanbever et al., 1998b Dujardin et al., 2002
LDV	Mild reversible increase in blood circulation	Vanbever et al., 1998b
Chromametry	Mild reversible erythema	Vanbever et al., 1998b

was observed. Patients submitted to electrochemotherapy tolerated the pulses well (Heller et al., 1999).

Toxicology of an electroporative pulse has been partially evaluated ex vivo in pig skin by scaling the degree of erythema, edema, and petechia (Rivière et al., 1995). The only skin alteration seen was a transient erythema, which tended to decrease within 4 h.

B. In Vivo Noninvasive Investigation of the Skin

The alterations of skin functions following the application of high-voltage pulses have been evaluated directly in vivo in hairless rat using noninvasive bioengineering methods (Vanbever et al., 1998b) (Fig. 7). In agreement with the visual evaluation, mild but transient erythema was measured by laser Doppler flowmetry and

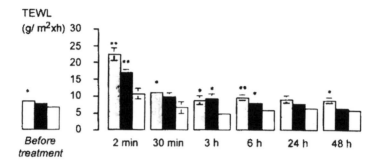

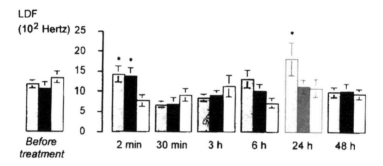

Figure 7 (A) Laser Doppler flowmetry measurement of skin blood flow. (B) Transepidermal water loss (TEWL) following electroporation (15 pulses of 100 V, voltage applied at the electrodes, and 500 ms applied at a rate of 1 pulse per minute) in hairless rat skin in vivo. White bars are measurements at the cathode, striped bars at the anode, and black bars at the control sites. Statistical significance: * $p < 0.05$ vs. control, ** $p < 0.01$ vs. control. (From Vanbever et al., 1998b, with permission.)

chromametry. Transepidermal water loss increased following electroporation of skin. However, the barrier function recovered control values within several hours.

C. Skin Electrical Properties

As the stratum corneum has a much higher electrical resistance than the other parts of the skin, an electric field applied to the skin will concentrate in the stratum corneum. In contrast, the field will be much lower in the viable tissues.

Skin resistance drops by orders of magnitude during high-voltage pulses and recovers either partially or fully to prepulse values. The recovery takes a few ms or up to several hours depending on pulse magnitude, length, number, rate (Pliquett et al., 1995a), and the experimental model (Gallo et al., 1997). For large pulses and/or multiple pulse protocols, recovery of the stratum corneum after each pulse is not complete; some of the newly-created pathways could not close completely and could be responsible for the persisting change in skin resistance (Pliquett et al., 1995b). Water and ion entrapment might also be involved (Vanbever et al., 1995). In vivo studies have confirmed that skin impedance recovers within several hours after skin electroporation (Dujardin et al., 2002).

Changes in skin capacitance are not pronounced, and the changes that do occur have a relatively short lifetime. Increases (up to 5 times) generally last for only about 1 s. After many or very large pulses, residual changes may remain (Pliquett et al., 1995b).

D. Biophysical Investigation of the Stratum Corneum Structure

Different biophysical methods have been used to study the stratum corneum following application of high-voltage pulses (Jadoul et al., 1997, 1998a, 1999) (Table 3). Hydration of the stratum corneum due to pulsing was observed by ATR-FTIR and measurement of total water content. Differential thermal analysis, x-ray diffraction, and freeze fracture electron microscopy studies revealed a dramatic perturbation of the lamellar ordering. The larger the pulses, the larger the perturbations. A disordering of the lipid lamellar stacking and of the lipid lateral packing (change from orthorhombic and hexagonal lateral lipid phase to liquid phase) were detected by SAXS and WAXS, respectively. Even though no change in phase transition temperature was detected, a decrease in enthalpy phase transition was measured after electroporation (Jadoul et al., 1999). No significant increase in lipid fluidity of the alkyl chain was detected by ATR-FTIR. No perturbation of the corneocyte content was evident.

In contrast to the heterogeneous permeabilization of the stratum corneum observed (Pliquett et al., 1996b; Prausnitz et al., 1996a), lipid disordering was generalized over the skin surface. However, the duration of these perturbations as well as their reversibility in vivo are still unknown.

E. Ultrastructure of the Skin

Light microscopy of animal skin exposed to electroporation in vivo or in vitro does not reveal any change in histological structure. A mild intraepidermal vacuolization was reported by Rivière et al. (1995).

Freeze fracture electron microscopy has been used to visualize the stratum corneum ultrastructure. A general perturbation of the intercellular lipid, i.e., a general loss or severe distortion of the lamellarity, was observed (Jadoul et al., 1998a). Spherical deformations, a network-like structure, and multilamellar vesicles were observed and could be related to the heating effect of the pulse (Gallo et al., 1999). Cross-fracture did not reveal any changes in corneocyte ultrastructure.

F. Sensation, Pain, and Muscle Stimulation

Sensation due to current applied to skin is often caused by direct electrical excitation of nerves (Reilly, 1992). Effects range from sensation of localized heat or cold, through tingling and itching, slight pricking, muscle contraction, to outright pain (Ledger, 1992; Prausnitz, 1996a). In general, increased current/charge, pulse rate, and pulse length all increase the levels of sensation. Because these effects will be strongly affected by the electrical field properties and distribution within the skin, careful design of electrodes and pulsing protocols will be essential for the reduction of unwanted side effects.

VI. POTENTIAL THERAPEUTIC APPLICATIONS

A. Transdermal Drug Delivery

All the literature data demonstrate that electroporation is an efficient method of enhancing the transdermal delivery of drugs, including macromolecules. (a) electroporation enhances transdermal transport in vitro by 1 to 4 orders of magnitude (Table 1) with a very short lag time (seconds to minutes depending on the experimental model); (b) electroporation quickly increases the plasma levels of drugs in vivo; (c) electroporation enhances transdermal delivery of macromolecules up to at least 40 kDa.

An important issue arising from these data is to determine how electroporation stands as compared with other enhancing methods for transdermal drug delivery. For conventional small molecular weight drugs, electroporation does not provide any major advantages over other physical methods. Molecular transport resulting from electroporation or iontophoresis is in the same range of magnitude (Prausnitz et al., 1993, 1995; Vanbever at al, 1994; Johnson et al., 1998). A larger fraction of skin surface area is, however, made available for transport using skin electroporation. The tolerance of skin to electroporation or iontophoresis is very similar (Jadoul et al., 1999). However, while iontophoresis is considered a safe procedure acceptable for clinical use, electroporation still needs to be made applicable in humans, e.g., by the development of an appropriate electrode configuration.

In contrast to conventional drugs, for which no significant advantage of electroporation over other enhancing methods can be found, macromolecule delivery using skin electroporation could be a promising alternative as a noninvasive delivery. Iontophoresis has been shown to enhance transdermal transport of macromolecules including peptides and oligonucleotides (Green, 1996; Brand and Iversen, 1996), with however a molecular weight cutoff of 10 kDa (Turner et al., 1997). In contrast, electroporation enhances transdermal delivery of up to at least 40 kDa molecules (Prausnitz et al., 1995; Banga and Prausnitz, 1999; Lombry et al., 2000). Low frequency sonophoresis also enhances protein delivery (Mitagroti et al., 1995).

The combination of electroporation and other enhancer methods might still open new perspectives, i.e., skin electroporation combined with a chemical enhancer (Weaver et al., 1997; Vanbever et al., 1997), iontophoresis (Bommannan et al., 1994), ultrasound (Kost et al., 1996), or pressure (Zhang et al., 1996).

In conclusion, electroporation is an efficient method of enhancing transdermal drug delivery and expands the range of compounds delivered transdermally. As compared to other enhancing methods, electroporation could be particularly interesting for transdermal delivery of macromolecules and fast and/or pulsatile transdermal delivery.

B. Topical Drug Delivery

Compared to chemical methods or other physical methods, e.g., iontophoresis and sonophoresis, which also enhance transdermal and topical drug delivery, electroporation could be a promising alternative for enhancing topical drug delivery. In contrast to these methods, skin electroporation has been shown to permeabilize the underlying tissue, including the keratinocytes (Mir and Orlowsky, 1999, Regnier and Préat, 1999). Hence if the keratinocytes or other epidermal cells are the target cells, electroporation could be particularly interesting and innovative for the topical delivery of macromolecules. Several reports support this hypothesis and confirm that electroporation enhances the topical delivery of macromolecules ranging from peptides to antigens to nucleic acids.

The immunosuppressive drug cyclosporine A, which is useful in treating psoriasis, would benefit from a topical rather than a systemic delivery. However, its molecular weight and lipophilicity limit the development of an efficient formulation. Wang et al. (1998b) have reported that electroporation enhanced by one order of magnitude the delivery of cyclosporine A formulated as a coevaporate to increase its water solubility.

Electroporation is a promising technique for the needle-free noninvasive delivery of antigen in the skin. The skin is increasingly recognized as an organ of immunity. Passive diffusion of antigen does not elicit strong immune response unless cholera toxin B is added as an adjuvant (Glenn et al., 1998). Electroporation delivery of several antigens, e.g., myristylated peptide, diphteria toxoid, or ovalbumin, elicited an antigen-specific serum IgG response (mainly Th2 response) (Misra et al., 1999; Dujardin et al., 2000).

Electroporation has also been investigated for the topical delivery of nucleic acids. Topical and transdermal transport of oligonucleotides (ODN) can be electrically enhanced (Zewert et al., 1995; Brand and Iversen, 1996; Regnier et al., 1998). Unlike iontophoresis, electroporation induced a rapid delivery of ODN in the keratinocytes (Regnier and Préat, 1999). Electroporation has also been shown to enhance the delivery of a plasmid DNA in the epidermis, inducing an intracellular delivery of the DNA in the keratinocytes of LTRs. The expression of a reporter gene plasmid was detected in the epidermis for at least 7 days. (Dujardin et al., 2001).

REFERENCES

A Badkar, G Betageri, G Hofmann, A Banga. Enhancement of transdermal iontophoretic delivery of a liposomal formulation of colchicine by electroporation. Drug Del 6:111–115, 1999.

A Banga, ed. Electrically-assisted transdermal and topical drug delivery. Bristol, PA: Taylor and Francis, 1998.

A Banga, M Prausnitz. Delivery of protein and gene-based drugs by skin electroporation. Trends Biotech 16:408–412, 1998.

A Banga, S Bose, T Ghosh. Iontophoresis and electroporation: comparisons and contrasts. Int J Pharm 179:1–19, 1999.

D Bommannan, J Tamada, L Leung, RO Potts. Effect of electroporation on transdermal iontophoretic delivery of luteinizing hormone releasing hormone (LHRH) in vitro. Pharm Res 11:1809–1814, 1994.

RM Brand and PL Iversen. Iontophoretic delivery of telomeric oligonucleotides. Pharm Res 13:851–854, 1996.

DC Chang, BR Chassy, JA Saunders, A Sowers, eds. Guide to Electroporation and Electrofusion. New York: Academic Press, 1992.

T Chen, EM Segall, R Langer, JC Weaver. Skin electroporation: rapid measurements of the transdermal voltage and flux of four fluorescent molecules show a transition to large fluxes near 50 V. J Pharm Sci 37:1368–1374, 1998a.

T Chen, R Langer, JC Weaver. Skin electroporation causes molecular transport across the stratum corneum through localized transport regions. J Inv Dermatol Symp Pro 3:159–165, 1998b.

T Chen, R Langer, JC Weaver. Charged microbeads are not transported across the human stratum corneum in vitro by short high-voltage pulses. Bioelectrochem Bioenerg 48:181–192, 1999.

YA Chizmadzhev, VG Zarnytsin, JC Weaver, RO Potts. Mechanism of electroinduced ionic species transport through a multilamellar lipid bilayer system. Biophys J 68:749–765, 1995.

YA Chizmadzhev, AV Indenbom, PI Kuzmin, SV Galichenko, JC Weaver, RO Potts. Electrical properties of skin at moderate voltages: contribution of appendageal macropores. Biophys J 74:843–856, 1998a.

Y Chizmadzhev, P Kuzmin, J Weaver, R Potts. Skin appendageal macropores as a possible pathway for electrical current. J Invest Dermatol Symp 3: 148–152, 1998b.

N Dujardin, P Van der Smissen, V Préat. Topical gene transfer into rat skin using electroporation. Pharm Res 18:61–66, 2001.

N Dujardin, E Staes, V Préat. Needle free immunization using skin electroporation. Proc Controlled Release of Bioactive Material 27, 2000.

N Dujardin, E Staes, Y Kalia, P Clarys, RH Guy, U Préat. In vivo assessment of skin electroporation using square wave pulses. J Control Release 79: 219–227, 2002.

DA Edwards, MR Prausnitz, R Langer, JC Weaver. Analysis of enhanced transdermal transport by skin electroporation. J Control Rel 34:211–221, 1995.

S Gallo, AR Oseroff, PG Johnson, SW Hui. Characterization of electric-pulse-induced permeabilization of porcine skin using surface electrodes. Biophys J 72:2805–2811, 1997.

S Gallo, A Sen, M Hensen, S Hui. Time-dependent ultrastructural changes to porcine stratum corneum following an electric pulse. Biophys J 76:2824–2832, 1999.

EA Gift, JC Weaver. Observation of extremely heterogenous electroporative molecular uptake by Saccharomyces cerivisiae which changes with electric field pulse amplitude. Biochim Biophys Acta 1234:52–62, 1995.

GM Glenn, GR Matyas, CR Alving. Skin immunization made possible by cholera toxin. Nature 391:351, 1998.

P Green. Iontophoretic delivery of peptides drugs. J Control Release 41:33–48, 1996.

T Gowrishankar, T Herndon, T Vaughan, J Weaver. Spatially constrained localized transport regions due to skin electroporation. J Control Rel 60:101–110, 1999.

RH Guy Current status and future prospects of transdermal drug delivery. Pharm Res 13:1765–1768, 1996.

J Hadgraft, RH Guy, eds. Transdermal Drug Delivery: Development Issues and Research Initiatives. New York, Marcel Dekker, 1989.

R Heller, R Gilbert, MJ Jaroszeski. Clinical applications of electrochemotherapy. Adv Drug Del Rev 35:119–129, 1999.

GA Hofmann, WV Rustrum, KS Suder. Electro-incorporation of microcarriers as a method for the transdermal delivery of large molecules. Bioelectrochem Bioenerg 38:209–222, 1995.

L Illic, T Gowrishankar, T Vaughan, T Herndon, J Weaver. Spatially constrained skin electroporation with sodium thiosulfate and urea creates transdermal microconduits. J Control Rel 61:185–202, 1999.

A Jadoul, V Préat. Electrically-enhanced transdermal delivery of domperidone. Int Pharm 154:229–234, 1997.

A Jadoul, V Regnier, J Doucet, D Durand, V Préat. X-ray scattering analysis of the stratum corneum treated by high voltage pulses. Phar Res 14:1275–1277, 1997.

\ Jadoul, H Tanajo, V Préat, F Spies, H Boddé. Electroperturbation of human stratum corneum fine structure by high voltage pulses: a freeze fracture electron microscopy and differential thermal analysis study. J Invest Dermatol Symp Proc 3:153–158, 1998a.

\ Jadoul, N Lecouturier, J Mesens, W Caers, V Préat. Electrically enhanced transdermal delivery of alnitidan. J Control Rel 54:265–272, 1998b.

\ Jadoul, J Bouwstra, V Préat. Effects of iontophoresis and electroporation on the stratum corneum. Review of the biophysical studies. Adv Drug Del Rev 35:89–105, 1999.

?G Johnson, SA Gallo, SW Hui, A Oseroff. A pulsed electric field enhances cutaneous delivery of methylene blue in excised full-thickness porcine skin. J Invest Dermatol 111:457–463, 1998.

f Kost, U Pliquett, S Mitragotri, A Yamamoto, R Langer, J Weaver. Synergistic effect of electric field and ultrasound on transdermal transport. Pharm Res 13:633–638, 1996.

S Mitragotri, D Blankschtein, R Langer. Ultrasound—mediated transdermal protein delivery. Science 269:850–853, 1995.

C Lombry, N Dujardin, V Préat. Transdermal delivery of macromolecules using skin electroporation. Pharm Res 17(1), 2000.

L Mir, S Orlowski. Mechanisms of electrochemotherapy. Adv Drug Del Rev 35:107–118, 1999.

A Misra, S Ganga, P Upadhay. Needle-free non-adjuvanted skin immunization by electroporation-enhanced transdermal delivery of diphtheria toxoid and a candidate peptide vaccine against hepatitis B virus. Vaccine 18:517–523, 1999.

S Mitragotri, D Blankschtein, R Langer. Ultrasound-mediated transdermal protein delivery. Science 269:850–853, 1995.

M Okino, H Tomie, H Kanesada, M Marumoto, K Esato, H Suzukiu. Optimal electric conditions in electrical impulse chemotherapy. Jpn J Cancer Res 83:1095–1101, 1992.

U Pliquett. Mechanistic studies of molecular transdermal transport due to skin electroporation. Adv Drug Deliv Rev 35:41–60, 1999.

U Pliquett, JC Weaver. Electroporation of human skin: simultaneous measurement of changes in the transport of two fluorescent molecules and in the passive electrical properties. Bioelectrochem Bioenerg 39:1–12, 1996a.

U Pliquett, JC Weaver. Transport of a charged molecule across the human epidermis due to electroporation. J Control Rel 38:1–10, 1996b.

U Pliquett, MR Prausnitz, YA Chizmadzhev, J Weaver. Measurement of rapid release kinetics for drug delivery. Pharm Res 12:549–555, 1995 (errata in Pharm Res 12:1244, 1995a).

U Pliquett, R Langer, JC Weaver. Changes in the passive electrical properties of human stratum corneum due to electroporation. Biochim Biophys Acta 1239:111–121, 1995b.

U Pliquett, EA Gift, JC Weaver. Determination of the electric field and anomalous heating caused by exponential pulses in electroporation experiments. Bioelectrochem Bioenerg 39:39–53, 1996a.

U Pliquett, TE Zewert, T Chen, R Langer, JC Weaver. Imaging of fluorescent molecule and small ion transport through human stratum corneum during high voltage pulsing: localized transport regions are involved. Biophys Chem 58:185–204, 1996b.

U Pliquett, R Vanbever, V Préat, JC Weaver. Local transport regions in human stratum corneum due to short and long high-voltage pulses. Bioelectro Bioener 47:151–161, 1998.

RO Potts, D Bommannan, O Wang, JA Tamada, JE Rivière, NA Monteiro-Rivière. Transdermal peptide delivery using electroporation. Pharm Biotechnol 10:213–238, 1997.

MR Prausnitz. The Effects of electric current applied to the skin: a review for transdermal drug delivery. Adv Drug Deliv Rev 18:395–425, 1996a.

MR Prausnitz. Do high-voltage pulses cause changes in skin structure? J Control Rel 40:321–326, 1996b.

MR Prausnitz. Reversible skin permeabilization for transdermal delivery of macromolecules. Crit Rev Ther Drug Carrier Syst 14:455–483, 1997.

M Prausnitz. A practical assessment of transdermal drug delivery by skin electroporation. Adv Drug Del Rev 35:61–76, 1999.

MR Prausnitz, VG Bose, R Langer, JC Weaver. Electroporation of mammalian skin: a mechanism to enhance transdermal drug delivery. Proc Natl Acad Sci USA 90:10504–10508, 1993a.

MR Prausnitz, DS Seddick, AA Kon, VG Bose, S Frankenburg, SN Klaus, R Langer, JC Weaver. Methods for in vivo tissue electroporation using surface electrodes. Drug Deliv 1:125–131, 1993b.

MR Prausnitz, U Pliquett, R Langer, JC Weaver. Rapid temporal control of transdermal drug delivery by electroporation. Pharm Res 11:1834–1837, 1994.

MR Prausnitz, ER Edelman, JA Gimm, R Langer, JC Weaver. Transdermal delivery of heparin by skin electroporation. Bio/Technology 13:1205–1209, 1995.

MR Prausnitz, JA Gimm, RH Guy, R Langer, JC Weaver, C Cullander. Imaging of transport pathways across human stratum corneum during high-voltage and low-voltage electrical exposures. J Pharm Sci 85:1363–1370, 1996a.

MR Prausnitz, CS Lee, CH Liu, JC Pang, T-P Singh, R Langer, JC Weaver.

Transdermal transport efficiency during skin electroporation and ionto-phoresis. J Control Rel 38:205–217, 1996b.

V Préat, R Vanbever, A Jadoul, V Regnier. Electrically enhanced transdermal drug delivery: iontophoresis vs. electroporation. In: Transdermal Drug Delivery. A Case Study, Iontophoresis. P Couvreur, D Duchêne, P Green and H Junginger, eds. Paris: Editions de la Santé, 1997, pp 58–67.

V Regnier, V Préat. Localisation of an FITC-labeled phosphorothioate oligo-deoxynucleotide in the skin after topical delivery by iontophoresis and electroporation. Pharm Res 15:1596–1602, 1998a.

V Regnier, V Préat. Mechanisms of a phosphorothioate oligonucleotide delivery by skin electroporation. Int J Pharm 184:147–156, 1999.

V Regnier, T Le Doan, V Préat. Parameters controlling topical delivery of oligo-nucleotides by electroporation. J Drug Target 5:275–289, 1998.

V Regnier, A Tahiri, André, M Lemaître, T Le Doan, V Préat. Electroporation mediated delivery of 3'-protected phosphodiester oligodeoxynucleotides to the skin. J Control Release, in press.

JE Rivière, NA Monteiro-Rivière, RA Rogers, D Bommannan, JA Tamada, RO Potts. Pulsatile transdermal delivery of LHRH using electroporation: drug delivery and skin toxicology. J Control Rel 36:229–233, 1995.

EW Smith, H Maibach, eds. Percutaneous Penetration Enhancers. Boca Raton, FL: CRC Press, 1995.

SI Sukharev, VA Klenchin, SM Serov, LV Chernomordik, YA Chizmadzhev. Electroporation and electrophoretic DNA transfer into cells. The effect of DNA interaction with electropores. Biophys J 63:1320–1327, 1992.

AV Titomirov, S Sukharev, E Kistanova. In vivo electroporation and stable trans-formation of skin cells of newborn mice by plasmid DNA. Biochim Bio-phys Acta 1088:131–134, 1991.

N Turner, L Ferry, M Price, C Cullander, RH Guy. Iontophoresis of poly-lysines: the role of molecular weight. Pharm Res 14:1322–1331, 1997.

R Vanbever, N Lecouturier, V Préat. Transdermal delivery of metoprolol by elec-troporation. Pharm Res 11:1657–1662, 1994.

R Vanbever, V Préat. Factors affecting transdermal delivery of metoprolol by electroporation. Bioelectrochem Bioenerg 38:223–228, 1995.

R Vanbever, V Préat. In vivo efficacy and safety of skin electroporation. Adv Drug Del Rev 35:77–88, 1999.

R Vanbever, E Le Boulangé, V Préat. Transdermal delivery of fentanyl by electro-poration. I. Influence of electrical factors. Pharm Res 13:559–565, 1996a.

R Vanbever, N De Morre, V Préat. Transdermal delivery of fentanyl by electro-poration. II. Mechanisms involved in drug transport. Pharm Res 13:1359–1365, 1996b.

R Vanbever, MR Prausnitz, V Préat. Macromolecules as novel transdermal trans-port enhancers for skin electroporation. Pharm Res 14:638–644, 1997.

R Vanbever, G Langers, S Montmayeur, V Préat. Transdermal delivery of fentanyl: rapid onset of analgesia using skin electroporation. J Control Rel 50: 225–235, 1998a.

R Vanbever, D Fouchard, A Jadoul, N De Morre, V Préat, J-P Marty. In vivo noninvasive evaluation of hairless rat skin after high-voltage pulse exposure. Skin Pharmacol Appl Skin Physiol 11:23–34, 1998b.

R Vanbever, M-A Leroy, V Préat. Transdermal permeation of neutral molecules by electroporation. J Control Rel 54:243–250, 1998c.

R Vanbever, UF Pliquett, V Préat, JC Weaver. Comparaison of the effects of short, high-voltage and long, medium voltage pulses on skin electrical and transport properties. J Control Rel 69:35–47, 1999.

KA Walters, J Hadgraft, eds. Pharmaceutical Skin Penetration Enhancement. New York: Marcel Dekker, 1993.

S Wang, M Kara, RR Krishnan. Transdermal delivery of cyclosporin-A using electroporation. J Control Rel 50:61–70, 1998a.

S Wang, M Kara, RR Krishnan. Topical delivery of cyclosporin-A coevaporate using electroporation. Technique Drug Dev. Ind. Pharm 23:657–998, 1998b.

JC Weaver. Electroporation: a general phenomenon for manipulating cells and tissues. J Cell Biochem 51:426–435, 1993.

JC Weaver. Electroporation theory: concepts and mechanisms. In: JA Nickoloff, ed. Molecular Biology: Methods, 1995.

JC Weaver, YA Chizmadzhev. Theory of electroporation: a review. Bioelectrochem Bioenerg 41:135–160, 1996.

JC Weaver, R Vanbever, TE Vaughan, MR Prausnitz. Heparin alters transdermal transport associated with electroporation. Biochem Biophys Res Com 234: 637–640, 1997.

JC Weaver, TE Vaughan, Y Chizmadzhev. Theory of skin electroporation: implications of straight-through aqueous pathway segments that connect adjacent corneocytes. J Invest Dermatol Symp Proc 3:143–147, 1998.

J Weaver, T Vaughan, Y Chizmadzhev. Theory of electrical creation of aqueous pathways across skin transport barriers. Adv. Drug Del Rev 35:21–39, 1999.

TE Zewert, UF Pliquett, R Langer, JC Weaver. Transdermal transport of DNA antisense oligonucleotides by electroporation. Biochem Biophys Res Com 212:286–292, 1995.

TE Zewert, U Pliquett, R Vanbever, T Chen, R Langer, JC Weaver. Creation of transdermal pathways for macromolecule transport by skin electroporation and a low toxicity pathway-enlarging molecule. Bioelectrochem Bioenerg. 49:11–20, 1999.

L Zhang, L Li, GA Hofmann, RM Hoffman. Depth-targeted efficient gene delivery and expression in the skin by pulsed electric fields: an approach to gene therapy of skin aging and other diseases. Biochem Biophys Res Comm 220:633–636, 1996.

L Zhang, L Li, ZL An, RM Hoffman, GA Hofmann. In vivo transdermal delivery of large molecules by pressure-mediated electroincorporation and electroporation: a novel method for drug and gene therapy. Bioelectrochem Bioenerg 42:283–292, 1997.

L Zhang, S Lerner, WV Rustrum, GA Hofmann. Electroporation mediated topical delivery of vitamin C for cosmetic applications. Bioelectrochem Bioenerg, 48:453–461, 1999.

7

Sonophoresis: Ultrasound-Enhanced Transdermal Drug Delivery

Victor Meidan
New Jersey Center for Biomaterials, Newark, New Jersey, U.S.A.

I. INTRODUCTION

Phonophoresis, or sonophoresis, is defined as the use of ultrasonic energy in order to enhance the topical or transdermal delivery of drugs. For some 40 years, physiotherapists have used the combination of ultrasound and steroid or analgesic to treat a diverse range of muscular and arthritic conditions. Unfortunately, most of this treatment has been conducted on a subjective and nonquantitative basis. Common deficiencies of the physiotherapeutic literature have included a lack of proper controls, incomplete accounts of the dosimetry and protocols employed, and the noncalibration of the ultrasound source. Consequently, much of the available data from these studies are inadequate (Meidan et al., 1995). Fortunately, over the last decade or so, drug delivery scientists have performed much more refined and quantitative studies. However, because individual groups have employed different ultrasound parameters (i.e., frequency, intensity, duration, and mode) as well as different drugs, skin types, and vehicles, a fully comprehensive theory of phonophoresis has not yet emerged, although much more is now known about this field than just a few years ago (Mitragotri et al., 2000c; Mitragotri and Kost, 2000).

The objective of this chapter is to discuss sequentially the physics of ultrasound propagation, the methods of phonophoresis application, and the biophysical interactions of ultrasound with tissue. This is followed by a review of the literature from the separate perspectives of animal studies, in vitro human skin studies, and clinical trials.

II. THE PHYSICS OF ULTRASOUND PROPAGATION

Ultrasound is an acoustic vibration propagating in the form of longtitudinal compression waves at frequencies beyond the human auditory range, i.e., 0.02 MHz. The intensity parameter, measured in W cm^{-2}, defines the quantity of energy conveyed by an ultrasonic wave as it passes through any given site, and this parameter is used to establish exposure. Therapeutic ultrasound is normally generated by a transducer that converts electricity to ultrasound by utilizing the piezoelectric principle. This effect describes the behavior of certain ceramic materials, which expand or contract when a voltage is applied across them. For continuous wave (c.w.) ultrasound production, an alternating voltage is applied to the transducer resulting in the continuous emission of ultrasound of the same frequency. To produce pulsed ultrasound, short bursts of alternating voltage are repeatedly applied to the transducer. The ultrasound emitted by the transducer propagates away from the front face in the direction in which the transducer is pointing. Once generated, ultrasound retains a specific waveform and constant frequency irrespective of the medium that the beam is passing through (this is strictly speaking not always true but nonlinear effects are negligible at phonophoretic intensities).

As an ultrasound beam propagates away from the transducer, it can be described as consisting of two regions (Fig. 1). These are the near field and the far field. The near field is a cylindrical beam of spatially fluctuating acoustic intensity caused by the constructive and destructive interference of ultrasonic waves. The far field is a diverging beam exhibiting a central acoustic intensity peak in the center of the beam, which smoothly falls off at either side. The boundary between these two fields occurs at a distance d from the transducer and is given by

$$d = \frac{r^2}{\lambda}$$

where r is the radius of the radiating surface of the transducer and λ is the wavelength. As long as r $>$ 5 λ, which usually holds true, the beam diverges in the far zone, with a divergence angle θ (in degrees) of

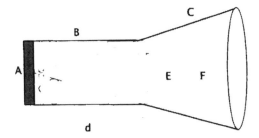

Figure 1 Diagrammatic representation of the output of a large transducer typically used in phonophoresis. (Adapted from Williams, 1983.) A, the transducer; B, cylindrical beam in the near field; C, diverging beam in the far field; E, F, acoustic energy contours such that E > F. Angle subtended by the divergent beam at the transducer surface is 2θ.

$$\theta = \frac{35 \, \lambda}{r}$$

As can be seen from Fig. 1, the intensity profile in the near field is complex and consists of many maxima and minima. Since acoustic intensity is not homogeneously distributed throughout the beam space (or in time in the case of pulsed ultrasound), several intensity parameters have been defined. The spatial-average temporal-average (SATA) intensity is the most commonly cited parameter. It is calculated by dividing the total power in an ultrasonic beam by the beam area and, in the case of pulsed ultrasound, averaging over the pulse repetition cycle. It is noteworthy that most researchers investigating phonophoresis have disregarded beam heterogeneity and have neglected to map out the ultrasonic intensities at different points within the beam. Therefore certain skin tissues have been subjected to intensities that are either much lower or much higher than the intensity (generally SATA) stated by the authors. In some studies, the transducer is put into continuous motion over the skin surface, and this, to an extent, overcomes this problem.

When an ultrasonic beam propagates through an interface between media exhibiting a mismatch in acoustic impedance, it can reflect back along its own path, interfering with that portion of itself that has not yet been reflected to form a standing wave. A standing wave field consists of a regular repeating pattern of nodes (where the displacement is zero) and antinodes (where the displacement varies from positive to negative at the same frequency as the incoming wave, but at twice its amplitude). A partial standing wave will develop if the reflector is less than 100% effective. In vivo, a standing wave field can form at a bone/soft tissue interface, and the extent to which this phenomenon occurs depends largely on the anatomy of the sonicated site. Standing waves can also form in vitro, within diffusion cells used to investigate phonophoresis. The potential development and influences of standing waves have been frequently disregarded in the literature. Nethertheless, it is possible to measure the acoustic dosimetry at the skin membrane. For instance, a hydrophone technique showed that the steel mesh supporting the skin sample in a Franz cell reflected back at least 75% of the energy to form a standing wave field within the skin and coupling gel (Meidan et al., 1998b).

Another interesting process that can occur at a bone/soft tissue interface is mode conversion. This is characterized by the partial conversion of the longitudinal ultrasonic waves into transverse waves. The transverse wave component is not readily propagated, and the energy is rapidly released as heat at the interface.

III. THE DESIGN OF PHONOPHORETIC SYSTEMS

Since air is a poor medium for ultrasound propagation, investigations into phonophoresis require the use of an ultrasonic coupling medium, which facilitates an efficient transfer of ultrasonic energy between the transducer and the skin surface. The medium is either a commercially available viscous gel or an aqueous solution. Generally, the therapeutic agent is homogeneously dispersed within the coupling medium. However, in some studies, the therapeutic agent has been deposited directly on the skin and the overlying coupling gel subsequently added (Hofmann and Moll, 1993; Meidan et al., 1998a, b). In terms of frequency, intensity, mode, and treatment time, individual research groups have employed different parameters. Generally, the ultrasonic frequency applied is between 0.020 and 3 MHz, although in an important contribution, Guy and coworkers used 10 and 16 MHz (Bommannan et al. 1992a, b). Intensity is rarely greater than 3 W cm^{-2}. In the historical physiotherapeutic setting, the intensity was often increased, stopping just short of the patient beginning to feel pain. Alternatively, the transducer was set into motion over the skin surface, thus reducing the energy deposition per unit area. Sometimes, a beam of pulsed ultrasound was used to achieve the same objective. The treatment time frequently did not exceed 10 minutes, though drug delivery researchers have used much longer sonication periods.

A limitation of many phonophoresis studies is that the ultrasound dosimetry in the system is ill-defined. Standing waves and beam heterogeneity have already been discussed. Another problem is that the ultrasonic generators used for phonophoresis can sometimes suffer from poor calibration, and the quantity of sound energy being emitted may not be truly reflected by the control dial on the machine (Williams, 1983). Therefore the machines used for ultrasound studies should ideally be calibrated. Such calibration was achieved in some reports by using radiation pressure force measurements (Ueda et al., 1996; Meidan et al., 1998b;

Meidan et al., 1999), calorimetric techniques (Hikima et al., 1998; Mitragotri et al., 2000c) or hydrophones (Johnson et al., 1996; Mitragotri et al., 1996).

One important factor is the ultrasound effect on drug stability. Although molecules are generally stable at the intensities employed in phonophoresis, this should be checked. For example it was verified that oligonucleotides in aqueous solution, exhibiting two distinct chemistries and two different chain lengths, were not degraded by c.w. 1.1 MHz ultrasound at 1.5 W cm^{-2} at physiological pH (Meidan et al., 1997).

Another crucial aspect relates to the use of the coupling gel. In some cases, topical creams or ointments containing the active drug have been employed as the coupling medium. The disadvantage of this approach is that, unlike contact media designed specifically for that purpose, topical pharmaceutical products are generally not formulated to optimize their efficiency as ultrasound couplants. Consequently, much of the ultrasound energy may be attenuated before it reaches the skin (Cameron and Monroe, 1992). The importance of vehicular effects was recently demonstrated in in vitro diffusion cell experiments where 5-fluorouracil flux through rat skin was measured under the influence of a 1.1 MHz c.w. standing wave field (Meidan et al., 1999). Sonication actually decreased percutaneous drug absorption due to diffusive loss of 5-fluorouracil, which is hydrophilic, from the skin surface into the overlying volume of aqueous coupling gel. In other studies, hairless mice were partially immersed in either lignocaine gel or aqueous lignocaine solution and exposed to 0.048 MHz ultrasound at 0.17 W cm^{-2} (Tachibana and Tachibana, 1993). Enhanced delivery was observed from the aqueous solution but not from the gel base. It can be concluded that formulation design of the donor vehicle/coupling gel may radically affect therapeutic efficacy in phonophoretic systems.

IV. BIOPHYSICAL INTERACTIONS OF ULTRASOUND

When ultrasound propagates through skin, several different partially interrelated biophysical interactions can occur. These are first-order forces, heating, radiation pressure, cavitational activity, and microstreaming. The flow chart in Fig. 2 illustrates the processes that can contribute to phonophoresis.

A. First-Order Forces

These forces are those directly associated with the alternating motion of the particles during wave propagation (these particles are strictly speaking not atoms or molecules but theoretical points within the medium). At the intensities and frequencies commonly employed for phonophoresis, the first-order forces involve small particle displacements, moderate particle velocities, but incredibly high

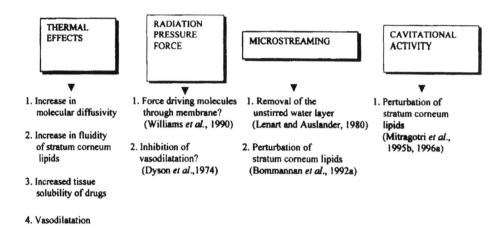

Figure 2   Potential mechanisms of phonophoresis.

values of instantaneous particle acceleration. For example, a 1 MHz, 1 W cm^{-2} ultrasonic field would induce particle displacements of 0.018 μm, particle velocities of 12 cm s^{-1}, and a maximum acceleration of 71000 g! However, for such forces to damage the skin, different portions of the same structure must be subjected to different forces so that it is twisted or torn (Williams, 1983). At 1 MHz, the wavelength is about 1.5 mm, and thus the distance between the zones of maximum opposite motion is half a wavelength or approximately 750 μm. Hence structures of this order or larger are subjected to cyclic strains equal to twice the maximum particle displacement. At an intensity of 1 W cm^{-2}, a 1 MHz beam stretches and compresses a structure 750 μm long by about 0.04 μm, which is only 0.005% of its length. Whole human skin is approximately 3 mm thick and will experience strains of this magnitude throughout its thickness. Although these strains are being applied a million times a second, the overall fatigue induced is still considered rather small compared to the strains normally encountered in strenuous exercise (Williams, 1983). Structures much smaller than a wavelength, such as individual corneocytes, are subjected to even smaller strains. This is because the whole cell and its immediate neighbors are displaced together in a synchronized fashion. It is therefore unlikely that the first-order forces mediate phonophoresis.

B.   Thermal Effects

As ultrasound propagates through tissue, its energy is progressively attenuated by the dual processes of scatter and absorption. Attenuation is commonly measured in terms of percentage energy loss per unit distance. For any given tissue, the intensity of the ultrasonic beam will decrease exponentially with tissue depth. Generally, the attenuation coefficient increases with increasing structural protein content and with decreasing water content. Consequently, tissues of high collagen content, such as bone and cartilage, undergo greater attenuation (96% cm^{-1} and 68% cm^{-1} respectively at 1 MHz), and therefore greater heating, than skin and subcutaneous fat (39% cm^{-1} and 13% cm^{-1} respectively at 1 MHz). However, attenuation also varies with frequency, with greater attenuation occurring at higher frequencies. In the 0.5 to 10 MHz frequency range, the attenuation coefficients for tissues other than bone are fairly well approximated by the equation

$$\alpha = \alpha_1 * f^{1.1}$$

where α_1 is the attenuation coefficient at 1 MHz, f is the frequency (in MHz), and α is the attenuation coefficient at the specified frequency (Goss et al., 1979). As previously explained, heat can also be generated by mode conversion at the soft tissue/bone interface. The region adjoining the bone is the periosteum and heating here may well be perceived as painful to the patient. Such periosteal pain limits the amount of ultrasonic exposure over bony anatomy. Heat will be dissipated over areas where there is a thick layer of absorbent tissue, and this will enable the patient to tolerate the heating. Another mismatch in acoustic impedance develops at the contact between transducer the and skin. Skin temperature increases will also occur at this interface.

It can be seen that it is extremely difficult to predict the amount of tissue heating that will develop during a specific sonication regimen although temperature elevations of several degrees centigrade are typical. The temperature rise will increase the fluidity of stratum corneum lipids as well as directly increase the molecular diffusivity of the permeant molecule through this layer. Both these effects will enhance drug transport in vitro. In the in vivo situation, ultrasonic energy will pass into deeper tissues, increasing the tissue solubility of drugs as well as initiating increased microvascular perfusion, i.e., hyperemia. The magnitude of the hyperemic reaction in human skin can be measured by using a water bath as a variable heat supply and a photometric sensor to quantify the extent of induced erythema. Figure 3 shows the response profile that is typically obtained.

All the effects mentioned here will tend to enhance drug flux, and early work has established that a 10°C rise in temperature will roughly double the in vivo skin absorption of many compounds (Scheuplein, 1978). However, if phonophoresis is mediated purely by heating then clearly the process is of no great therapeutic interest, since the same enhancement could then be obtained by applying other heating sources such as microwaves or simple conductive heating.

C. Radiation Pressure Force

Any structure that absorbs a component of a beam of energy is subjected to a force, termed the radiation pressure force, which tends to push that object in the

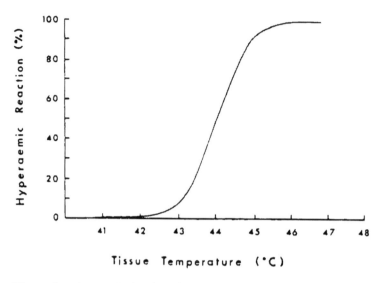

Figure 3 The magnitude of the hyperemic response in human skin as a function of temperature. (From Williams, 1983.)

direction of wave propagation (Williams, 1983). One hypothesis is that an induced radiation pressure, acting on the penetrant molecules, can "push" the drug through the skin (Saxena et al., 1993). Early work demonstrated that radiation pressure indeed played a role in the observed ultrasound-accelerated transport of electrolytes through cellophane membranes (Lenart and Auslander, 1980). A model system of the diffusion of molecules through a water-cooled agar gel demonstrated differences in the phonophoretic effect of ultrasound (Williams et al., 1990). The relatively large molecule sodium diethylene triamine pentacetic acid (DTPA) underwent ultrasound-accelerated diffusion while the smaller sodium pertechnetate molecule did not. The authors proposed that the larger DTPA molecule absorbs a greater proportion of the ultrasound than sodium pertechnetate and is consequently subjected to a greater radiation pressure force, which increases its diffusion rate through gel. This conclusion was derived even though charge and other differences between the two molecules were also apparent. In any case, some physicists have rejected the pushed-penetrant hypothesis, claiming that at the atomic level of description, the particles merely oscillate about a fixed point according to the characteristics of the acoustic field. Consequently, the penetrant molecules would be subjected to zero radiation pressure (Simonin, 1995). Even if the hypothesis were valid, phonophoresis is often associated with standing waves, and so the molecules could be driven in different directions at different locations within the exposed skin.

Another proposal is that ultrasound exerts a radiation force on the stratum corneum, thus perturbing its barrier properties. One author considered a 1 MHz, 1 W cm^{-2} beam and computed from theory that the stratum corneum would be subjected to a maximum pressure equivalent to that produced by 5 mg of weight distributed over a 1 cm^2 area of its surface (Simonin, 1995). This is insufficient to modify the microstructure of the cornified layer. However, although the primary radiation pressure forces are very weak, these forces are magnified if a standing wave field is generated above the soft tissue/bone interface. In this situation, radiation pressure forces act over a distance of 0.4 mm in soft tissue at 1 MHz, so that unanchored particles that are denser than their suspending medium are pushed to the zones of maximum acoustic pressure. This process has been observed within small blood vessels where blood cells have amassed together in discrete motionless bands separated by clear plasma (Dyson et al., 1974). Such an impairment in microvascular blood perfusion may result in less efficient heat removal from sonicated tissues, thus enhancing any thermal phonophoretic effect.

D. Microstreaming

When a structure within an ultrasonic field experiences an unequal distribution of radiation pressure forces across its length, it is subjected to a force known as the acoustic torque. In a liquid or semiliquid medium, the torque can produce microscopic currents known as acoustic microstreaming. At a frequency of 0.025 MHz, the dimensions of the streaming patterns are determined by the boundary layer thickness, the size of which is estimated to be approximately 3.6 μm in water. These small dimensions mean that the velocity gradients, i.e., the rate of change of velocity with distance, and hence the hydrodynamic shear stresses are large even though the actual streaming velocities that produce them are relatively low—of the order of a few cm s^{-1}. Microstreaming can also develop secondarily to cavitation.

In research with cellophane membranes, it was shown that microcurrents can decrease the thickness of the unstirred water layer in the vicinity of the membrane. This effect was the major factor responsible for the enhanced diffusion of solutes under the influence of 1 MHz ultrasound (Lenart and Auslander, 1980). This type of process may be relevant to clinical phonophoresis. It is also possible that phonophoresis may be mediated by microstreaming occurring within the skin. In in vivo work with the hairless guinea pig model (Bommannan et al., 1992a), 0.2 W cm^{-2} c.w. ultrasound was applied at frequencies of 10 and 16 MHz for 5 or 20 minutes. Transmission electron microscopy was used to track the skin permeation of the electron-dense tracer colloidal lanthanum hydroxide. Under control conditions, the tracer was found not to penetrate the stratum corneum. However, with high-frequency ultrasound, the penetration of tracer by an intercellular route through the epidermis to the upper dermis was observed. Since surface skin temperature increases of not more than 1°C were measured, ultrasonic heating was not responsible. Cavitation can also be ruled out at such intensity and frequency. It is probable that microstreaming induces disordering of the intercellular lipids of the stratum corneum, thus permeabilizing this pathway to drug delivery.

E. Cavitational Activity

Cavitation is defined as acoustically induced bubble activity (Apfel, 1997). It includes diverse types of behavior, ranging from the relatively gentle volumetric pulsations of gas-filled bodies (stable cavitation) to the destructive and free radical–generating formation and implosion of vapor-filled voids (transient cavitation). Cavitation is powered by the energy from the incident ultrasonic beam, about 10% of which may be reradiated from the bubble as an outgoing spherical wave. The remainder may be lost as heat, or may induce shock waves that can disrupt biological tissues. The intensity threshold for the onset of cavitation increases with increasing beam frequency as well as decreasing dissolved gas content in the irradiated medium. Frequency dependence is due to the fact that, at higher frequencies, there is insufficient time for gas molecules to diffuse into cavities during the extremely brief rarefaction phase of the acoustic cycle. Figure 4 shows the commonly quoted threshold intensities that have to be attained before transient cavitation will develop in aqueous media in vitro.

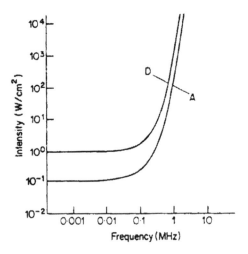

Figure 4   The threshold intensity for the development of transient cavitation in aqueous media, as a function of frequency. (A) represents an aerated medium and (D) represents degassed medium. (From Williams, 1983.)

There is now evidence that cavitation can develop in human epidermis (Wu et al., 1998). Following sonication with c.w. 0.168 MHz ultrasound at 1.2 W cm^{-2}, human epidermis samples were immersed in saline solutions containing a fluorescent lipid probe. Epifluorescence imaging of the cornified layer revealed the presence of voids of dimensions of 20 μm—sufficiently wide to allow the passage of very large molecules. The authors stated that the formation of these defects could be explained by cavitation forcing the lipid molecules out of the stratum corneum.

Langer and colleagues used a 1 MHz c.w. beam at 2 W cm^{-2} in order to enhance the penetration of a wide range of compounds through human epidermis in vitro (Mitragotri et al., 1995b). Large improvements in transdermal delivery were obtained for three out of the seven investigated compounds. Powerful evidence that the enhanced delivery was due to cavitation was derived from the fact that drug penetration was radically reduced by raising the frequency to 3 MHz, or by using deaerated or compressed epidermis. Furthermore, confocal microscopy of fluorescein-stained keratinocytes showed that following sonication at 1 MHz, there was bleaching of fluorescein in the keratinocytes but not in the intercellular lipids. Fluorescein is bleached in the presence of hydrogen peroxide, which is formed from the free radicals generated by transient cavitation. Significantly, 3 MHz ultrasound did not cause any bleaching. This data indicates that transient cavitation occurring in the keratinocytes somehow permeabilizes the stratum corneum. In sequential in vitro work (Mitragotri et al., 1996), this group found that employing very low frequency ultrasound (0.02 MHz) induced greater cavitational activity within the stratum corneum, thus facilitating even more efficacious phonophoresis. Electrical resistance measurements, indicative of the integrity of the stratum corneum lipid bilayers, decreased 25-fold as result of sonication. From this and other theoretical considerations, the authors proposed that cavitation disorders the intercellular lipid bilayers of the stratum corneum, allowing aqueous channels to form within the lipid domains. Together with keratinocytes, the perturbed bilayers constitute a new transcellular transport pathway that permits the rapid ingress of permeant molecules. The mechanism is illustrated in Fig. 5.

V. PHONOPHORESIS IN ANIMAL SKIN

A. Steroids

Kost and coworker documented successful hydrocortisone phonophoresis in vitro through whole sections of hairless mouse skin (Machluf and Kost, 1993). The authors reported a duration-responsive effect when a 1 MHz pulsed beam was operated at an intensity of 3 W cm^{-2}. While 30 minutes of sonication did not affect hydrocortisone flux, sonication for 1.5 hours or more produced significant

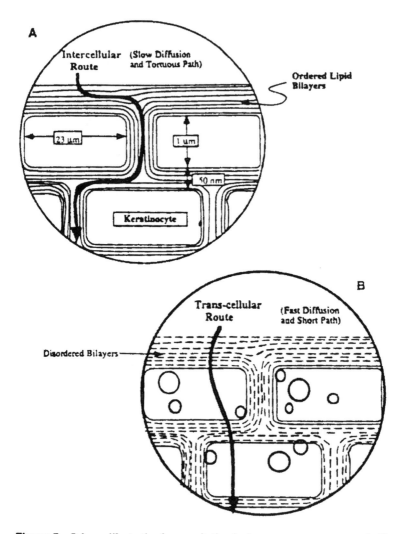

Figure 5 Scheme illustrating how cavitation in the stratum corneum may facilitate phonophoresis. The scheme shows the drug transport pathway during (A) control conditions and (B) during low frequency phonophoresis. (From Mitragotri et al., 1996.)

flux enhancement, which persisted even after the energy had been switched off. DSC and FTIR analysis failed to reveal any perturbations of stratum corneum structure. In other in vitro experiments with whole hairless mouse skin (Machet et al., 1998), it was shown that 20 min of sonication with c.w. 1.1 MHz ultrasound at 1.2 W cm^{-2} did not enhance either hydrocortisone or estradiol flux. However, the diffusion apparatus incorporated a cooling coil that largely removed thermal effects as an artifact of ultrasound exposure. In in vitro work with whole Wistar rat skin, it was determined that c.w. 1.1 MHz ultrasound alone (2.25 W cm^{-2}) did not significantly enhance hydrocortisone penetration, although a significant synergistic effect was observed when ultrasound was combined with azone (Meidan et al., 1998b). The possibility of in vivo hydrocortisone phonophoresis was examined in a study involving 24 purebred greyhounds (Muir et al., 1990). Aliquots of coupling gel containing 10% hydrocortisone were deposited at skin surface sites that were then irradiated with a 1 MHz beam conveying an intensity of 2.75 W cm^{-2}. In this model, sonication did not significantly improve drug delivery.

One group recently analyzed the effect of 0.02 MHz continuous ultrasonic waves on the in vitro transport of clobetasol 17-propionate across hairless mouse skin (Fang et al., 1999). The sound, which was applied for 4 hours at an intensity of 0.1 W cm^{-2}, produced a 55% increase in steroid flux and a 69% increase in the amount of drug accumulated in the skin. Although the authors claimed that cavitation was the most likely mechanism of enhancement, the effect could also have been induced by ultrasonic heating, as sonication produced a 5°C rise in skin temperature. Interestingly, when the hairless mouse skin was replaced by Wistar rat skin, the workers measured a more pronounced flux enhancement of 210%. The authors therefore proposed that phonophoresis was conveyed by appendageal routes of which there are more in Wistar rat skin, although other differences between the two types of tissue are also apparent.

B. Indomethacin

In vitro investigations into indomethacin phonophoresis were performed using 15 minutes of 0.48 W cm^{-2} c.w. ultrasound at 1 MHz (Al-Suwayeh and Hikal, 1991). Although ultrasonic heating was minimized by maintaining both donor and receptor solutions at 37°C, it was nethertheless found that sonication approximately doubled indomethacin permeation as compared to drug flux under control conditions. Indomethacin studies were also carried using an in vivo Wistar rat model (Miyazaki et al., 1992b). The protocol involved applying indomethacin ointment to shaved areas of the anesthetized animals and directing a 1 MHz c.w. ultrasound beam at the treated area for between 5 and 20 minutes with a range of intensities (0.25, 0.5, 0.75, and 1 W cm^{-2}). Blood samples were taken in the hours following phonophoresis, and indomethacin levels were quantified by

HPLC. It was found that 0.75 W cm^{-2} was the intensity that induced the greatest absorption enhancement, while 10 minutes constituted the optimal duration. However, the authors suggested that 0.5 W cm^{-2} was the preferred intensity, since its application for 10 minutes did not cause marked skin temperature elevations or any significant skin tissue damage. With this sonication regimen, indomethacin delivery was enhanced 2.7 times relative to the control. In further work with the same model (Miyazaki et al., 1992a), the effect of 1:3 pulsed output was examined, again at different intensities and durations. Pulsed ultrasound was found to be a weaker enhancer than the same energy delivered in c.w. form, perhaps because a pulsed beam generates less tissue heating.

C. Insulin

In 1992, it was reported that significant transdermal insulin transfer could be induced through the skin of anesthetized diabetic rabbits by using a 0.105 MHz pulsed beam for 90 minutes (Tachibana, 1992). Drug absorption, as evaluated from both blood insulin and blood glucose concentrations, was elevated in the 6 hour period following sonication. Crucially, skin biopsy studies revealed that the treated statum corneum remained histologically intact and unaffected by ultrasound. It is noteworthy that both the transducer and the insulin reservoir were maintained at 4°C, thus indicating that nonthermal acoustic phenomena were probably mediating the observed transport enhancement. A few years later, Langer and coworkers also used low-frequency ultrasound in their in vivo experiments on hairless rats (Mitragotri et al., 1995a). They reported that they could transdermally deliver therapeutic quantities of insulin by using a 0.02 MHz source at 0.225 W cm^{-2}. Again, sonication caused no skin damage. Also, a control treatment of ultrasound without drug did not affect blood glucose levels, thus demonstrating that ultrasonic waves were not directly inducing changes in the pancreas.

D. Other Drugs

Ultrasound-promoted delivery of various other drugs has been documented in the literature reports. For instance, it was found that 3 hours of sonication with a 1 MHz, 3 W cm^{-2} pulsed beam almost doubled caffeine permeation through full thickness hairless mouse skin (Machluf and Kost, 1993). Additionally, an intensity-drug flux response was demonstrated for in vitro digoxin penetration through hairless mouse skin under the influence of 10 minutes 3.3 MHz c.w. ultrasound (Pinton et al., 1991; Machet et al., 1996). Successful in vivo phonophoresis of salicylic acid has also been documented using both 10 MHz and 16 MHz ultrasound in hairless guinea pigs (Bommannan et al., 1992b) and 0.02 MHz ultrasound in hairless rats (Mitragotri et al., 1996). It is likely that enhancement was mediated by different mechanisms in each of these two studies. The effectiveness of the 0.02 MHz frequency was again demonstrated in recent in vitro mannitol experiments performed on full thickness pig skin (Mitragotri et al., 2000c). The dependence of permeability enhancement on the ultrasound parameters was similar to that of cavitation as measured in a model system of aluminum foil pitting. This indicates that cavitation was almost certainly permeabilizing the skin.

The problem with analyzing all these results is that since each group used different methods and model systems, it is difficult to derive any underlying patterns from the collated data. In a more systematic approach, one team (Ueda et al., 1995) assessed the partition coefficient dependency of phonophoresis by selecting nine candidate molecules exhibiting different octanol–water partition coefficients (log Ko/w ranging between -1.7 and 3.86) but relatively similar molecular weights (130 to 254 Da). In vitro diffusion studies were then performed using whole hairless rat skin and 0.15 MHz c.w. ultrasound at 2 W cm^{-2} for 1 hour. Flux was measured in the 3 hour period before sonication, during sonication, as well as in the 4 hours following sonication. While the penetration of all the molecules was enhanced during the sonication period, enhancement for hydrophilic drugs was greater than enhancement for lipophilic drugs. For example, mean 5-fluorouracil (log $Ko/w = -1.7$) flux underwent more than a sevenfold enhancement while mean flurbiprofen (log $Ko/w = 3.86$) flux only doubled. After the ultrasound was switched off, flux of the hydrophilic molecules remained high, decreasing only marginally, while flux of the lipophilic molecules decreased more substantially toward their presonication values. One limitation of this important study was that histological analysis of the skin was not undertaken in order to assess barrier integrity. In sequential work with a similar system (Ueda et al., 1996), it was found that ultrasound significantly increased the leaching out of sterols and ceramides from the skin, reducing their total amount by some 3%. This strongly suggests that irradiation produces a defect in lipid packing in the stratum corneum, thus permeabilizing it.

E.   The Limitations of Animal Skin

Despite the interesting findings described above, there are indications that animal skin may be a poor model for human skin in terms of its biophysical interactions with ultrasound. In a comparative morphological study (Yamashita et al., 1997), samples of both hairless mouse and human skin were immersed in water and subjected to 0.048 MHz ultrasound at 0.5 W cm^{-2} for 5 min. Scanning electron microscopy of mouse skin showed that the cells of the cornified layer were almost totally detached while the polygonal cells of the stratum spinosum and basal cells were exposed. Furthermore, large craterlike pores some 100 microns across were formed. In contrast, the human skin samples were much more resistant to the effects of sonication, showing only slight removal of the keratinocytes around the hair follicles. In other in vitro work (Meidan et al., 1998a), it was shown that the application of a 1.1 MHz ultrasonic free field (i.e. no standing waves) at 0.1 W cm^{-2} to Wistar rat skin caused sebum to be discharged from the sebaceous glands, thus filling much of the hair follicle shafts. The discharge caused the transfollicular pathway to be blocked for hydrophilic permeants that penetrate via this route. Hence, sonication actually reduced drug transport. Importantly, sebum discharge was not induced in guinea pig skin, and it is not known whether it can occur in human skin. It can be concluded that in order for results to be really meaningful, phonophoresis studies should be performed with human skin.

VI. PHONOPHORESIS ACROSS HUMAN SKIN IN VITRO

A. Ibuprofen

Successful ibuprofen phonophoresis across human epidermis was reported in a study in which epidermis samples were sonicated for 30 minutes at both time zero and at 6 hours (Brucks et al., 1989). 1 MHz c.w. ultrasound was used at an intensity of 1 W cm^{-2}. In order to control for ultrasonic heating, two controls were employed: one with no ultrasound and another where a comparative heat alone source was applied. The flux of [^{14}C]-ibuprofen was determined by scintillation counting of samples taken from the receptor solution. It was found that ultrasound enhanced ibuprofen permeation to a significantly greater extent than the heat alone treatment. This suggests that the enhancent was due to a mechanical artifact of the beam—either cavitation, microstreaming, or radiation pressure. Furthermore, the enhancement was reversible as drug flux returned to control levels some 10 min after the ultrasound was switched off. Another positive feature was that the ultrasound-induced skin temperature increases were not of a magnitude that should be destructive to tissue, and there was no visual damage to the skin. In related mechanistic investigations, human epidermis samples were subjected to 2 W cm^{-2}, 1 MHz c.w. ultrasound for 30 minutes (Nanavaty et al., 1989). Attenuated total reflectance Fourier transform infrared spectroscopy (ATR-FTIR) was used to compare sonicated epidermis samples with control samples. Although ultrasound induced no major irreversible changes in the skin, it did induce minor conformational perturbations in the lipids and/or proteins of the stratum corneum.

B. Steroids

One team (Machet et al., 1998) recently investigated the possibility of hydrocortisone phonophoresis in whole human skin. The ultrasonic energy was applied for

20 min in the form of a 1.1 MHz c.w. beam conveying an intensity of 1.2 W cm^{-2}. In a key step, heating was largely eliminated as an artifact of sonication by a cooling coil that maintained the donor solution–skin surface interface temperature 2°C cooler than the temperature of the receiver compartment. Under these conditions, ultrasound did not significantly affect transdermal hydrocortisone flux. In other whole skin experiments (Fang et al., 1999), a 0.02 MHz, 0.1 W cm^{-2} c.w. beam was used in an an attempt to improve the transdermal absorption of the potent steroid clobetasol 17-propionate. Sonication, which lasted for 4 hours, produced a 5°C elevation in skin temperature. This heating was probably responsible for the significant 20% increase in steroid flux that was reported. A limitation of this important study was that no histological work was undertaken in order to assess skin status.

C. Proteins

In a potentially exciting development, Langer and coworkers (Mitragotri et al., 1995a) reported that they could transdermally deliver insulin, erythropoeitin, and interferon-γ at a therapeutically relevant rate by using low-frequency ultrasound (0.02 MHz). Human epidermis samples were mounted in diffusion cells and pulsed (100 ms per second) ultrasound was applied for 4 h at intensities of up to 0.225 W cm^{-2}. Human skin permeability to insulin was dependent upon intensity as can be seen from Fig. 6. Importantly, the phonophoresis was reversible. For instance, 3 hours after the energy had been switching off, the insulin permeation rate had returned to zero. Furthermore, transepidermal water loss values increased 100-fold during sonication but decreased back down to control levels

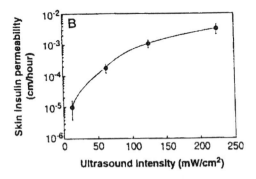

Figure 6 The variation of in vitro transdermal insulin permeability as a function of ultrasound intensity. The skin is impermeable to insulin at zero intensity. (From Mitragotri et al., 1995a.)

within 15 h after sonication. Clearly, these highly promising studies need to be performed within a clinical setting.

D. Other Drugs

One team evaluated the effect of 3.3 MHz c.w. ultrasound on the permeability of digoxin across whole human skin (Machet et al., 1996). Drug flux was monitored over the 48 h period following 10 minutes of sonication at either 1, 2, or 3 W cm^{-2}. The results were negative in that sonication did not accelerate digoxin penetration at any of the selected intensities. Negative results were also reported regarding attempts to phonophoretically improve the absorption of the virustatic agent zidovudine through whole human skin (Pelucio-Lopes et al., 1993). A 1.1 MHz c.w. source was employed for 20 min at an intensity of 1.5 W cm^{-2}. Notably, the thermal effects of ultrasound were suppressed by circulating cold water through specially jacketted donor cells.

In a systematic study, Mitragotri and coworkers chose seven different test

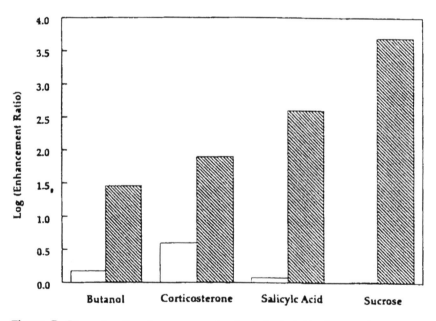

Figure 7 Phonophoretic enhancement ratios for 1 MHz (white bars) and 0.02 MHz (hatched bars) ultrasound for four different test molecules. The enhancement ratio is the ratio between the phonophoretic and the passive molecular permeability. (From Mitragotri et al., 1996.)

compounds exhibiting a wide range of octanol–water partition coefficients and investigated their susceptibility to phonophoresis across human epidermis (Mitragotri et al., 1995b). The compounds were benzene, butanol, caffeine, corticosterone, estradiol, progesterone, testosterone. The workers used c.w. 1 MHz ultrasound at 2 W cm^{-2} and monitored drug permeation over a 20 h period. Ultrasound improved the penetration of three out of the seven molecules. Even though sonication induced a 7°C increase in diffusion cell temperature, it was convincingly demonstrated that cavitation was the principal mechanism involved.

In sequential work with a slightly different set of test molecules, the authors applied 0.02 MHz ultrasound—a much lower frequency—at 0.125 W in a pulsed (1:9 on:off ratio) mode for 1 hour. Remarkably, low-frequency ultrasound enhanced the penetration of all seven investigated molecules through human epidermis by between 3 to 3000 times. Crucially, after the sound was switched off, barrier properties returned to normal as evaluated by water permeability measurements. As explained previously, cavitation-induced disordering of the intercellular stratum corneum lipids is believed to cause this tremendous yet reversible enhancement effect. Figure 7 compares the enhancement ratio induced by 1 MHz ultrasound (c.w., 2 W cm^{-2}) with that induced by 0.02 MHz ultrasound (pulsed,

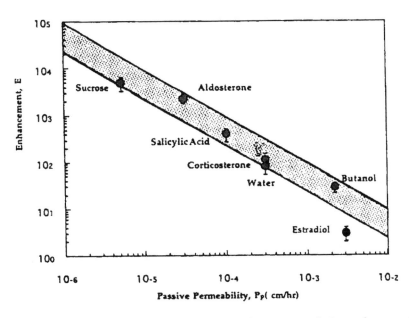

Figure 8 Comparison of predicted transport enhancement and observed transport enhancement for low-frequency phonophoresis. The lines represent the limits of the predicted range. (From Mitragotri et al., 1996.)

0.125 W cm^{-2}) in the case of butanol, corticosterone, salicylic acid, and sucrose. It can be seen that the low frequency beam was more efficacious by up to three orders of magnitude, and this is almost certainly due to its induction of greater cavitation within the stratum corneum. From their derived kinetic data, the group developed a mathematical model for explaining low frequency ultrasound phonophoresis. The model predicts that transport enhancement depends most directly on the passive skin permeability of the penetrant. Hydrophilic or larger molecules that passively permeate relatively slowly through the skin are preferentially accelerated by low-frequency ultrasound. Figure 8 plots phonophoretic enhancement as a function of passive skin permeability, and it can be seen that the observed enhancement values correlated well with the calculated enhancement values.

VII. CLINICAL STUDIES

Surprisingly, investigations into phonophoresis have included relatively few well-controlled clinical trials. The studies that have been performed have tended to focus on the use of steroids, nicotinate esters, analgesics, and anesthetics.

A. Nicotinate Esters

Hadgraft and colleagues (Benson et al., 1991) investigated the influence of 5 minutes of 1 W cm^{-2}, 3 MHz c.w. ultrasound on the percutaneous absorption of ethyl nicotinate, methyl nicotinate, and hexyl nicotinate across human forearm skin. The group used laser Doppler velocimetry in order to quantify the nicotinate absorption. All three nicotinate esters were found to undergo ultrasound-enhanced delivery, though for hexyl nicotinate the difference was not statistically significant. In another study, Hadgraft and coworker compared a method of sonication followed by drug application to drug application followed by sonication (Murphy and Hadgraft, 1990). For the more lipophilic hexyl nicotinate, the rate-limiting step was assumed to be partitioning from the lipid-rich stratum corneum. This process was not affected by ultrasound pretreatment but was enhanced by ultrasound posttreatment. The authors proposed that ultrasonic energy can enhance the rate of drug partitioning out of the intercellular lipids once a reservoir had formed. For methyl nicotinate, the main barrier to penetration were the structured stratum corneum lipids. Since both ultrasound pretreatment and posttreatment resulted in improving absorption, the authors suggested that for this drug, successful phonophoresis was due to ultrasound fluidizing the stratum corneum lipids. In another trial, it was shown that ultrasound can enhance benzyl nicotinate penetration through human skin (Hofmann and Moll, 1993). A reflection photometry method was used to quantify the skin reddening effect induced by benzyl nicotinate absorption. The lag time of drug delivery was significantly reduced by the application of a 0.5 W cm^{-2} c.w. beam. Unfortunately, the authors did not state the ultrasonic frequency of their equipment.

B. Anesthetics

In a particularly early study, the possibility of lignocaine phonophoresis through human forearm skin was examined by using 5 min of pulsed 0.87 MHz ultrasound at 2 W cm^{-2} (McElnay et al., 1985b). Compared to nonsonicated applications of drug, a statistically insignificant increase in anesthesia onset rate was reported when ultrasound energy was applied. In sequential work with a cream containing both lignocaine and prilocaine (Benson et al., 1988), the researchers identified a significant elevation in anesthesia duration. In these trials, ultrasonic heating was not monitored or controlled thus yielding no data on the mechanisms involved in any observed phonophoresis. In contrast, Williams (Williams, 1990) placed a water path maintained at 37°C between the transducer and skin in order to minimize the thermal effects of sonication. He used a sensory perception method in order to quantify anesthetic absorption following 50 min of 1.1 MHz c.w. ultrasound treatment at 0.25 W cm^{-2}. The creams investigated were Americaine, containing 20% benzocaine and 0.1% benzethonium; Emla cream, containing 2.5% lignocaine and 2.5% prilocaine; and Nupercainal, containing 1% dibucaine. The results were negative in that ultrasound induced no detectable effects upon the penetration rates of any of the agents tested.

C. Other Drugs

In a relatively early trial, 3 g of 0.025% fluocinolone acetonide gel was applied to the flexor surface of the forearms of 12 human volunteers (McElnay et al., 1987). The treated site was then irradiated for 5 minutes with 0.87 MHz pulsed ultrasound at 2 W cm^{-2}. The study was of a double-blind, sham ultrasound controlled, crossover nature. Drug absorption data was derived from skin blanching tests. Although sonication significantly enhanced steroid penetration, the magnitude of the effect was deemed too small to result in greater therapeutic efficacy within the clinical setting. Other workers have evaluated the effect of ultrasound at different modes, intensities, and frequencies on benzydamine absorption (Benson et al., 1986; Benson et al., 1989). The trials were of a double-blind, placebo-controlled nature. It was found that sonication did not improve benzydamine absorption across human flexor surface skin. Negative findings were also documented following attempts to enhance ultrasonically the delivery of trolamine salicylate (Oziomek et al., 1991).

VIII. ULTRASOUND FOR THE TRANSDERMAL EXTRACTION OF ANALYTES

Apart from its drug delivery applications, ultrasonic permeabilization of the skin could conceivably be used to extract transdermally clinically relevant analytes from the interstitial fluid. Such "reverse phonophoresis" would constitute a potentially noninvasive diagnostic tool, for example as a method of monitoring blood glucose concentrations in diabetics.

The idea has been taken up by Langer and colleagues (Mitragotri et al., 2000b) who placed a chamber on the backs of anesthetized Sprague Dawley rats and filled it with 1% sodium lauryl sulfate solution. A 0.02 MHz transducer was then immersed in the solution and switched on for a few minutes to produce a pulsed 7 W cm^{-2} beam. Subsequently, a vacuum was applied to the sonicated sites for 15 minutes in order to extract the analytes. The treatment led to a 100-fold elevation in glucose permeability, which persisted for at least 3 hours. Other ultrasonically extracted analytes were albumin, calcium, urea, triglycerides, lactate, and dextran. Crucially, transdermally extracted fluid had a composition similar to that of interstitial fluid. The efficacy of low-frequency ultrasound—surfactant combination was proven in other rat studies that employed similar though not identical protocols (Kost et al., 2000; Mitragotri et al., 2000a). In particular, these studies demonstrated that the concentration of any given analyte in the transdermal extractions correlated well with its plasma concentration.

Seven volunteers with type 1 diabetes mellitus have already participated in a glucose extraction clinical trial (Kost et al., 2000). Prehydrated skin sites on the forearm were irradiated for 2 minutes with a 0.02 MHz, 5 W cm^{-2} pulsed beam. The ultrasound propagated through a chamber filled with 1% sodium lauryl sulfate in saline. After sonication, the surfactant solution was replaced by pure saline. A vacuum was applied for 5 minutes, after which the saline was removed for glucose quantification. Vacuum extractions were performed every 30 minutes over a 4 hour period. Crucially, there was a good correlation between the ultrasonically extracted glucose concentrations and directly measured blood glucose concentrations. However, the extraction profile lagged behind the direct measurement profile by 30 minutes—representing the time it takes for glucose in the capillaries to diffuse to the skin surface. Although the calibration factors varied greatly between different individuals, the factor remained steady for a fixed treatment site on any given patient. Of great importance, the skin sites regained normal permeability status by 24 hours after sonication, while the patients reported no skin irritation. Even so, further studies evaluating the long-term safety of the technique are clearly warranted.

IX. CONCLUSIONS

Table 1 presents the results obtained from various literature reports in which ultrasound was used in an attempt to enhance the absorption of different molecules through human skin. For each study, penetration magnitude was quantified from a distinct measurement parameter, and these are listed in the table. The enhancement ratios were calculated by dividing the ultrasonic drug absorption value by the control drug absorption value as evaluated by the corresponding measurement parameter for that study. Hence an enhancement ratio of 2 indicates that ultrasound treatment doubled drug absorption. Where sonication did not significantly affect drug transport, the enhancement ratio value is listed as 1.

It can be seen from Table 1 that at medium to high frequencies (0.75 to 3 MHz), many molecules undergo modest but significant transport enhancement, and this is probably due to ultrasonic heating. A skin surface temperature increase of 10°C will improve the penetration of many compounds by between 1.4 and 3 times in vivo (Scheuplein, 1978), and ultrasonic heating on a smaller magnitude largely accounts for many of the modest improvements in drug delivery documented in the literature. The extent of heating is influenced by a host of factors including the sonication time, wave parameters, transducer motion, anatomical site, and quantity of coupling gel. Variations in these conditions explain the sometimes conflicting results obtained by different groups working with the same test molecule. It is noteworthy that when heating was eliminated as an artifact of phonophoresis, the penetration of five different anesthetic molecules (Williams, 1990), two steroids (Machet et al., 1998), mannitol (Machet et al., 1998) and an antiviral agent (Pelucio-Lopes et al., 1993) were not affected by ultrasound.

However, ultrasound can also permeabilize the skin by nonthermal mechanisms. In particular, there is now convincing proof that low-frequency ultrasound (0.02 MHz) induces cavitational disordering within the stratum corneum, which facilitates massive improvements in transdermal drug delivery—see Table 1. There is good evidence that these changes in the stratum corneum are reversible, and hence the process is potentially useful within the clinical setting (Mitragotri et al., 1996). More studies need to be performed in order to identify the role of the various parameters that govern the effectiveness of low-frequency phonophoresis so that the process can be optimized. An example of the type of quantitative biophysical research required was provided in a recent in vitro study (Mitragotri et al., 2000c). Here it was shown that the enhancement effect depended upon the total energy density delivered to the skin regardless of the particular values of intensity, exposure time, or duty cycle, provided an energy density threshold of 222 J cm^{-2} threshold was reached. Future research should also be

Table 1 Ultrasonically Enhanced Drug Delivery Through Human Skin—A Summary of the Literature Findings

Molecule	Molecular mass (Da)	f(MHz)	Intensity (W cm^{-2})	Enhancement ratio	Measurement parameter	Reference
Aldosterone	360	0.02	0.125 (P)	1400	Mean flux	Mitragotri et al., 1996
Benzene	78	1	2	1	Mean flux	Mitragotri et al., 1995b
Benzocaine/benzethonium chloride	165.2/448.1	1.1	0.25*	1	Electrical sensory perception	Williams, 1990
Benzydamine HCl	345.9	0.75	1.5	1	Residual amounts in vehicle	Benson et al., 1989
		1.5	1.5	1		
		3	1.5	1		
			1 (P)	1		
Butanol	74	0.02	0.125 (P)	29	Mean flux	Mitragotri et al., 1996
		1	2	1	Mean flux	Mitragotri et al., 1995b
Caffeine	194	1	3 (P)	1.92	Mean flux	Machluf and Kost, 1993
		1	2	1	Mean flux	Mitragotri et al., 1995b
Clobetasol-17 propionate		0.02	0.1	1.2	Mean flux	Fang et al., 1999
Corticosterone	346	0.02	0.125 (P)	80	Mean flux	Mitragotri et al., 1996
		1	2	4	Mean flux	Mitragotri et al., 1995b
Dibucaine	379.9	1.1	0.25*	1	Electrical sensory perception	Williams, 1990
Digoxin		3.3	3	1	Mean flux	Machet et al., 1996
Estradiol	272	0.02	0.125 (P)	3	Mean flux	Mitragotri et al., 1996
		1	2	13	Mean flux	Mitragotri et al., 1995b
		1.1	1.5	1	Mean flux	Machet et al., 1996

Fluocinolone acetonide	425.5	0.87	2 (P)	1.32	AUC from skin blanching	McElnay et al., 1987
Hexyl nicotinate	207.3	3	1	1	AUC vasodilation from LDV	Benson et al., 1991
Hydrocortisone	362.5	1.1	1.5*	1	Mean flux	Machet et al., 1998
		1	3 (P)	1.17	Mean flux	Machluf and Kost, 1993
Ibuprofen	206.3	1	1	1.82	Flux rate over 30 min	Brucks et al., 1989
Lignocaine	288.8	0.87	2 (P)	1	Time to onset of an- aesthesia	McElnay et al., 1985b
Lignocaine/prilocaine	288.8/257	1.1	0.25*	1	Electrical sensory per- ception	Williams, 1990
		1.5	1 (P)	1.56	Time to total sensation recovery	Benson et al., 1988
		3	1 (P)	2.01	Time to partial sensa- tion recovery	Benson et al., 1988
Mannitol	182.2	1.1	1.5*	1	Mean flux	Machet et al., 1998
Methyl nicotinate	137.1	3	1	1.59	AUC vasodilation from LDV	Benson et al., 1991
		3	1	2.04	AUC vasodilation from LDV	McElnay et al., 1993
Progesterone	274	1	2	1	Mean flux	Mitragotri et al., 1995b
Salicylic acid	138	0.02	0.125 (P)	400	Mean flux	Mitragotri et al., 1996
Sucrose	342	0.02	0.125 (P)	5000	Mean flux	Mitragotri et al., 1996
Testosterone	288	1	2	5	Mean flux	Mitragotri et al., 1995b
Zidovudine	267.2	1.14	1.5*	1	Mean flux	Pelucio-Lopes et al., 1993

* Denotes that controlling for thermal effects was conducted by minimizing ultrasonic heating. P denotes pulsed ultrasound otherwise c.w. mode ultrasound.

directed towards addressing the possible bioeffects and safety of low-frequency ultrasound.

Another delivery strategy involves employing a combination of ultrasound and chemical enhancers. When human epidermis samples were treated with a linoleic acid/ethanol mixture and then sonicated with a 1 MHz, 1.4 W cm^{-2} beam, the in vitro penetration of five different molecules was greater than that measured following the application of either ultrasound or the chemicals alone (Johnson et al., 1996). Of specific interest was the fact that the molecules exhibiting the largest molecular weight underwent more enhancement than smaller molecules with both ethanol/linoleic acid alone as well as with combination treatment, suggesting that both modalities permeabilize the stratum corneum *via* similar mechanisms. It has recently been shown that the efficacy of low-frequency ultrasound in enhancing transdermal penetration can be improved by its combination with sodium lauryl sulphate (Mitragotri et al., 2000d). Synergism can also be achieved by using ultrasound in conjunction with electrical energy. The combination of electroporesis and ultrasound has been shown to be more efficacious in permeabilizing the human epidermis in vitro than either modality alone (Kost et al., 1996). Similarly, the iontophoretic transfer of both sodium benzoate and D$_2$O was improved when the skin samples were presonicated (Ueda et al., 1996). Clearly, more work is required in order to elucidate the nature of these promising synergistic approaches.

REFERENCES

S Al-Suwayeh, AH Hikal. Influence of ultrasound on in vitro diffusion of indomethacin through hairless mouse skin. Pharm Res 8(10)(suppl):S-140, 1991.

RE Apfel. A tutorial on acoustic cavitation. J Acoust Soc Am 101:1227–1237, 1997.

HAE Benson, JC McElnay, J Whiteman, R Harland. Lack of effect of ultrasound on the percutaneous absorption of benzydamine. J Pharm Pharmacol 38(suppl):73P, 1986.

HAE Benson, JC McElnay, R Harland. Phonophoresis of lignocaine and prilocaine from Emla cream. Int J Pharm 44:65–69, 1988.

HAE Benson, JC McElnay, R Harland. Use of ultrasound to enhance percutaneous absorption of benzydamine. Phys Ther 69(2):113–118, 1989.

HAE Benson, JC McElnay, R Harland, J Hadgraft. Influence of ultrasound on the percutaneous absorption of nicotinate esters. Pharm Res 8(2):204–209, 1991.

D Bommannan, GK Menon, H Okuyuma, PM Elias, RH Guy. Sonophoresis 2.

Examination of the mechanism(s) of ultrasound enhanced transdermal drug delivery. Pharm Res 9(8):1043–1047, 1992a.

D Bommannan, H Okuyama, P Stauffer, RH Guy. Sonophoresis 1: The use of high frequency ultrasound to enhance transdermal drug delivery. Pharm Res 9(4):559–564, 1992b.

R Brucks, M Nanavaty, D Jung, F Siegel. The effect of ultrasound on the in vitro penetration of ibuprofen through human epidermis. Pharm Res 6(8):697–701, 1989.

MH Cameron, LG Monroe. Relative transmission of ultrasound by media customarily used for phonophoresis. Phys Ther 72(2):142–148, 1992.

M Dyson, JB Pond, B Woodward, J Broadbent. The production of red blood cell stasis and endothelial damage in the blood vessels of chick embryo heated with ultrasound as a stationary wave field. Ultrasound Med Biol 1:133–148, 1974.

JY Fang, CL Fang, KC Sung, HY Chen. Effect of low frequency ultrasound on the in vitro percutaneous absorption of clobetasol 17-propionate. Int J Pharm 191:33–42, 1999.

SA Goss, LA Frizzel, F Dunn. Ultrasonic absorption and attenuation in mammalian tissues. Ultrasound Med Biol 63:181–186, 1979.

T Hikima, Y Hirai, K Tojo. Effect of ultrasound application on skin metabolism of prednisolone 21-acetate. Pharm Res 15:1680–1683, 1998.

D Hofmann, F Moll. The effect of ultrasound and in vivo penetration of benzyl nicotinate. J Control Rel 27:185–192, 1993.

ME Johnson, S Mitragotri, A Patel, D Blankschtein, R Langer. Synergistic effects of chemical enhancers and therapeutic ultrasound on transdermal drug delivery. J Pharm Sci 85:670–679, 1996.

J Kost, U Pliquett, S Mitragotri, A Yamamoto, R Langer, J Weaver. Synergistic effect of electric field and ultrasound on transdermal transport. Pharm Res 13:633–638, 1996.

J Kost, S Mitragotri, AR Gabbay, M Pishko, R Langer. Transdermal monitoring of glucose and other analytes using ultrasound. Nature Med 6:347–350, 2000.

I Lenart, D Auslander. The effect of ultrasound on diffusion through membranes. Ultrasonics 18:216–218, 1980.

L Machet, J Pinton, F Patat, B Arbeille, L Pourcelot, L Vaillant. In vitro phonophoresis of mannitol, oestradiol and hydrocortisone across human and hairless mouse skin. Int J Pharm 133:39–45, 1996.

L Machet, N Cochelin, F Patat, B Arbeille, MC Machet, G Lorette, L Vaillant. In vitro phonophoresis of mannitol, oestradiol and hydrocortisone across human and hairless mouse skin. Int J Pharm 165:169–174, 1998.

M Machluf, J Kost. Ultrasonically enhanced transdermal drug delivery. Experimental approaches to elucidate the mechanism. J Biomater Sci Polymer Edn 5:147–156, 1993.

JC McElnay, TA Kennedy, R Harland. Enhancement of the percutaneous absorption of fluocinolone using ultrasound. Br J Clin Pharmac 21:609–610P, 1985.

JC McElnay, TA Kennedy, R Harland. The influence of ultrasound on the percutaneous absorption of fluocinolone acetonide. Int J Pharm 40:105–110, 1987.

JC McElnay, MP Matthews, R Harland, DF McCafferty. The effect of ultrasound on the percutaneous absorption of lignocaine. Br J Clin Pharmac 20:421–424, 1985.

VM Meidan, AD Walmsley, WJ Irwin. Phonophoresis—is it a reality? Int J Pharm 118:129–145, 1995.

VM Meidan, D Dunnion, WJ Irwin, S Akhtar. The effect of ultrasound on the stability of oligodeoxynucleotides in vitro, Int J Pharm 152:121–125, 1997.

VM Meidan, MF Docker, AD Walmsley, WJ Irwin. Low intensity ultrasound as a probe to elucidate the relative follicular contribution to total transdermal absorption. Pharm Res 15(1):85–92, 1998a.

VM Meidan, MF Docker, AD Walmsley, WJ Irwin. Phonophoresis of hydrocortisone with enhancers: an acoustically defined model. Int J Pharm 170:157–168, 1998b.

VM Meidan, AD Walmsley, M Docker, WJ Irwin. Ultrasound-enhanced diffusion into coupling gel during phonophoresis of 5-fluorouracil. Int J Pharm 185:205–213, 1999.

S Mitragotri, J Kost. Low-frequency sonophoresis: a noninvasive method of drug delivery and diagnostics. Biotechnol Prog 16(3):488–492, 2000.

S Mitragotri, D Blankschtein, R Langer. Ultrasound-mediated transdermal protein delivery. Science 269:850–853, 1995a.

S Mitragotri, DA Edwards, D Blankschtein, R Langer. A mechanistic study of ultrasonically-enhanced transdermal drug delivery. J Pharm Sci 84(6):697–705, 1995b.

S Mitragotri, D Blankschtein, and R Langer. Transdermal drug delivery using low-frequency sonophoresis. Pharm Res 13:411–420, 1996.

S Mitragotri, M Coleman, J Kost, R Langer. Analysis of ultrasonically extracted interstitial fluid as a predictor of blood glucose levels. J Appl Physiol 89: 961–966, 2000a.

S Mitragotri, M Coleman, J Kost, R Langer. Transdermal extraction of analytes using low frequency ultrasound. Pharm Res 17:466–470, 2000b.

S Mitragotri, J Farrell, H Tang, T Terahara, J Kost, R Langer. Determination of threshold energy dose for ultrasound-induced transdermal drug transport. J Contr Rel 631:41–52, 2000c.

S Mitragotri, D Ray, J Farrell, H Tang, B Yu, J Kost, R Langer. Synergistic effect of low frequency ultrasound and sodium lauryl sulphate on transdermal transport. J Pharm Sci 89:892–900, 2000d.

S Miyazaki, Y Kohata, M Takada. Effect of ultrasound on the transdermal absorption of indomethacin-continuous mode and pulsed mode. Yakuzaigaku 52(4):264–271, 1992a.

S Miyazaki, H Mizuoka, Y Kohata, M Takada. External control of drug release and penetration. 6. Enhancing effect of ultrasound on the transdermal absorption of indomethecin from an ointment in rats. Chem Pharm Bull 40(10):2826–2830, 1992b.

WS Muir, FP Magee, JA Longo, RR Karpman, PR Finley. Comparison of ultrasonically applied vs intra-articular injected hydrocortisone levels in canine knees. Orthopaed Rev 19:351–356, 1990.

TM Murphy, J Hadgraft. A physico-chemical interpretation of phonophoresis in skin penetration enhancement. In: Prediction of Percutaneous Penetration: Methods, Measurements and Modelling. RC Scott, RH Guy, and Hadgraft, J eds. London: IBC Technical Services, 1990, pp 333–336.

M Nanavaty, R Brucks, H Grimes, FP Siegel. An ATR-FTIR approach to study the effect of ultrasound on human skin. Proceed Intern Symp Control Rel Bioact Mater 16:310–311, 1989.

RS Oziomek, DH Perrin, DA Herold, CR Craig. Effect of phonophoresis on serum salicylate levels. Med Sci Sports Med 23(4):397–401, 1991.

C Pelucio-Lopes, L Machet, L Vaillant, F Patat, M Lethieq, Y Furet, L Pourcelot, G Lorette. Phonophoresis of azidothymidine (AZT). Int J Pharm 96:249–252, 1993.

J Pinton, L Valliant, C Machet, F Patat, M Lethiecq, L Pourcellot, D Guilloteau, G Lorette. Does phonophoresis increase the ex-vivo percutaneous absorption? J Invest Dermatol 96(4):650 abstracts, 1991.

J Saxena, N Sharma, MC Makoid, UV Banakar. Ultrasonically mediated drug delivery. J Biomater Appl 7(3):277–296, 1993.

R Scheuplein. The skin as a barrier. In: The Physiology and Pathophysiology of Skin. (Jarret, A) London: ed. Academic Press, 1978:1669–1692.

JP Simonin. On the mechanisms of in vitro and in vivo phonophoresis. J Control Rel 33:125–141, 1995.

K Tachibana. Transdermal delivery of insulin to alloxan-diabetic rabbits by ultrasound exposure. Pharm Res 9:952–954, 1992.

K Tachibana, S Tachibana. Use of ultrasound to enhance the local anesthetic effect of topically applied aqueous lidocaine. Anesthesiology 78:1091–1096, 1993.

H Ueda, K Sugibayashi, Y Morimoto. Skin penetration–enhancing effects of drugs by phonophoresis. J Control Rel 37:291–297, 1995.

H Ueda, M Ogihara, K Sugibayashi, Y Morimoto. Change in the electrochemical properties of skin and the lipid packing in stratum corneum by ultrasonic irradiation. Int J Pharm 85:217–224, 1996.

AR Williams. Ultrasound: Biological Effects and Potential Hazards. (Williams, A.R.) London: Academic Press, 1983.

AR Williams. Phonophoresis: an in vivo evaluation using three topical anaesthetic preparations. Ultrasonics 28:137–140, 1990.

AR Williams, EH Rosenfield, KA Williams. Gel-sectioning technique to evaluate phonophoresis in vitro. Ultrasonics 28:132–136, 1990.

J Wu, J Chappelow, J Yang, L Wiemann. Defects generated in human stratum corneum specimens by ultrasound. Ultrasound Med Biol 24:705–710, 1998.

N Yamashita, K Tachibana, K Ogawa, N Tsujita, and A Tomita. Scanning electron microscopic evaluation of the skin surface after ultrasound exposure. Anat Rec 247:455–461, 1997.

8

Metabolic Approach to Transdermal Drug Delivery

Peter M. Elias and Kenneth R. Feingold
University of California, San Francisco,
San Francisco, California, U.S.A.

Janice Tsai
University of California, San Francisco, San Francisco, California,
U.S.A.; National Cheng Kung University, Taiwan; and Cellegy
Pharmaceuticals, Inc., Foster City, California, U.S.A.

Carl Thornfeldt
Cellegy Pharmaceuticals, Inc., Foster City, California, U.S.A.

Gopinathan Menon
Avon Products, Inc., Suffern, New York, U.S.A.

I. INTRODUCTION

The traditional view of the stratum corneum (SC) regards the outer layer of the epidermis to be relatively impermeable, highly resilient, and analogous to plastic wrap. This passive model, which holds that permeation is governed solely by the physical-chemical properties of the SC (1), still dominates strategies for transdermal drug delivery. Based on this view, the permeability characteristics of topical drugs can be predicted either from in vitro studies with isolated stratum corneum sheets, or by modeling methods that predict permeability based solely upon the permeants' molecular structures. Hence, the widespread utilization of cadaver-derived, frozen-thawed, or freshly obtained skin for penetration and drug delivery studies.

The SC also displays site-related variations in the number of cell layers, the extent of hydration, and the overall quantities and distribution of its principal lipid species, differences that again can be integrated into the characteristics of the plastic wrap model, as can the contribution of vehicle components in predicting transdermal delivery. Yet, while the plastic wrap model is able to explain the role of the SC as a component physical-chemical barrier to the external environment, it inadequately explains (a) the variable absorption of hydrophilic, amphiphilic, and hydrophobic molecules; (b) the large variations in permeability of different topographic sites, such as palms/soles vs. facial skin; or (c) individual variations in the susceptibility to toxic insults or irritants.

II. TWO-COMPARTMENT MODEL

Most of the latter observations are better explained by the unique structural and biochemical heterogeneity of the SC (2). In this so-called two-compartment model, the lipids of the SC are segregated within intercellular domains, while the lipid-depleted corneocytes act primarily as spacers, while providing protection from mechanical insults.

The permeability barrier forms coincident with the secretion of lamellar body (LB) contents at the stratum granulosum (SG)–SC interface. The delivery of LB contents to the SC interstices, a key event in epidermal terminal differentiation, ultimately results in the formation of a hydrophobic, extracellular lamellar membrane matrix (3). Variations in intercellular lipid content and membrane structure provide a structural basis for the wide variations in permeability of different human skin sites (e.g., palms/soles, leg, abdomen, face). Whereas there are some variations in SC lipid composition over different skin sites, regional differences in total lipid content alone can account for the differences in the water vapor–permeability of these sites (4). It follows then that the lipid matrix in the SC interstices comprises the permeability barrier to transdermal drug delivery, as well. Thus the SC interstices appear to subserve both the putative intercellular transport route as well as the "reservoir function" of the SC (5).

III. EPIDERMAL LIPID SYNTHESIS

Numerous studies have described the changes in lipid composition that accompany SC formation in murine, porcine, and human epidermis (6). These studies demonstrate the loss of phospholipids and the emergence of ceramides, cholesterol, and free fatty acids during barrier formation. Moreover, a further gradient in lipid transformation occurs during the final stages of terminal differentiation and barrier formation; e.g., small amounts of glucosylceramides and phospholipids persist in the lower SC (i.e., stratum compactum) but disappear from the outer SC (i.e., stratum disjunctum) (2). In addition, cholesterol sulfate content is maximal in the SG and stratum compactum, decreasing in the stratum disjunctum (2,7). Eventually, the three major classes of lipids in the SC are present as an approximately equimolar mixture of cholesterol, free fatty acids, and ceramides. These biochemical transformations are paralleled by changes in extracellular membrane structure, leading to the generation of mature lamellar membrane structures (Fig. 1). Life in a terrestrial environment requires the complete biochemical and structural sequence described above.

Epidermal lipid synthesis is both highly active and largely autonomous from systemic influences (8,9). The concept that lipid synthesis in the epidermis is relatively autonomous from circulating influences is supported by two further observations (8): (a) epidermal lipid synthesis is not altered in animals with either very high or very low serum lipids, reflecting its paucity of LDL receptors (10), an important evolutionary adaptation to survival in a terrestrial environment. (2) Only very small amounts of systemically administered lipids are incorporated into the epidermis, even when the barrier is disrupted (8,9). The synthesis of specific epidermal lipids, particularly of the three key SC lipids, i.e., ceramides, free fatty acids, and cholesterol, is discussed below.

Maturation of Stratum Corneum Lamellar Membrane Structures

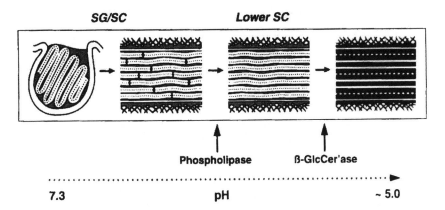

Figure 1 Diagram of major structural changes after lamellar body secretion. Proposed orchestration of these changes by selected hydrolytic enzymes. The pH gradient is also shown. (Modified from Ref. 42, with permission.)

Abundant cholesterol is synthesized in the epidermis (accounting for about 30% of total cutaneous sterologenesis) (11). Whereas most of this synthetic activity is localized to the basal layer, all of the nucleated epidermal cell layers retain the capacity to synthesize abundant cholesterol (11). While being autonomous from the influence of circulating lipids, epidermal cholesterol synthesis nevertheless is regulatable by external influences, i.e., changes in permeability barrier status, required for barrier homeostasis (cited in 8). Moreover, permeability barrier disruption upregulates LDL receptor expression, suggesting that increased transport of lipids into or within the epidermis must occur (cited in 8). Acute barrier disruption provokes an increase in cholesterol synthesis, attributable both to a prior increase in mRNA for HMGCoA reductase and other key enzymes of cholesterol synthesis (12), and to changes in the phosphorylation (activation) state of HMGCoA reductase. The changes in mRNA for the key enzymes are in turn regulated by increased expression of sterol regulatory element binding protein-2 (SREBP-2), a transcription factor for both cholesterol and fatty acid synthesis (see also below) (12).

The free fatty acids (FFA) in the SC comprise both essential and nonessential species, the latter comprising the bulk of the FFA in the intercellular lamellae (4,6). The essential FA, linoleic acid, comprises only a small proportion of the FFA in the SC, where it is largely ω-esterified to N-acyl fatty acids in Cer 1 and 4 (13,14). Nonessential free fatty acids are synthesized in all epidermal cell layers, and as with cholesterol and ceramide synthesis, FFA synthesis is regulated by barrier requirements (15). Both of the key enzymes of FFA synthesis, FA synthase and acetyl CoA carboxylase (ACC), are upregulated. Moreover, as with the enzymes of cholesterol biosynthesis, their expression appears to be transcriptionally regulated by SREBP-2 (12).

Finally, as with transport mechanisms for cholesterol, keratinocytes also possess specific mechanisms to take up and bind the essential FFA, linoleic acid, and the prostaglandin precursor, arachidonic acid, from the blood (6). Whether one or more of these transport mechanism(s), is (are) in turn, regulated by barrier requirements remain(s) to be determined.

The ceramides in mammalian SC, which account for almost 50% of lipid mass, are a surprisingly well-conserved family of seven different species, derived in large part from glucosylceramides, and presumably to a lesser extent from sphingomyelin, which are secreted from lamellar bodies into the SC interstices (2). These molecules demonstrate minor species-to-species variations in sphingoid base structure, N-acyl chain length, and α/ω hydroxylation (6,16). Moreover, the two ω-hydroxylated species, Cer 1 and Cer 4, contain an ω-esterified linoleate moiety (=acylceramide) (14). Acylceramides have two putative functions: (a) to form "molecular rivets," proposed to link adjacent intercellular membrane structures in the SC (16); and (b) to generate the deesterified ω-hydroxyceramides, which are proposed to bind covalently to amino acids in the outer portion of the cornified envelope (17). The other major ceramide species are a mixture of cera-mides with or without α-hydroxylated N-acyl groups, which interact with choles-terol and free fatty acids to form the bulk of the intercellular lamellae. Recent evidence suggests that two of these Cer species, i.e., Cer 2 and Cer 5, derive in large part from epidermal sphingomyelin rather than glucosylceramides.

Like sterologenesis, ceramide synthesis also proceeds in all epidermal cell layers (8), while glucosylceramide synthesis, catalyzed by the enzyme glucosyl-ceramide synthase, GC synthase), peaks in the SG, coincident with the formation of lamellar bodies (18). As with cholesterol and FA synthesis, ceramide synthesis is regulated by barrier requirements. Barrier-induced increases in ceramide syn-thesis are explicable by antecedent changes in both the activity and the mRNA levels for serine palmitoyl transferase (SPT), the rate-limiting enzyme for cera-mide synthesis (12). In contrast, GC synthase activity is not regulated by alter-ations in barrier function (18). How SPT activity and ceramide synthesis are regulated remains unknown. In fact, SREBP-2 does not regulate SPT, but it could regulate the pool of FA that is available for sphingoid base and N-acyl group production (12).

IV. STRUCTURAL BASIS FOR THE STRATUM CORNEUM BARRIER

A decrease in the concentrations of any of three critical lipid species affects the barrier integrity (19). Several excellent papers and reviews on the biophysical aspects of barrier lipids have been published (20,21), which have contributed to our current understanding of the mechanics of barrier function. One recent model of barrier lipid organization is the domain-mosaic model (22), which considers the SC lipid matrix as a two-phase system with a discontinuous lamellar-crystalline structure embedded in a continuous liquid-crystalline structure. If the crystalline areas are considered impermeable, while the liquid-crystalline areas remain per-meable, and if water is located primarily in the liquid-crystalline areas, the diffu-sion of both hydrophilic and hydrophobic compounds would be restricted to the liquid-crystalline domain. As a result, permeating molecules would perform a "random walk" in liquid-crystalline channels separated by two or more crystalline domains. Molecules would move up to the next layer within the SC when two such channels fuse or cross. In addition, the diffusing molecule has to follow a highly tortuous pathway in the vertical dimension, due to the extended diffusional pathlength in the lateral direction, which further limits bulk diffusion. Thus the domain-mosaic model advocates a meandering polar pathway for water and hy-drophilic molecular movement through the grain boundaries of the lipid mosaic, thus adding another level of complexity to the tortuous lipid pathway of the brick and mortar model.

The morphological basis of the aqueous pore pathway has been debated extensively, based on theoretical calculations of the molecular weight of compounds that traverse the SC and their respective activation energies (23). Based on TEM studies of tracer permeation under various permeabilization strategies, Ref. 24 identified the lacunar domains embedded within the lipid bilayers as the likely morphological basis of the pore pathway. These lacunae, to a large extent, correspond to sites of desmosome degradation. Under basal anhydrous conditions, they comprise scattered and discontinuous aqueous domains within the extracellular matrix, which can become transiently interconnected to form a continuous pore pathway under appropriate conditions of permeabilization (e.g., sonophoresis, iontophoresis, prolonged hydration). The pore pathway reverts back to its original discontinuous state once the permeabilizing stimulus is stopped or reversed (as in the case of sonophoresis) or no longer exists (as in hydration). It should be noted that such a lacunar system does not correspond to the grain boundaries of the domain-mosaic model; instead it forms an extended macrodomain within the SC (24).

V. LAMELLAR BODY SECRETION

As noted above, the unique two-compartment organization of the SC is attributable to the secretion of lamellar body (LB)–derived lipids and colocalized hydrolases at the SG–SC interface (2). Epidermal LB bodies first appear in the spinous layers as 0.2 to 0.5 micrometer ovoid structures that contain parallel arrays of lipid-enriched disks enclosed by a trilaminar membrane. In near-perfect cross sections, each lamella shows a major electron-dense band that is shared by electron-lucent material divided centrally by a minor electron dense band (25). Above the spinous layer, i.e., within the SG, LB are interspersed with electron-dense and irregular-to-stellate shaped granules that are composed of profilaggrin and loricrin. The filaggrin subunits of profilaggrin play the role of matrix molecules that aggregate and align keratin filaments. The quantities of LB reach their highest density in the uppermost SG cells, where they occupy about 20% of the cell cytosol, and, as seen from electron micrographs of oblique sections, they are highly polarized in the apical cytosol of the upper SG. Under basal conditions, the rate of LB secretion appears to be slow, yet it is sufficient to provide for barrier integrity in the absence of stress. LB secretion requires active metabolism, since neither new organelle formation nor secretion occur at 4°C. Moreover, calcium is an important regulator of LB secretion, with the epidermal calcium gradient restricting LB secretion to low maintenance levels under basal conditions (26,27). Finally, the secretion inhibitors, monensin and brefeldin A, block both LB formation and secretion after barrier disruption, leading to inhibition of barrier recovery (28).

Following acute barrier disruption, much of the preformed pool of LB in the outermost SG cell is quickly secreted (29,30). The outermost SG cell contains a number of structural specializations that allow it to secrete lamellar bodies rapidly through its highly keratinized matrix (30). These include a widely disbursed trans-Golgi network in the apical cytosol, which results in oganelle formation near the cellular periphery as well as deep invaginations of the apical SG plasma membrane, which allow LB fusion/secretion from deep within the cytosol (30).

VI. EXTRACELLULAR PROCESSING (ECP)

During the final stages of epidermal differentiation, a sequence of membrane transitions occurs within the SC extracellular domains (Fig. 1). Extrusion of LB contents at the SG–SC interface is followed sequentially by unfurling, elongation, and processing of secreted contents into mature lamellar membrane unit structures (2,29). Since occlusion inhibits each of these steps, the entire LB secretory response therefore is linked to barrier homeostasis. The lamellar basic unit bilayer pattern typically consists of a series of six electron-lucent lamellae alternating with five electron-dense lamellae (Fig. 2). In skin regions where the SC displays a high lipid content, double and triple basic units occur frequently. These basic unit structures persist throughout the SC, until contamination with sebum, other topical contaminants, and/or friction results in loss of the cohesive bilayer arrays. As noted above, marked alterations in lipid composition occur, including depletion of glucosylceramides and phospholipids, with accumulation of ceramides and FFA in the SC. These alternatives can be attributed to the fact that LB deliver not only glucosylceramides, phospholipids, and cholesterol but also a family of hydrolytic enzymes to the SC interstices (2). Abundant evidence links these hydrolases to the extracellular processing (ECP) steps that are critical for barrier homeostasis. For example, the hydrolases that have been studied (glycosidase, phospholipase, steroid sulfatase, and neutral/acidic lipases) are localized specifically within the outer epidermis and SC membrane domains (2). Moreover, in some cases, the activities and mRNA levels of these enzymes are regulated by barrier requirements; and in some cases (e.g., 31) they have been shown to be required for normal barrier homeostasis.

The most compelling direct evidence for the central role of ECP in barrier formation comes from studies on glucosylceramide-to-ceramide processing. As with other lipid hydrolases, β-glucocerebrosidase (β-GlcCer'ase) is concentrated in the outer epidermis, the highest levels being in the SC (31), where it is localized to membrane domains. In contrast, endogenous β-glycosidase activity is low in the SC of mucosal epithelia (32), which is up to an order of magnitude more permeable than epidermis. Moreover, glycosylceramides [type(s) unspecified]

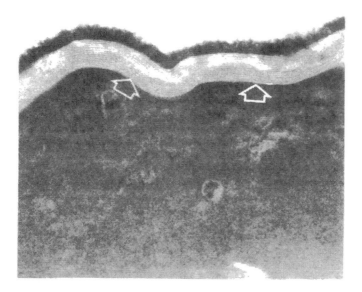

Figure 2 Ultrastructure of stratum corneum lamellar bilayers (arrows) in normal human skin, visualized following ruthenium tetroxide post-fixation. ×145,000.

predominate over ceramides; untransformed LB contents persist into the outer SC, i.e., mature lamellar membrane structures do not form in mucosal epithelia. Recently, these correlative observations have been supported by direct evidence for the role of β-GlcCer'ase in the ECP of glucosylceramides-to-ceramides: (a) applications of specific conduritol-type inhibitors of β-GlcCer'ase both delay barrier recovery after acute perturbations and produce a progressive abnormality in barrier function when applied to intact skin (33). (b) In a transgenic murine model of Gaucher disease (GD), produced by targeted disruption of the β-GlcCer'ase gene, homozygous animals are born with an ichthyosiform dermatosis and a severe barrier abnormality (34). (c) In the severe type 2 neuronopathic form of GD, infants present with a similar clinical phenotype, including an ichthyosiform erythroderma (35). In all three situations (inhibitor, transgenic murine, type 2 GD), the functional barrier deficit is accompanied by an accumulation of glucosylceramides and the persistence of immature LB-derived membrane structures within the SC interstices. Together, these studies point to a critical role for glucosylceramide-to-ceramide processing in barrier homeostasis.

Phospholipids, which are integral components of all cell membranes, harbor within their structures bioactive moieties, such as arachidonic acid at the sn-2 position, whose generation is rate limiting for the production of eicosanoids and other lipid mediators. Moreover, the FFA and lysophospholipids resulting from phospholipase A_2 (PLA_2) action can act as second messengers themselves or serve as structural intermediates in membrane remodeling and phospholipid degradation. While the types of PLA_2 present in mammalian epidermis have not been extensively studied, recent studies have shown that both the $cPLA_2$ (85 kD cytosolic) and type 2 $sPLA_2$ (14 kD secretory) are constitutively expressed in epidermis under basal conditions (36,37). Since we have shown that bromphenacyl bromide (BPB) and MJ-33 (type 1 $sPLA_2$ inhibitors), but not MJ-45 (a type 2 $sPLA_2$ inhibitor) modulate barrier function, this suggests a role for the type 1 rather than the type 2 isoenzyme in barrier function (36,37). That epidermal barrier function appears to require an $sPLA_2$ with characteristics of the type 1 form, fits well with its putative role in barrier homeostasis.

The lamellar body also delivers considerable sphingomyelin (SM) to the SC interstices, and an additional pool of SM becomes available potentially with the disappearance of the plasma membrane. Both sources of SM are possible precursors of either bulk Cer or specific Cer species in the SC interstices. Recent studies have shown that processing of SM by acid sphingomyelinase, an enzyme present in abundance in LB, is required for normal barrier homeostasis. Moreover, based upon an extensive comparison of Cer compositions in human and mouse epidermis, SM accounts for much, if not all, of the Cer 2 and Cer 5 in the SC.

Just as with glucosylceramides and phospholipids, cholesterol sulfate content increases with epidermal differentiation and then decreases quantitatively during passage from the inner to the outer SC (38). Both cholesterol sulfate and steroid sulfatase are concentrated in SC membrane domains, but not in LB (38). That cholesterol sulfate plays a critical role in desquamation is demonstrated definitively by the accumulation of cholesterol sulfate in SC membranes in recessive X-linked ichthyosis (RXLI), and by the ability of topical cholesterol sulfate to induce hyperkeratosis (39).

Recent studies have demonstrated a barrier abnormality in RXLI (40). Activity of steroid sulfatase, a close relative of arylsulfatase C, is absent in RXLI (39). This microsomal enzyme, which is expressed in high levels in the outer epidermis of mammals (38), hydrolyzes the sulfate moiety from the 3-β-hydroxy group on a variety of sterols and steroids. Cholesterol sulfate, in addition to its apparent roles in desquamation and differentiation, also plays an important negative role in barrier homeostasis. Recent studies have shown accumulation of cholesterol sulfate detrimental to the barrier, and that cholesterol sulfate degradation does not contribute significantly to the cholesterol pool required for the barrier (40). Thus the role of steroid sulfatase as a processing enzyme is primarily to prevent an accumulation of cholesterol sulfate in the SC interstices.

VII. ORCHESTRATION OF EXTRACELLULAR PROCESSING BY THE STRATUM CORNEUM pH GRADIENT

That the SC displays an acidic external pH ("acid mantle") is well documented, and such acidic conditions are considered crucial for the resistance to microbial invasion. However, the origin of the acid mantle is not known—passive mechanisms, including net catabolic processes, sebaceous gland-derived FFA, microbial metabolism, and apical proton flux due to the high electrical resistance of the SC have been proposed. Alternatively, protons could also be generated actively, e.g., by ion pumps inserted into the plasma membrane. If the limiting membrane of the LB contained such pumps, as is suggested by inhibitor studies (41), then acidification of the extracellular space (ECS) could begin with insertion of such pumps coincident with LB secretion. Ongoing proton secretion at the SG–SC interface, followed by one or more passive mechanisms, could explain the pH gradient across the SC. The concept that acidification is required for sequential ECP is supported by the observation that barrier recovery is delayed when acutely perturbed skin is immersed in neutral pH buffers (42) (Fig. 3). Simultaneously, the activity of at least one ECP enzyme with an acid pH optimum, β-glucocerebrosidase, is inhibited. Thus in addition to its acknowledged antimicrobial function, the pH gradient is important for permeability barrier homeostasis, presumably through the sequential regulation of ECP enzymes. Both β-glucocerebrosidase and sphingomyelinase possess acidic pH options, while secretory phospholipase A_2 and steroid sulfatase are most active at a neutral pH. One could expect sequential processing then to occur first, for phospholipids and cholesterol sulfate, followed by glucosylceramides and SM (Fig. 2).

By a variety of independent methods, we and others have demonstrated a Ca^{2+} gradient in the epidermis, with the highest levels in the intercellular spaces and cytosol of the SG, followed by a rapid decline in Ca^{2+} from the SC (27). Whereas the epidermis displays a similar K^+ gradient, different gradients exist for other ions, including Na^+ and Cl^- (43). Both acute and chronic barrier disruption dramatically alter the Ca^{2+} gradient, displacing Ca^{2+} outward through the SC interstices (43,27). Finally, exposure to high Ca^{2+} (and K^+) delays barrier recovery following acute perturbations, in a manner that is independent of pH (Fig. 3) and reversible by L-type Ca^{2+} channel inhibitor (26).

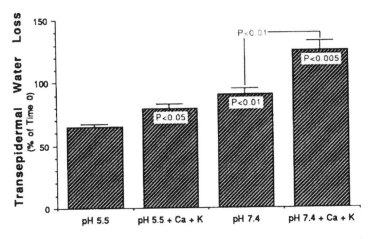

Figure 3 The effects of pH and ions on barrier recovery are unrelated and additive. (Reproduced from Ref. 42, with permission.)

VIII. METABOLIC BASIS FOR ENHANCEMENT OF TRANSDERMAL DRUG DELIVERY

Several approaches and strategies to overcome the barrier have been devised to enhance transdermal drug delivery, with varying degrees of success. Briefly, these have been classified as physical (mechanical), chemical, and metabolic approaches. Combinations of these strategies can also be employed either to further increase efficacy (44,45) or to extend the time available for transdermal delivery. These techniques vary from straightforward approaches (e.g., occlusion, tape stripping) to highly technical procedures that use sophisticated instrumentation and miniaturization, e.g., iontophoresis, electroporation. In this review, we will focus solely on metabolic approaches.

The concept of a biochemical approach for enhancing cutaneous permeability derives from the dynamic nature of the response to acute barrier disruption. The repair response, regardless of the manner of primary disruption, occurs quickly (over hours). A number of cellular and metabolic responses are required (see above), and if one or more are frustrated, the repair process is delayed (Tables 1 and 2); thus the enhanced window of opportunity for transdermal drug delivery. Through the use of pharmacological agents aimed at inhibiting epidermal lipid synthesis, LB secretion, ECP, or alteration of lamellar membrane composition, the repair response can be modulated. Regardless of the biochemical target, all of these methods alter lamellar bilayer structure by modifying the critical molar

Table 1   Implications for Transdermal Drug Delivery

1. Regardless of effectiveness of enhancer, metabolic response to barrier perturbations will result in rapid normalization of barrier function.
2. Strategies that interfere with metabolic response may prolong patency of barrier.
3. In order to work, such strategies require access to epidermal metabolic apparatus (concept of 1° and 2° enhancers).
4. Some 2° enhancers are themselves potential candidates for transdermal drug delivery.

ratio of the three key SC lipids. Enzyme inhibition is accompanied by morphological and physiological changes in SC lamellar bilayers (Fig. 4), leading to increased transepidermal water loss (e.g., 15,18,46–48) as well as enhanced permeation of drugs through the SC (44).

The first pharmacological study to support this particular concept came from experiments in adult hairless mice in which topical HMGCoA reductase inhibitors, such as lovastatin and fluvastatin, caused both a delay in the kinetics of barrier recovery, and a barrier defect following repeated applications to intact skin (cited in 15,18,28,46,48). Since the ability of the inhibitors to alter barrier homeostasis could be reversed by coapplications of either mevalonate (the immediate product of HMGCoA reductase) or cholesterol (a distal product), the inhibitor effect could not be ascribed to nonspecific toxicity. Likewise, application of specific pharmacologic inhibitors of ACC (15), SPT (47), and GC synthase (18), key enzymes of FA and ceramide synthesis, also provoked a delay in barrier recovery, as measured by increased TEWL rates. Yet while the applications of some inhibitors are additive in their capacity to alter barrier recovery, coapplic-

Table 2   Examples of Lipid Synthetic and Processing Inhibitors That Modulate Barrier Homeostasis

Required lipid	Enzyme targets	Example
Cholesterol	HMGCoA Reductase	Lovastatin, fluvastatin
Free fatty acids	Acetyl CoA carboxylase	TOFA
	Secretory phospholipase A$_2$	MJ33, BPB
Ceramides/ glucosylceramides	Serine palmitoyl transferase	β-choloralanine
	Glucosylceramide synthase	Morpholino agents (e.g., P4)
	β-Glucocerebrosidase	Conduritols

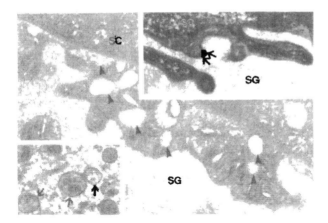

Figure 4 Effects of applications of HMGCoA reductase inhibitors on lamellar body contents (A: lovastatin; B: fluvastatin; C: vehicle). (A) Note that most organelle contents are empty (solid arrow). (B) Note focal abnormalities at stratum granulosum (SG)–stratum corneum (SC) interface (open arrows). (C) Replete, compact secreted contents in vehicle-treated control. B and C, ruthenium tetroxide postfixation. (A) ×85,000; (B) ×125,000; (C) ×105,000.

ations of other inhibitors can paradoxically normalize the kinetics of barrier recovery (49). It should be noted that these studies were designed to determine which of the three key lipids is (are) required for permeability barrier homeostasis (8). Yet these inhibitors also cause the "window to remain open longer," and/or they "open the window de novo," providing an opportunity for enhanced drug delivery (Table 1). Thus all of the pharmacological "knockout" studies support the concept that interference with the biosynthesis, secretion, or ECP of epidermal lipids can lead not only to variable increases in TEWL but also to enhanced bioavailability of topical drugs for transdermal delivery.

The effects of these inhibitors on skin barrier function are explained in part by ultrastructural studies, which demonstrate that each of the pharmacologic inhibitors produces abnormalities of both LB contents (Fig. 4) and SC extracellular lamellae. In addition to abnormal extracellular lamellae, these agents often produce the formation of separate lamellar and nonlamellar domains within the SC interstices. The basis for such domain separation presumably relates to changes in the critical mole ratio, with deletion or excess of any one of the three key lipids, i.e., a portion of the excess species presumably no longer remains organized in a lamellar phase. For example, a 50% reduction in cholesterol would result in an excess of both ceramides and free fatty acids, with a portion of this excess forming a putative nonlamellar phase. The result of phase separation is more permeable SC interstices due not only to the deletion of a key hydrophobic lipid but also to the creation of additional interphase penetration pathways, distinct from the primary lamellar route.

Whereas vehicles, hydration, and conventional chemical enhancers also can produce domain separation, they do not significantly modify the structure or spacing of the extracellular lamellae themselves. Instead, the primary effect of standard chemical enhancers, vehicles, and physical methods appears to be expansion of the pore pathway, as well as fluid–lipid domains in the SC interstices (24). Hence they display effects on barrier function of less than one order of magnitude. In contrast, and in summary, strategies that interfere with the synthesis, delivery, activation, or assembly/disassembly of the extracellular lamellar membrane interfere with permeability barrier homeostasis.

Finally, these biochemical approaches also can be viewed vectorially (Table 3), i.e., as operative within different layers of the epidermis. For example, most lipid synthesis occurs within the basal layer, while lamellar body formation, acidification, and secretion occur in suprabasal nucleated cell layers, while ECP,

Table 3 Mechanistic and Vectorial Classification of Various Biochemical Enhancers

Localization	Types
Basal layer	*Lipid synthesis inhibitors*
	HMGCoA reductase
	Acetyl CoA carboxylase
	Serine palmitoyl transferase
	Glucosylceramide synthase
	Lipid secretion inhibitors
	Organellogenesis
	Golgi processing
	Extracellular acidification inhibitors
	Protonophores
	Proton pumps
	Extracellular processing
	β-Glucocerebrosidase
	Secretory phospholipase A_2
	Acid sphingomyelinase
	Ion pump inhibitors, e.g., bafilamycin
Stratum corneum	*Membrane biochemical alterations*
	Lipid analogues[a]
	Other lipid hydrolases, e.g., acid lipase
	Complex lipids[b]

[a] Nonphysiological, e.g., epicholesterol, transvaccenic acid.
[b] Some (e.g., cholesterol esters), but not all, are effective.

membrane assembly, and membrane turnover in turn occur within the SC interstices. Therefore, in theory, strategies also could be developed to exploit the selective localization of the steps leading to the generation of SC extracellular lamellae in order potentially to enhance transdermal drug delivery.

IX. ASSESSMENT OF EFFICACY OF BIOCHEMICAL APPROACHES (TABLE 4)

To date, the effectiveness of these biochemical approaches for transdermal drug delivery has been assessed mainly in adult hairless mouse epidermis (44,45). In our initial studies, caffeine and lidocaine were used as model permeants to assess whether their penetration characteristics parallel changes in TEWL measurements. The biochemical approaches here consisted of the topical application of either drug plus a cholesterol and/or fatty acid lipid synthesis inhibitor in two different, conventional enhancer/vehicle systems, i.e., either dimethylsulfoxide or propylene glycol:ethanol (7:3 vol), followed by assessment of both TEWL and drug delivery. Results from these studies showed that these biochemical modulators enhance total body absorption of lidocaine and caffeine across living skin at levels above those achieved with either of the two conventional enhancer-vehicle systems (44), i.e., the net delivery of both drugs increased several times over either drug in the conventional enhancer system (Fig. 5). Moreover, changes in TEWL correlated linearly with transdermal delivery of both drugs (Fig. 6). This study was the first to show that biochemical enhancers can increase transdermal drug delivery in a widely employed animal model. Additional work is needed to explore whether TEWL serves as a universal, accurate, and reproducible predictor for transdermal delivery of drugs, with a broad range of physical–chemical properties. Although important regulatory issues remain (Table 4), it is important to note that a number of the most effective biochemical enhancers are already

Table 4 Potential Issues with Biochemical/metabolic Enhancers

1. Which classes of drugs are enhanced, and which are not?
2. What are the size (mass), charge, solubility, ionic limits for various drug classes?
3. Will these methods be additive, synergistic among themselves, and/or useful with physical methods, i.e., sonophoresis, iontophoresis?
4. To what extent is irritation a consequence of enhanced permeability, and what strategies can be coemployed to decrease the likelihood of irritation?
5. What are the regulatory issues associated with use of this class of agents?
   a. Drug plus a drug enhancer?
   b. Creating a larger and prolonged "window" through the skin?

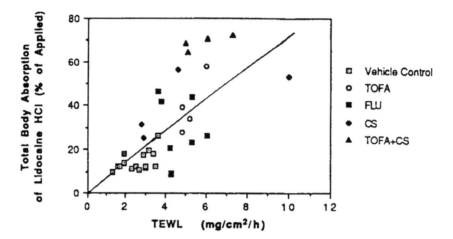

Figure 5 Correlation of lidocaine absorption, from 4 to 6 hours after lipid synthesis inhibitors were applied to acetone-treated hairless mice for 4 hours, with the changes in TEWL ($r = 0.698$). (Reproduced from Ref. 44, with permission.)

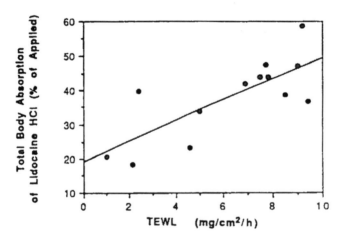

Figure 6 Correlation of lidocaine HCl absorption in consecutive two hour periods following barrier disruption by acetone treatment, with the changes in TEWL ($r = 0.768$). (Reproduced from Ref. 44, with permission.)

Table 5 Some Biochemical Enhancers That Are Approved Agents

Lipid synthesis inhibitors
HMGCoA reductase inhibitors (lovastatin)
Fatty acid synthesis inhibitors (TOFA[a])
Lipid secretion inhibitors
Protonophores (monensin[veterinary], chloroquine)
Extracellular processing
Sphingomyelinase inhibitor (desipramine)
Ion pump/channel inhibitors
L-type calcium channel (verapamil, nifedipine)

[a] TOFA = 5-(tetradecyloxy)-2-furancarboxylic acid.

approved drugs (Table 5), and that some could be useful as therapeutic products for skin disease themselves. In summary, these preliminary studies suggest that great potential exists for the biochemical enhancer approach to increase transdermal drug delivery.

REFERENCES

1. RJ Scheuplein, IH Blank. Permeability of the skin. Physiol Rev 51:702–747, 1971.
2. PM Elias, GK Menon. Structural and lipid biochemical correlates of the epidermal permeability barrier. Adv Lipid Res 24:1–26, 1991.
3. PM Elias. Epidermal lipids, barrier function, and desquamation. J Invest Dermatol 80:44s–49s, 1983.
4. MA Lampe, AL Burlingame, J Whitney, ML Williams, BE Brown, E Roitman, PM Elias. Human stratum corneum lipids: characterization and regional variations. J Lipid Res 24:120–130, 1983.
5. MK Nemanic, PM Elias. In situ precipitation: a novel cytochemical technique for visualization of permeability pathways in mammalian stratum corneum. J Histochem Cytochem 28:573–578, 1980.
6. NY Schurer, PM Elias. The biochemistry and function of stratum corneum lipids. Adv Lipid Res 24:27–56, 1991.
7. AW Ramasinghe, PW Wertz, DT Downing, IC MacKenzie. Lipid composition of cohesive and desquamated corneocytes from mouse ears. J Invest Dermatol 86:187–190, 1986.
8. KR Feingold. The regulation and role of epidermal lipid synthesis. Adv Lipid Res 24:57–82, 1991.

9.   KR Feingold. Permeability barrier homeostasis: its biochemical basis and regulation. Cosm and Toiletr 112:49–59, 1997.

10. M Ponec, L Hawkes, J Kampanaar, BJ Vermeer. Cultured human skin fibroblasts and keratinocytes: differences in the regulation of cholesterol synthesis. J Invest Dermatol 81:125–130, 1983.

11. KR Feingold, BE Brown, SR Lear, AH Moser, PM Elias. Localization of de novo sterologenesis in mammalian skin. J Invest Dermatol 81:365–369, 1983.

12. IR Harris, AM Farrell, WM Holleran, S Jackson, C Grunfeld, PM Elias, KR Feingold. Parallel regulation of sterol regulatory element binding protein-2 and the enzymes of cholesterol and fatty acid synthesis but not ceramide synthesis in cultured human keratinocytes and murine epidermis. J Lipid Res 39:412–422, 1998.

13. PN Wertz, DL Downing. Ceramides of pig epidermis: structure determination. J Lipid Res 24:759–765, 1983.

14. PW Wertz, DT Downing, RK Feinkel, TN Traczyk. Sphingolipids of the stratum corneum and lamellar granules of fetal rat epidermis. J Invest Dermatol 83:193–195, 1984.

15. M Mao-Qiang, PM Elias, KR Feingold. Fatty acids are required for epidermal permeability barrier function. J Clin Invest 92:791–798, 1993.

16. DT Downing. Lipid and protein structures in the permeability barrier of mammmalian epidermis. J Lipid Res 33:301–313, 1992.

17. DC Swartzendruber, PW Wertz, DJ Kitko, KC Madison, DT Downing. Evidence that the corneocyte has a chemically bound lipid envelope. J Invest Dermatol 88:709–713, 1987.

18. CSN Chujor, WM Holleran, KR Feingold, PM Elias. Glucosylceramide synthase activity in murine epidermis: quantitation, localization, regulation, and requirement for barrier homeostasis. J Lipid Res 39:277–288, 1998.

19. GK Menon, R Ghadially. Morphology of lipid alterations in the epidermis: a review. Microsc Res Tech 37:180–192, 1997.

20. JA Bouwstra, GS Gooris, K Cheng, A Weerheim, W Bras, M Ponec. Phase behavior of isolated skin lipids. J Lipid Res 37:999–1011, 1996.

21. N Kitson, J Thewalt, M Lafleur, M Bloom. A model membrane approach to the epidermal permeability barrier. Biochemistry 33:6707–6715, 1994.

22. B Forslind. A domain mosaic model of the skin barrier. Acta Dermatovenereol 74:1–6, 1994.

23. H Schaefer, TE Redelmeier. Skin Barrier. Principles of Percutaneous Absorption. Basel: Karger, 1996.

24. GK Menon, PM Elias. Morphologic basis for a pore-pathway in mammalian stratum corneum. Skin Pharmacol 10:235–246, 1997.

25. L Landmann. The epidermal permeability barrier. Anat Embryol 178:1–13, 1988.

26. SH Lee, PM Elias, E Proksch, GK Menon, M Mao-Qiang, KR Feingold. Calcium and potassium are important regulators of barrier homeostasis in murine epidermis. J Clin Invest 89:530–538, 1992.

27. GK Menon, PM Elias, KR Feingold. Integrity of the permeability barrier is crucial for maintenance of the epidermal calcium gradient. Br J Dermatol 130:139–147, 1994.

28. M-Q Man, BE Brown, S Wu-Pong, KR Feingold, PM Elias. Exogenous nonphysiologic vs. physiologic lipids: divergent mechanisms for correction of permeability barrier dysfunction. Arch Dermatol 131:809–816, 1995.

29. GK Menon, KR Feingold, PM Elias. The lamellar body secretory response to barrier disruption. J Invest Dermatol 98:279–289, 1992.

30. PM Elias, C Cullander, T Mauro, U Rassner, L Komuves, B Brown, GK Menon. The secretory granular cell: the outermost granular cell as a specialized secretory cell. J Invest Dermatol Symp Proc 3:87–100, 1998.

31. WM Holleran, Y Takagi, G Imokawa, S Jackson, JM Lee, PM Elias. β-Glucocerebrosidase activity in murine epidermis: characterization and localization in relationship to differentiation. J Lipid Res 33:1201–1209, 1992.

32. F Chang, PW Wertz, SM Squier. Comparison of glycosidase activities in epidermis, palatal epithelium, and buccal epithelium. Comp Biochem Biophys [B] 150:137–139, 1991.

33. WM Holleran, Y Takagi, KR Feingold, GK Menon, G Legler, PM Elias. Processing of epidermal glucosylceramides is required for optimal mammalian permeability barrier function. J Clin Invest 91:1656–1664, 1993.

34. WM Holleran, E Sidransky, GK Menon, M Fartasch, J-U Grundmann, EI Ginns, PM Elias. Consequences of β-glucocerebrosidase deficiency in epidermis: ultrastructure and permeability barrier alterations in Gaucher disease. J Clin Invest 93:1756–1764, 1994.

35. E Sidransky, M Fartasch, RE Lee, LA Metlay, S Abella, A Zimran, W Gao, PM Elias, EI Ginns, WM Holleran. Epidermal abnormalities may distinguish Type 2 from Type 1 and Type 3 of Gaucher disease. Pediatr Res 39:134–141, 1996.

36. M Mao-Qiang, KR Feingold, M Jain, PM Elias. Extracellular processing of phospholipids is required for permeability barrier homeostasis. J Lipid Res 36:1925–1935, 1995.

37. M Mao-Qiang, M Jain, KR Feingold, PM Elias. Secretory phospholipase A$_2$ activity is required for permeability barrier homeostasis. J Invest Dermatol 106:57–63, 1996.

38. PM Elias, ML Williams, ME Maloney, JA Bonifas, BE Brown, S Grayson, EH Epstein Jr. Stratum corneum lipids in disorders of cornification: steroid sulfatase and cholesterol sulfate in normal desquamation and the pathogenesis of recessive X-linked ichthyosis. J Clin Invest 74:1414–1421, 1984.

39. ML Williams. Lipids in normal and pathological desquamation. Adv Lipid Res 24:211–262, 1991.

40. E Zettersten, M Mao-Qiang, J Sato, M Denda, A Farrell, R Ghadially, ML Williams, KR Feingold, PM Elias. Recessive X-linked ichthyosis: role of cholesterol-sulfate accumulation in the barrier abnormality. J. Invest. Dermatol 111:784–790, 1998.

41. SJ Chapman, A Walsh. Membrane-coating granules are acidic organelles which possess proton pumps. J Invest Dermatol 93:466–470, 1989.

42. T Mauro, S Greyson, WN Gao, M-Q Man, E Kriehuber, M Behne, KR Feingold, PM Elias. Barrier recovery is impeded at neutral pH, independent of ionic effects: implications for extracellular lipid processing. Arch Derm Res: 290:211–215, 1998.

43. M Mao-Qiang, T Mauro, G Bench, R Warren, PM Elias, KR Feingold. Calcium and potassium inhibit barrier recovery after disruption, independent of the type of insult in hairless mice. Exp Dermatol 6:36–40, 1997.

44. J-C Tsai, RH Guy, CR Thornfeldt, KR Feingold, PM Elias. Metabolic approaches to enhance transdermal drug delivery. I. Effect of lipid synthesis inhibitors. J Pharm Sci 85:643–648, 1996.

45. EH Choi, SH Lee, SK Ahn, SM Hwang. The pretreatment effect of chemical skin penetration enhancers in transdermal drug delivery using iontophoresis. Skin Pharmacol Appl Skin Physiol 12:326–335, 1999.

46. KR Feingold, M Mao-Qiang, E Proksch, GK Menon, B Brown, PM Elias. The lovastatin-treated rodent: a new model of barrier disruption and epidermal hyperplasia. J Invest Dermatol 96:201–209, 1991.

47. WM Holleran, M Mao-Qiang, WN Gao, GK Menon, PM Elias, KR Feingold. Sphingolipids are required for mammalian barrier function: inhibition of sphingolipid synthesis delays barrier recovery after acute perturbation. J Clin Invest 88:1338–1345, 1991.

48. GK Menon, KR Feingold, M-Q Man, M Schaude, PM Elias. Structural basis for the barrier abnormality following inhibition of HMG CoA reductase in murine epidermis. J Invest Dermatol 98:209–219, 1992.

49. M Mao-Qiang, KR Feingold, PM Elias. Inhibition of cholesterol and sphingolipid synthesis causes paradoxical effects on permeability barrier homeostasis. J Invest Dermatol 101:185–190, 1993.

9

The Application of Supersaturated Systems to Percutaneous Drug Delivery

Mark Pellett
Wyeth Consumer Healthcare, Havant, England

S. Lakshmi Raghavan and Jonathan Hadgraft
NRI, University of Greenwich, Chatham, England

Adrian Davis
GlaxoSmithKline Consumer Healthcare, Weybridge, England

I. INTRODUCTION

As discussed in previous chapters, the stratum corneum forms the main barrier to percutaneous absorption, and a number of methods are available to enhance the diffusion of compounds into and across it. Many of these techniques, for example iontophoresis or chemical penetration enhancers, can give rise to local intolerance either directly or through increased local delivery of other vehicle excipients. Supersaturation provides an alternative mechanism of penetration enhancement that is specific to the drug and without the need to perturb the skin barrier.

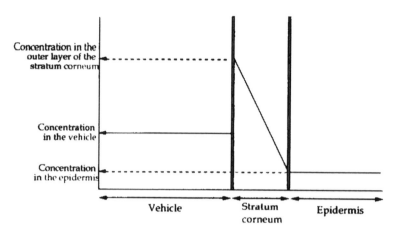

Figure 1   Schematic profile of steady-state diffusion across the skin.

The concept and an understanding of supersaturated systems evolved from crystallization theory, in which a solution must first of all be supersaturated in order for crystals to form. The potential benefit to topical and transdermal delivery was first recognized by Higuchi (1960). He considered a system in which the vehicle did not alter the barrier properties of the skin, and the rate-controlling step to percutaneous absorption was in the outer layers of the stratum corneum (Fig. 1). Consequently, from the definition of the partition coefficient, and when a vehicle is saturated with the drug, the stratum corneum must also be saturated. Similarly, when a supersaturated solution is placed on the surface of the skin, the concentration of the drug in the stratum corneum will become supersaturated. When this happens, penetration enhancement proportional to the degree of supersaturation will occur.

This can also be explained in terms of chemical potential where flux through a membrane is driven by the chemical potential gradient. In supersaturated systems the chemical potential is substantially higher in the vehicle than that for saturated solutions, depending upon the degree of supersaturation. Equilibrium would then require a similar rise in the chemical potential in the outer layers of the stratum corneum at least transiently. If the skin lipids act as antinucleant agents, and permit the permeant to remain in a higher activity state, supersaturation will also occur in the stratum corneum.

The theory that supersaturated systems could offer greater penetration was tested by Coldman et al. (1969), who introduced the steroid fluocinolone acetonide and its acetate ester onto human skin in a volatile:nonvolatile cosolvent vehicle (isopropanol: isopropyl myristate or isopropanol:propylene glycol). By

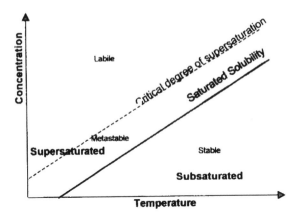

Figure 2 Diagram showing the critical degree of supersaturation and the different stability states of supersaturated systems.

comparing results of occluded formulations with unoccluded formulations, it was shown that the unoccluded formulations gave an enhanced penetration. It was concluded that, as the volatile component evaporated, the solution became more concentrated, and eventually a state of supersaturation was reached, giving rise to an enhanced penetration.

The importance of choosing the correct and optimum vehicle was highlighted by Poulsen (1968). In a series of experiments investigating the release of fluocinolone acetonide and its acetate ester from gelled propylene glycol: water vehicles into isopropyl myristate, release was dependent on the solubility of the steroid in the vehicle, its concentration in the vehicle, and the partition coefficient between the vehicle and the isopropyl myristate.

Supersaturated systems are defined as systems in which the concentration in solution exceeds that of a saturated solution. They are thermodynamically unstable, and in many cases crystals form spontaneously (labile state). In a few examples, the compounds remain in solution in a so-called metastable state, but if foreign particles are introduced, or if the system is subjected to external forces such as ultrasound, nucleation and crystallization occur. The boundary between the metastable and labile states is called the critical degree of supersaturation (Fig. 2).

The application of supersaturated systems to topical and transdermal drug delivery has been used for many years, but control and the intentional creation of stable supersaturated states has been the subject of more intense research in the last decade.

II. MATHEMATICAL MODELING

As explained in previous chapters, Fick's first law of diffusion describes the steady-state transport of a compound across a membrane:

$$J = \frac{DAK}{h} (C_v - C_r)$$

where J is the flux, A is the area of diffusion, K is the membrane–vehicle partition coefficient, D is the diffusion coefficient, C_v is the drug concentration in the vehicle, C_r is the concentration in the receptor phase, and h is the diffusional pathlength. C_r is usually very small, and under sink conditions, $(C_v - C_r)$ is generally approximated to C_v.

The flux of any given compound across a membrane from a saturated solution, irrespective of its concentration, is constant, provided that there are no interactions between the membrane and the components of the formulation. This was demonstrated by Twist and Zatz (1986), who showed that the flux of methyl paraben from saturated solutions of different solvents was constant. Therefore, under normal circumstances, the flux of a drug is limited by its saturated solubility, which in turn can also limit its bioavailability. The preparation of stable supersaturated systems not only circumvents some of the safety issues that are associated with other mechanisms of enhancement but can also lead to increased bioavailability.

III. PREPARATION OF SUPERSATURATED SOLUTIONS AND THEIR APPLICATION TO PERCUTANEOUS ABSORPTION

A number of methods can be used to prepare supersaturated systems (Khamskii, 1969):

1. Heating and then cooling
2. The removal of a solvent, i.e., evaporation
3. The reaction of two or more solutes to produce a new compound that is less soluble in solution than the original starting solutes
4. Addition of a substance to a solution that reduces the solubility of the solute, for example biphasic mixing

The following review of the application of supersaturated systems to topical and transdermal delivery is divided into the sections used to prepare supersaturated systems.

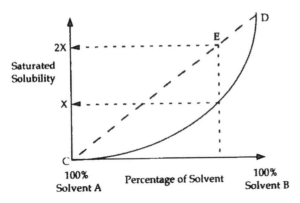

Figure 3 Graph showing the solubility of a drug in a binary cosolvent system and the method used to obtain a supersaturated solution.

A. Binary Mixtures

This technique has received a great deal of attention in the last decade and unlike other techniques, it is easier to control the desired degree of saturation, and manufacturing and packaging issues are more easily resolved. This technique involves constructing a curve for the saturated solubility of the drug in a binary cosolvent system where the drug is more soluble in one of the solvents than the other (Fig. 3). By preparing a saturated solution of the drug in solvent B and diluting with solvent A, a point, E, is obtained along the line CD. This is a supersaturated solution, and its degree of saturation is calculated by dividing the amount of drug in solution by its saturated solubility in the same cosolvent mixture. In this example, the solution E has two degrees of saturation. The problem of crystallization is usually overcome by adding an antinucleant polymer to solvent A so that crystallization is either inhibited or retarded.

Davis and Hadgraft (1991a, 1991b) investigated the stability of eight-times supersaturated solutions of hydrocortisone acetate and the flux from a range of subsaturated, saturated, and supersaturated solutions across a silicone membrane soaked in isopropyl myristate. Supersaturated solutions were prepared by the biphasic mixing of water, a desolubilizer, to a solution of the drug in propylene glycol, a solubilizer; the antinucleant polymers polyacrylate, PVP, and HPMC were investigated for their ability to maintain eight-times supersaturated solutions of the drug. HPMC was demonstrated to be the most effective antinucleant polymer, as it inhibited crystal growth for a period of 1 month, whereas crystallization started within the first few hours with the other two polymers. Furthermore, a study performed on 16 volunteers showed that a 0.02% supersaturated gel of

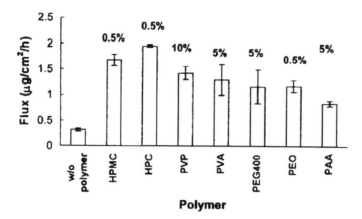

Figure 4 The flux of hydrocortisone acetate (from a 4.8 × saturated solution) across a silicone membrane. The figure shows the amount of polymer required to produce the maximum flux; w/o polymer was the control where no polymer was used.

hydrocortisone acetate was bioequivalent to a 1% cream in reducing a surfactant-induced erythema (Davis, 1995).

Raghavan et al. (2001a) studied the effect of a number of polymers on the stability of supersaturated systems and permeation enhancement of hydrocortisone acetate. Flux enhancement was observed with all the polymers they used, but the extent of enhancement depended on the nature and amount of polymer. The flux increased with the amount of polymer, reached a maximum, and decreased at higher concentrations. While the increase of flux with increasing polymer concentration was attributed to increased inhibition of crystallization by the polymers, the decrease at high polymer concentrations was suggested to be due to changes in the microviscosity of the applied solution and solubility change induced by the presence of the polymer.

The amount of polymer needed for maximum permeation enhancement was different for each polymer (Fig. 4). The studies indicate that cellulose polymers were most effective, whereas polyacrylamide was the least effective. They also found that polyethylene oxide (PEO) and polyethylene glycol 400 (PEG400), which have very similar structures, had different effects on the permeation enhancement. While only 0.5% of the former was sufficient to produce maximum enhancement, 5% of the latter was required for the same effect. They have suggested this variation to be due to the difference in their molecular weights, which has also been found by Megrab et al. (1995), who obtained a variation of flux of estradiol with variation in the molecular weight of polyvinyl pyrrolidone.

Iervolino et al. (2000, 2001) studied the influence of hydroxypropyl β-cyclodextrins (hp-βCD) on the stability of supersaturated solutions. Cyclodextrin, a well-known solubilizer of poorly soluble drugs (through the formation of inclusion complexes), has also been used to supersaturate pancratistatin, an anticancer drug for parenteral use (Torres-Labandeira et al., 1990). It has also been shown to inhibit crystallization of amorphous nifedipine in spray-dried powders (Uekema et al., 1992). Although Iervolino et al. (2000) found that HPMC was more effective in enhancing the stability of supersaturated solutions than hp-βCD, the latter was found to be effective at low degrees of supersaturation. Different mechanisms of stabilization of supersaturation by HPMC and hp-βCD were proposed; HPMC by inhibiting nucleation and hp-βCD by changing the solubility. The permeation of ibuprofen across silicone membrane, as well as across skin, also increased proportionally with degree of supersaturation. However, it was found to be slightly higher than expected. This was attributed to the possibility of contribution from hp-βCD acting as a penetration enhancer. It is also interesting to note from their work that whenever the supersaturated system was unstable, the standard deviation obtained from the permeation experiments was large. This is a reflection of the uncertainty in the degree of saturation for these unstable systems. The surface of skin appeared to create more nucleation possibilities than the silicone.

Dias et al. (2002) also studied the effect of using both hp-βCD and hydroxypropyl methylcellulose for permeation enhancement of diclofenac in supersaturated solutions across silicone membranes. In accordance with the work of Iervolino et al. (2000), they found that the antinucleant HPMC was required to stabilize supersaturated solutions by inhibiting the crystallization process.

Moser et al. investigated the role of supersaturation in the dermal delivery of a lipophilic drug. Supersaturation was shown to be a useful method to increase the skin penetration of a Lavendustin derivative (LAP), a highly lipophilic drug (Moser, 2001a,b,c,d,e). The uptake and permeation of LAP increased with increase in supersaturation up to five times the saturated solution. At higher degrees of supersaturation, the systems were found to be unstable. They also found that neither the apparent diffusion coefficient nor the partition coefficient were affected by changes in degrees of supersaturation. Attempts to use several polymers, such as hydroxypropyl methyl cellulose, polyvinyl pyrrolidone, Aerosil 200, poloxamer 188, and Eudragit L155, to stabilize supersaturation were unsuccessful, although sodium carboxymethyl cellulose was found to have some stabilizing effect. Their efforts to find a correlation between the stabilization and solubility parameters were not successful. It is also interesting to note that the authors found that stability is primarily a question of the degree of supersaturation rather than the actual concentration.

Megrab et al. (1995) achieved 18 degrees of saturation with estradiol giving a tenfold increase in flux over a 6 hour period when compared with a saturated solution in the same cosolvent mixture. A flux of 18 times that of the saturated solution would have been expected, but it is probable that some crystallization of the drug occurred in the donor phase or within superficial layers of the stratum corneum, which reduced chemical potential. Within 10 minutes of exposure to a supersaturated solution, the investigators found an 18-fold increase in the concentration of the drug in the stratum corneum.

Kemken et al. (1992) demonstrated the use of microemulsions in the development of supersaturated preparations of bupranolol. Water-free microemulsion bases were applied to the skin under occlusive patches. Water uptake from the skin led to the production of a microemulsion and a decrease in the solubility of the drug, resulting in a supersaturated preparation. Crystal formation was not observed until after a 2- or 3-day period, and an equilibrium was reached after about 10 to 14 days. The effect of the microemulsion formulations on reducing an induced tachycardia in male albino New Zealand rabbits was monitored. Effect versus time curves were plotted for each microemulsion and other test preparations, and the different rates of the therapeutic effect of bupranolol were compared. Some preparations reduced tachycardia faster than others; this was related to preparation/water solubility curves and the formation of supersaturated states presenting enhanced absorption rates. Interacting effects on the skin of the individual components of the preparations were not considered.

Pellett et al. (1997a) investigated the in vitro delivery of piroxicam across silicone and human skin. Solutions with degrees of saturation up to 5.3-fold were prepared, but those above a 4-fold degree of saturation were unstable, and crystallization occurred within 8 h of preparation. The solutions up to 4 degrees of saturation were stable for 16 h, and transport across silicone and human skin increased linearly, i.e., delivery of piroxicam from a 4-fold supersaturated solution was four times greater than delivery from a saturated solution in the same vehicle (see Fig. 5).

In a later study, Pellett et al. (1997b) determined the amount of permeant in the stratum corneum using a tape stripping technique. As in the previous study, supersaturated solutions of piroxicam up to 4 degrees of saturation were investigated, and the amount of piroxicam in the viable layers of the skin was shown to increase with increasing degree of saturation. The study clearly demonstrated that the stratum corneum was capable of supporting supersaturated systems, and it was suggested that the intercellular lipids of the stratum corneum possessed an antinucleant capacity. The solubility of a permeant in the stratum corneum and the creation of supersaturated states within this membrane is therefore an important consideration when attempting to enhance delivery across the skin. This effect has also been observed by other investigators and used to explain the failure of iontophoresis to provide the anticipated fluxes for the delivery of CQA 206–291, a dopamine agonist (Hager et al., 1993).

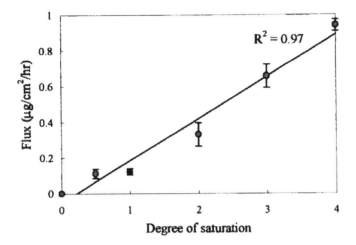

Figure 5 The relationship between the degree of saturation and the steady-state flux of piroxicam across human skin in vitro.

The likelihood of success with which stable supersaturated systems are formed is not fully known, but it is likely that there is a limit to the amount of drug that can be held in solution; so the higher the concentration of the drug in solution, the greater its instability. This was demonstrated by Iervolino et al. (2000) where high concentrations of supersaturated solutions of ibuprofen were found to be unstable when compared to similar degrees of saturation at low concentrations. In this study, a linear relationship between flux and degree of saturation using silicone membranes was observed up to 4 degrees of saturation without an antinucleant polymer, and in the presence of HPMC enhancement was observed, although not linearly, at up to 7 degrees of saturation.

Schwarb et al. (1999) studied the effect of supersaturation on the delivery of fluocinonide across silicone membranes using in vitro diffusion cells and a pharmacodynamic response (skin blanching assay) from in vivo applications. Supersaturated solutions up to 3.8 degrees of saturation were prepared using PVP and a vehicle composition of ethanol, water, glycerol, and propylene glycol. The stability of the solutions was tested under worst case conditions by constantly stirring the solutions and providing a nucleating surface. For the supersaturated solutions with a low concentration, a solution with a theoretical degree of saturation of 3.76 and a fluocinonide concentration of 100 µg/mL generated a stable solution of up to approximately 3.2 degrees of saturation. It is probable that this discrepancy of approximately 0.56 degrees of saturation is caused by an initial crystallization of the drug that is then halted by the presence of the PVP. It was also shown that fluocinonide permeability across silicone membranes was linearly related to degree of saturation. An increased pharmacodynamic response was anticipated in the in vivo study, but the results did not support this expectation. It was proposed that this might have been caused by vehicle effects on the skin, for example, irritation or drug binding.

B. Evaporation

An increase in thermodynamic activity can also be achieved by evaporation of the volatile components of a formulation. For example, a compound may be present in a formulation in a subsaturated state, but when it is applied to the skin surface, the volatile components may evaporate, resulting in a decreased solubility in the residual components on the skin surface. It is then possible that the solubility of the compound on the skin surface is reduced to the extent that a supersaturated state exists. In fact, it is almost inevitable that the application of some products on the market, particularly lotions, are susceptible to the formation of supersaturated states and may infer penetration enhancement by this means. A number of investigators have studied this effect.

In the first of a series of publications, the transdermal delivery of nifedipine centered around the production of supersaturated solutions using volatile:nonvolatile vehicles with antinucleant polymers to stabilize the supersaturated solutions (Kondo and Sugimoto, 1987). At first, the diffusion of nifedipine from a range of combinations of IPM, PG, and acetone vehicles across an ethylene–vinyl acetate copolymer membrane was monitored, and initial flux values were found to be greater than the final steady-state fluxes. This was explained by the initial evaporation of the volatile components of the vehicle, giving rise to a supersaturated solution, after which the drug crystallized out of solution. The incorporation of polymers into the IPM:PG:acetone vehicle gave rise to up to five-times enhancement in flux.

In the second paper (Kondo et al., 1987a), the diffusion of nifedipine from ethanol:diethyl sebacate vehicles was investigated in vitro across ethylene–vinyl acetate copolymer membranes and excised male Wistar rat skin. In vivo studies were also performed on male Wistar rats. With both in vitro and in vivo experiments, greatest diffusion was seen from 75:25 ethanol:diethyl sebacate combinations, which possessed 1.75 degrees of saturation after complete evaporation of the ethanol.

The third and final paper in this series examined the in vivo penetration of nifedipine from a range of combinations of propylene glycol, isopropyl myristate, and acetone vehicles across male Wistar rat skin in vivo (Kondo et al., 1987b). The initial plasma concentrations from the rats dosed with the binary cosolvent vehicles of propylene glycol:acetone and isopropyl myristate:acetone were greater than those from respective propylene glycol and isopropyl myristate vehicles, but final plasma concentrations dropped to the levels seen with the saturated propylene glycol and isopropyl myristate vehicles, respectively. An even greater degree of penetration was observed from a propylene glycol:isopropyl myristate: acetone vehicle. In vitro diffusion studies on excised rat skin showed that propylene glycol was diffusing through the skin to a greater extent in the ternary system than in the binary system. This was explained in terms of phase separation of the isopropyl myristate and propylene glycol solvents after the evaporation of the acetone and a supersaturation of the isopropyl myristate phase with respect to the propylene glycol, resulting in a greater thermodynamic activity of the propylene glycol and hence greater penetration. Nifedipine is more soluble in propylene glycol than in isopropyl myristate, and so the loss of propylene glycol from the system can also contribute to the increased thermodynamic activity of the drug. After the initial plasma concentrations, the levels decreased due to precipitation of the drug at the application site, but upon addition of antinucleant polymers to the preparations, greater plasma levels were observed. Each polymer varied in its ability to maintain the elevated plasma levels, presumably owing to their ability to maintain the supersaturated states in the preparations.

Chiang et al. (1989) used volatile solvents to increase the flux of minoxidil across human cadaver skin. Initially, a fixed ratio of propylene glycol/ethanol/ water vehicles were used and, at room temperature, evaporation of the volatile components of the preparations occurred largely within the first 30 min. Flux values were determined from 8 to 24 h time periods and increased with increasing concentrations up to a point where the drug started to crystallize out. The preparations were demonstrated to be supersaturated by seeding and observing crystallization under a microscope. The flux of minoxidil from isothermodynamic states was also investigated by varying the propylene glycol content of the vehicle at the expense of the ethanol and maintaining the same concentration of drug in the propylene glycol. There was no statistically significant difference in the fluxes from such preparations. Similar observations of an enhanced flux from unoccluded formulations have been made by other workers when volatile solvents are present in the vehicle (Mathias et al., 1985; Tanaka et al., 1985).

The use of a volatile/nonvolatile cosolvent system to give supersaturation and enhanced penetration was also used by Theeuwes et al. (1976). Hydrocortisone was dissolved in a mixture of acetone and water, and the volatile component was evaporated. The diffusion of the resultant supersaturated solution was monitored across an ethylene–vinyl acetate copolymer membrane. Up to 3.9 degrees of saturation was achieved, giving a proportional increase in flux.

Fang et al. (1999) investigated the delivery of sodium nonivamide acetate across rat skin from supersaturated solutions prepared by the evaporation of the volatile components of an aqueous ethanolic vehicle applied to the skin surface.

A range of potential antinucleant polymers was tested, and it was shown that methylcellulose and hydroxypropyl cellulose had the most stabilizing effect. Diffusion experiments were conducted over 72 h, where delivery over the first 12 h increased rapidly followed by a plateau. It was proposed that the plateau was caused by the precipitation of the drug on the skin surface, but as discussed above, it is probable that crystallization within the skin membrane itself also limited continued enhancement.

C. Heating and Cooling

Saturated solubilities generally increase with increasing temperature (Fig. 2), and consequently when a fixed concentration of a solid dispersion is heated, the dispersant completely dissolves in the solvent. When the solution is cooled to a temperature below that of the saturated solubility, the solution may crystallize, or a stable supersaturated system may result.

An indomethacin gel (Henmi et al., 1994) was prepared by heating a mixture of the drug with hydrogenated soybean phospholipid and liquid paraffin to 95°C, cooling to room temperature, placing it in an air incubator at 40°C for three days, and then storing at room temperature. The permeation of the drug from this formulation was compared to that prepared at room temperature in the same vehicle components. There was a 10-fold increase in the permeation of the heated and cooled formulation when compared to that prepared at room temperature. Formulations of other drugs were prepared in a similar manner by heating and cooling, and their permeation rates were compared to suspensions of the drugs prepared at 37°C. Flurbiprofen and ketoprofen showed an increase in the permeation rates from the heated and cooled formulations, but there was no difference in the rates for ibuprofen formulations. Hydrogenated soybean phospholipid was considered not to be a penetration enhancer, and the enhancement was attributed to the formation of supersaturated states on cooling the formulations.

In the production of some transdermal patches, hot melt technology is used. The active can be added to the molten polymer adhesive at an elevated temperature. The solution is then extruded in a molten state onto a backing membrane. On cooling, it is possible that the active is present above its solubility limit. During storage, the supersaturated state is thermodynamically unstable, and crystallization can occur. This gives rise to characteristic crystals in the adhesive matrix.

D. Reactions

In some instances, it is possible to attain supersaturated states when two reactants give rise to a product that has a lower solubility than the reactants in the reaction

Table 1 Flux Values and Enhancement Ratios for Penetration of Flurbiprofen Across Human Stratum Corneum

Pre-treatment (donor phase)	Flux ($\mu g/cm^2/h$) (Mean $\pm$ SE, n $\geq$ 5)	Enhancement ratio
EtOH (saturated soln)	3.2 $\pm$ 0.3	1.0
2.8% OA in EtOH (saturated soln)	6.6 $\pm$ 0.7	2.1
EtOH (6-fold supersaturated soln)	14.3 $\pm$ 0.8	4.5
2.8% OA in EtOH (6-fold supersaturated soln)	31.8 $\pm$ 2.8	9.9

media. No specific examples of this technique could be found for topical and transdermal drug delivery.

IV. APPLICATION OF SUPERSATURATION TO ACHIEVE SYNERGISTIC METHODS OF DRUG DELIVERY

Pellett et al. (1998) studied the synergistic effects on the enhancement of flurbiprofen permeation across human skin using supersaturated solutions and oleic acid (OA). Oleic acid is thought to exhibit an enhancing effect by increasing the diffusivity of a drug in the stratum corneum (Mak et al., 1990), whereas supersaturated systems increase the concentration in the vehicle. Therefore increasing both these parameters would be expected to result in a multiplicative effect on flux. Such a result would also imply independent mechanisms of action for the two methods of enhancement. Multiplicative, synergistic enhancement of flurbiprofen across human skin using these two types of penetration enhancement (supersaturation and chemical enhancers) was investigated by pretreating stratum corneum for 1 h with a 2.8% ethanolic solution of oleic acid and then applying a supersaturated solution with 6 degrees of saturation. Flux values were determined using Fick's first law, and enhancement ratios were calculated (Table 1).

The enhancement ratio for the 6-fold supersaturated solution after pretreatment with ethanol was 4.5, and the enhancement ratio observed after pretreatment with OA was 2.1. Therefore a multiplicative effect would be expected to result in a 9.5-fold increase (4.5 $\times$ 2.1) in penetration. In fact, a 9.9-fold increase was observed.

The fact that multiplicative effects on flux were observed suggested that the mechanisms of enhancement for supersaturation and oleic acid were independent of each other. In other situations with other methods of enhancement, it is possible that the mechanisms of action are not as mutually exclusive as those found here.

Megrab et al. (1995) also demonstrated a synergistic effect between supersaturation and increased partitioning into the skin for estradiol. In the absence of vehicle solvent interactions, flux across membranes from saturated solutions is constant, but this study showed that as the proportion of propylene glycol was increased in the vehicle of saturated solutions, the uptake of the drug into the stratum corneum increased with increasing proportion of propylene glycol. The same effect was also observed when supersaturated solutions were applied. Therefore as propylene glycol is thought to increase partitioning into the membrane, and supersaturation increases the concentration in the vehicle beyond that of the normally limited saturated solubility, synergy between these two mechanisms of enhancement were demonstrated.

These studies showed that the enhancement of the permeation of flurbiprofen across membranes can be achieved using synergistic methods of enhancement, namely supersaturation and chemical enhancers. The combination of the mechanisms of action for these two methods resulted in a multiplicative increase in flux, as predicted from Fick's laws of diffusion.

V. ANTINUCLEANT POLYMERS

The precise mechanism of nucleation and subsequent crystal growth is not fully understood. Frank (1949) suggests that if a molecule diffuses to the flat surface of a crystal it only has one binding surface, but if it arrives at a kink or step, it has more than one binding surface with which to anchor itself to the crystal. With screw dislocations, the crystal can continue to grow in a "spiral staircase" manner. Four models predicting crystal growth of theophylline monohydrate were presented by Rodriguez-Hornedo and Hsui-Jean (1991):

1. Solute transport within the bulk solution to the crystal surface
2. Attachment of growth units to a surface that is "rough" on a molecular scale
3. Nucleation of two-dimensional clusters to the surface, which expand and merge to form new layers
4. Spreading of layers from a screw lattice dislocation, which acts as a continuous source of steps

They showed that the crystals grew according to the screw dislocation theory. In a similar manner, the mechanism of action of antinucleant polymers is also largely unknown, although in a review by Pellett et al. (1997a), and with subsequent articles, a number of potential theories have been developed.

Mehta et al. (1970) exposed crystals to supersaturated solutions and measured growth rates through a microscope as a function of stirring rates, degrees of supersaturation, and different solvents. It was shown that crystal growth occurred as a result of either surface-controlled or diffusion-controlled mechanisms, depending on the crystal and solvent used. Furthermore, crystal growth only occurred above certain critical degrees of saturation that also varied according to the crystal and the solvent. For sulfamethiazole, the higher the alcohol polarity, the lower the critical degree of supersaturation. In this study, only two compounds were investigated, namely sulfathiazole and methylprednisolone, but similar observations of a critical degree of supersaturation are well documented and have been observed with other compounds (Mullin, 1961).

HPMC and PVP have been shown to inhibit the crystal growth of paracetamol from supersaturated solutions (Femi-Oyewo and Spring, 1994), and after dissolution of nifedipine, spironolactone, and griseofulvin in the presence of certain polymers, they were shown to be supersaturated, but only in one of the two solvents tested (Hasegawa et al., 1985).

Holder and Thorne (1979) demonstrated the inhibition of the nucleation of *n*-paraffins in xylene by structurally similar synthetic polymers. Three possible theories for their mechanism of action were proposed, and it was concluded that all three factors could be influencing the nucleation process:

1. The synthetic polymers increased the solubility of the paraffin
2. The prevention of seeds from growing beyond a critical size, above which nucleation and crystallization occur
3. The disruption of the normal structure of the precipitating crystal (*n*-paraffin existing in several forms with different solubilities)

Simonelli et al. (1970) investigated the inhibition of sulfamethiazole crystals by PVP and suggested that PVP formed a netlike structure over a growing crystal face with fingerlike crystal growths occurring between the pores of the PVP network. Owing to the curvature of these protrusions, a higher degree of supersaturation was required to continue crystal growth, which was thus halted. Sekikawa et al. (1978) demonstrated that coprecipitates (a coprecipitate is formed upon complete evaporation of a solvent, leaving the crystal structure of the drug encapsulated in the polymer matrix) were formed between various drugs and PVP, but only in those drugs that formed stable supersaturated solutions. This has been supported by the behavior of a similar nature with other drugs (Corrigan and Timoney, 1975).

More recently, Raghavan et al. (2001b) proposed a model for the influence of polymers on the nucleation as well as crystal growth of hydrocortisone acetate based on the experimental investigations of crystallization and morphology of

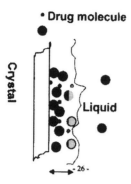

Figure 6 A schematic showing how the presence of an impurity (the antinucleant polymer) impedes the adsorption of the active to the crystal surface.

the drug in the presence of a number of polymers. According to their model (Fig. 6), crystallization is inhibited by the association of the polymer onto the drug, and crystal growth is inhibited by the adsorption of the polymer onto the growing crystal surface through hydrogen bonding. Additionally, accumulation of the polymer in the hydrodynamic layer surrounding the crystal increases the resistance to diffusion to the crystal surface. In crystal growth, it is well known that there is a stagnant boundary layer at the interface between the growing crystal face and the bulk solution. In agitated systems, this layer is very thin or almost nonexistent. In a stagnant solution, as in this method, this layer can be very thick, especially in the presence of the polymer, and will provide a diffusional barrier for the diffusion of the drug molecules.

The adsorption of the polymer onto the crystal surface, according to the model, is due to hydrogen bonding between the drug and the polymer. The extent to which adsorption occurs will depend on the nature and strength of hydrogen bonding between the drug and the polymer. The presence of hydrogen bonding was established from infrared spectroscopic studies on solid dispersions of hydrocortisone acetate and polyvinyl pyrrolidone. In this example, Raghavan et al. (2001a) found a distinct development of the drug–polymer IR band, which replaced the two hydroxyl bands of hydrocortisone acetate. Such an existence of hydrogen bonding was also observed between ibuprofen and hydroxypropyl methylcellulose by Iervolino et al. (2001).

Both Raghavan et al. (2001b) and Iervolino et al. (2001) also found that both a crystal's size and its morphology were modified in the presence of polymers. The particle sizes were found to be smaller when the polymer was present. Moreover, the morphologies were different in the presence of different polymers, indicating that the drug–polymer interactions are dependent on the nature of the polymer. The formation of smaller and differently shaped crystals in the presence of polymers supports the model proposed by Raghavan et al. (2001a). The polymers inhibit crystal growth, whereas the habit modification indicates that the extent of polymer adsorption in each face is different, the extent depending on the strength of hydrogen bonding at each face.

A number of papers have reported the inhibition of polymorphic transitions of drugs to more stable forms (Miyazaki et al., 1976; Ebian et al., 1975) in the presence of polymer additives. Davis and Hadgraft (1993) postulated that different polymorphs of a compound can have different saturated solubilities in a solvent and therefore one can be supersaturated with respect to the other. Furthermore, the choice of solvent and the crystallization conditions dictate the type of crystals (e.g., hydrate, alcoholate) and polymorphs formed (Khoshkhoo and Anwar, 1991; Corrigan and Timoney, 1974). Solvate forms of drugs, sometimes called pseudopolymorphs, can also have different solubilities and dissolution rates from their corresponding nonsolvate forms (De Smidt et al., 1986; Shefter and Higuchi, 1963). Moreover, as with the polymorphs, an anhydrous form of a drug can be supersaturated with respect to its hydrate form. Pellett et al. (1997a) suggested that the inhibition of the formation of different hydrate forms of piroxicam were responsible for the preparation of stable supersaturated solutions.

Other methods of crystal growth inhibition have been investigated by other workers, such as the inhibitory effect of viscous systems on the polymorphic transition of chlortetracycline (Miyazaki et al., 1976) and the crystal growth inhibition of chlorpropamide (Abd El-Bary et al., 1990). The ability of 2-hydroxypropyl-β-cyclodextrin (hp-βCD) to inhibit crystallization of amorphous nifedipine in spray-dried powders (Uekama et al., 1992) and the inhibition of crystal growth of isosorbide 5-mononitrate from tablets or powders (Uekama et al., 1985) by β-cyclodextrins may have some implications in the preparation of stable supersaturated systems for topical drug delivery. The use of hydroxypropyl cyclodextrins to supersaturate pancratistatin has also been reported (Torres-Labandeira et al., 1990).

Recently, Iervolino et al. (2001) investigated the crystal morphology of ibuprofen in the presence of hp-βCD and HPMC after crystals were formed from supersaturated solutions. It was found that HPMC modified crystal growth and structure, whereas hp-βCD had no influence on crystal growth. Differences in enhancement brought about by the inclusion of hp-βCD in the solutions were thought to have been due to the polymer acting as a penetration enhancer at certain drug/polymer ratios.

VI. CONCLUSIONS

Supersaturation can be useful in dermal and transdermal drug delivery, but producing stable supersaturated states is not easy. There is increased knowledge of the mechanisms of supersaturation, but there is still much to be learned, particularly at the molecular level. Supersaturation can lead to the application of lower doses with a concomitant reduction in possible adverse drug reactions. It can be used in combination with other enhancement strategies to produce synergistic effects. The knowledge base associated with the production of supersaturated states and their stabilization has also led to progress in understanding instability difficulties in transdermal patches.

REFERENCES

A Abd El-Bary, MAA Kassem, NST Foda, SS Badawi. Controlled crystallization of chlorpropamide from surfactant and polymer solutions. Drug Dev Ind Pharm 16:1649–1660, 1990.

C Chiang, GL Flynn, ND Weiner, GJ Szpunar. Bioavailability assessment of topical delivery systems: effect of vehicle evaporation upon in vitro delivery of minoxidil from solution formulations. Int J Pharm 55:229–236, 1989.

MF Coldman, BJ Poulsen, T Higuchi. Enhancement of percutaneous absorption by the use of volatile: nonvolatile systems as vehicles. J Pharm Sci 58: 1098–1102, 1969.

OI Corrigan, RF Timoney. Anomalous behaviour of some hydroflumethazide crystal samples. J Pharm Pharmacol 26:838–840, 1974.

OI Corrigan, RF Timoney. The influence of polyvinylpyrrolidone on the dissolution properties of hydroflumethazide. J Pharm Pharmacol 27:759–764, 1975.

AF Davis, J Hadgraft. Effect of supersaturation on membrane transport: 1. Hydrocortisone acetate. Int J Pharm 76, 1–8, 1991a.

AF Davis, J Hadgraft. The use of supersaturation in topical drug delivery. In: Prediction of Percutaneous Penetration: Methods, Measurements, Modelling (RC Scott, RH Guy, J Hadgraft, HE Boddé, eds.). London: IBC Technical Services, 1991b, Vol. 2, pp 279–287.

AF Davis, J Hadgraft. Supersaturated solutions as topical drug delivery systems. In: Pharmaceutical Skin Penetration Enhancement (KA Walters, J Hadgraft, eds.). New York: Marcel Dekker, 1993, Vol. 59, pp 243–267.

AF Davis. Novel formulations for improved topical therapy. Thesis, University of Wales, 1995.

JH De Smidt, JG Fokkens, H Grijseels, DJA Crommelin. Dissolution of theophylline monohydrate and anhydrous theophylline in buffer solutions. J Pharm Sci 75:497–501, 1986.

MMR Dias, SL Raghavan, MA Pellett, WJ Pugh, J Hadgraft. The effect of β-cyclodextrins on the permeation of diclofenac from supersaturated solutions. Int J Pharm 2002.

AR Ebian, MA Moustafa, SA Khalil, MM Motawi. Succinylsulfathiazole crystal forms II: Effects of additives on kinetics of interconversion. J Pharm Sci 64:1481–1484, 1975.

J-Y Fang, C-T Kuo, Y-B Huang, P-C Wu, Y-H Tsai. Transdermal delivery of sodium nonivamide acetate from volatile vehicles: effects of polymers. Int J Pharm 176:157–167, 1999.

MN Femi-Oyewo, MS Spring. Studies on paracetamol crystals produced by growth in aqueous solutions. Int J Pharm 112:17–28, 1994.

FC Frank. The influence of dislocations on crystal growth. Discussions of the Faraday Society 5:48–54, 1949.

DF Hager, MJ Laubach, JW Sharkey, JR Siverly. In vitro iontophoretic delivery of CQA 206–291: influence of ethanol. J Cont Rel 23:175–182, 1993.

A Hasegawa, R Kawamura, H Nakagawa, I Sugimoto. Physical properties of solid dispersions of poorly water-soluble drugs with enteric coating agents. Chem Pharm Bull 33:3429–3435, 1985.

T Henmi, M Fujii, K Kikuchi, M Matsumoto. Application of an oily gel formed by hydrogenated soybean phospholipids as percutaneous-type enhancer. Chem Pharm Bull 42:651–655, 1994.

T Higuchi. Physical chemical analysis of percutaneous absorption process from creams and ointments. J Soc Cosmet Chem 11:85–97, 1960.

GA Holder, J Thorne. Inhibition of crystallisation by polymers. Polymer Preprints Am Chem Soc Div Poly Chem 20:766–769, 1979.

M Iervolino, SL Raghavan, J Hadgraft. Membrane penetration enhancement of ibuprofen using supersaturation. Int J Pharm 198:229–238, 2000.

M Iervolino, BM Capello, SL Raghavan, J Hadgraft. Penetration enhancement of ibuprofen from supersaturated solutions through human skin. Int J Pharm 212:131–141, 2001.

J Kemken, A Ziegler, BW Muller. Influence of supersaturation on the pharmacodynamic effect of bupranolol after dermal administration using microemulsions as vehicle. Pharm Res 9:554–558, 1992.

EV Khamskii. Crystallisation from Solutions. New York: Consultants Bureau, 1969.

S Khoshkhoo, J Anwar. Crystallization of polymorphs: effects of supersaturation and solvent. J Pharm Pharmacol 43:36, 1991.

S Kondo, I Sugimoto. Enhancement of transdermal delivery by superfluous thermodynamic potential. I. Thermodynamic analysis of nifedipine transport across the lipoidal barrier. J Pharmacobio-Dyn 10:587–594, 1987.

S Kondo, H Yamasaki-Konishi, I Sugimoto. Enhancement of transdermal delivery by superfluous thermodynamic potential. II. In vitro–in vivo correlation of percutaneous nifedipine transport. J Pharmacobio-Dyn 10:662–668, 1987a.

S Kondo, K Yamanaka, I Sugimoto. Enhancement of transdermal delivery by superfluous thermodynamic potential. III. Percutaneous absorption of nifedipine in rats. J Pharmacobio-Dyn 10:743–749, 1987b.

BC Lippold. How to optimize drug penetration through the skin. Pharm Acta Helv 67:294–300, 1992.

VHW Mak, RO Potts, RH Guy. Oleic acid concentration and effect in human stratum corneum: non-invasive determination by attenuated total reflectance infrared spectroscopy in vivo. J Cont Rel 12:67–75, 1990.

CGT Mathias, RH Guy, RS Hinz, HI Maibach. Percutaneous penetration enhancement of solvent-deposited solids by the volatile solvent. J Invest Derm 84:360, 1985.

NA Megrab, AC Williams, BW Barry. Oestradiol permeation through human skin and silastic membrane: effects of propylene glycol and supersaturation. J Cont Rel, 36:277–294, 1995.

SC Mehta, PD Bernardo, WI Higuchi, AP Simonelli. Rate of crystal growth of sulfathiazole and methylprednisolone. J Pharm Sci 59:638–644, 1970.

K Moser, K Kriwet, C Froehlich, YN Kalia, RH Guy. Supersaturation: enhancement of skin penetration and permeation of a lipophilic drug. Pharmaceutical Research 18(7):1006–1011, 2001a.

K Moser, K Kriwet, C Froehlich, A Naik, YN Kalia, RH Guy. Permeation enhancement of a highly lipophilic drug using supersaturated systems. Journal of Pharmaceutical Sciences 90:607–616, 2001b.

K Moser, K Kriwet, YN Kalia, RH Guy. Enhanced skin permeation of a lipophilic drug using supersaturated formulations. Journal of Controlled Release 73:245–253, 2001c.

K Moser, K Kriwet, YN Kalia, RH Guy. Stabilization of supersaturated solutions of a lipophilic drug for dermal delivery. International Journal of Pharmaceutics 224:169–176, 2001d.

K Moser, K Kriwet, A Naik, YN Kalia, RH Guy. Passive skin penetration enhancement and its quantification in vitro. European Journal of Pharmaceutics and Biopharmaceutics 52:103–112, 2001e.

S Miyazaki, M Nakano, T Arita. Effect of additives on the polymorphic transformation of chlortetracycline hydrochloride crystals. Chem Pharm Bull 24:2094–2101, 1976.

JW Mullin. Crystallization, London: Butterworths, 1961, pp 26–29.

MA Pellett, S Castellano, J Hadgraft, A Davis. The penetration of supersaturated solutions of piroxicam across silicone membranes and human skin in vitro. J Cont Rel 46:205–214, 1997a.

MA Pellett, MS Roberts, J Hadgraft. Supersaturated solutions evaluated with an in vitro stratum corneum tape stripping technique. Int J Pharm 151:91–98, 1997b.

MA Pellett, AC Watkinson, KR Brain, J Hadgraft. Synergism between super-saturation and chemical enhancement in the permeation of flurbiprofen through human skin. In: Perspectives in Percutaneous Penetration (KR Brain, VJ James, KA Walters, eds.). Cardiff: STS Publishing, 1998, 173–176.

BJ Poulsen, E Young, V Coquilla, M Katz. Effect of topical vehicle composition on the in vitro release of fluocinolone acetonide and its acetate ester. J Pharm Sci 57:928–933, 1968.

SL Raghavan, A Trividic, AF Davis, J Hadgraft. Effect of cellulose polymers on supersaturation and in vitro membrane transport of hydrocortisone acetate. Int J Pharm 193:231–237, 2000.

SL Raghavan, B Kiepfer, AF Davis, SG Kazarian, J Hadgraft. Membrane transport of hydrocortisone acetate from supersaturated solutions: influence of polymers. Int J Pharm 221:95–105, 2001a.

SL Raghavan, A Trividic, AF Davis, J Hadgraft. Crystallization of hydrocortisone acetate: influence of polymers. Int J Pharm 212:213–221, 2001b.

N Rodriguez-Hornedo, W Hsui-Jean. Crystal growth kinetics of theophylline monohydrate. Pharm Res 8:643–648, 1991.

FP Schwarb, G Imanidis, EW Smith, JM Haigh, C Surber. Effect of concentration and degree of saturation of topical fluocinonide formulations and in vitro membrane transport and in vivo availability on human skin. Pharm Res 16: 909–915, 1999.

H Sekikawa, M Nakano, T Arita. Inhibitory effect of polyvinylpyrrolidone on the crystallization of drugs. Chem Pharm Bull 26:118–126, 1978.

AP Simonelli, SC Mehta, WI Higuchi. Inhibition of sulfathiazole crystal growth. J Pharm Sci 59:633–638, 1970.

E Shefter, T Higuchi. Dissolution behaviour of crystalline solvated and nonsolvated forms of some pharmaceuticals. J Pharm Sci 52:781–791, 1963.

S Tanaka, Y Takashima, H Murayama, S Tsuchiya. Studies on drug release from ointments. V. Release of hydrocortisone butyrate propionate from topical dosage forms to silicone rubber. Int J Pharm 27:29–38, 1985.

F Theeuwes, RM Gale, RW Baker. Transference: a comprehensive parameter governing permeation of solutes through membranes. J Mem Sci 1:3–16, 1976.

JJ Torres-Labandeira, P Davignon, J Pitha. Oversaturated solutions of drug in hydroxypropylcyclodextrins: parenteral preparation of pancratistatin. J Pharm Sci 80:384–386, 1990.

JN Twist, JL Zatz. Influence of solvents on paraben permeation through idealized skin model membranes. J Soc Cosmet Chem 37:429–444, 1986.

K Uekama, K Oh, T Irie, M Otagiri, Y Nishimiya, T Nara. Stabilization of isosorbide 5-mononitrate in solid state by B-cyclodextrin complexation. Int J Pharm 25:339–346, 1985.

K Uekama, K Ikegami, Z Wang, Y Horiuchi, F Hirayama. Inhibitory effect of 2-hydroxypropyl-B-cyclodextrin on crystal growth of nifedipine during storage: superior dissolution and oral bioavailability compared with poly-vinylpyrrolidone k-30. J Pharm Pharmacol 44:73–78, 1992.

10

Minimally Invasive Systems for Transdermal Drug Delivery

James A. Down and Noel G. Harvey
BD Technologies, Research Triangle Park, North Carolina, U.S.A.

I. INTRODUCTION

A. The Concept of Minimally Invasive Drug Delivery

A discussion of minimally invasive drug delivery must begin with a consideration of what invasive delivery means. Administration of drugs via needles and syringes has been with us for more than a hundred years. For example, the first all-glass syringe patent was licensed to Becton Dickinson & Co. from Luer in 1898 (1). Metal cannula needles on piston syringes have become the most prevalent ethical* device-based drug delivery modality in existence, with multiple billions being used each year in many health care applications. Conventional needles, whether on syringes or catheters, represent the preeminent invasive delivery mode in existence. They are, however, also the most efficient and cost-effective device-based system for administering agents into the systemic circulation and are presently the general method for delivering polypeptide agents, which are otherwise proteolyzed by the oral route. Without syringes and needles, unknown millions of diabetics would have been unable to survive in this century despite Banting and Best's discovery of insulin because syringe-based delivery was an unconditionally necessary component of diabetic therapy.

---

* We are disregarding cigarettes in this discussion, which are a much more widely used form of drug delivery "device" than syringes.

The reason, as discovered more than a century ago, is that a thin, sharp sterile metal pipe is an ideal way to breach the stratum corneum and deliver agents past the skin barrier into the microvascularization of the dermis or lower tissues and thence into the systemic circulation. The principle has been proven billions upon billions of times. Despite this, conventional needle-based delivery suffers many well-recognized drawbacks, not the least of which is the negative psychosocial connotation of drug administration via needles. Other problems include the pain of administration; safety concerns over the possibility of transmission of blood-borne pathogens; the lack of compliance, the inability or dislike of patients to self-administer via needles; and the lack of ease of use, especially for younger or elderly patients.

To address these needs, a plethora of new technologies have arisen or are in development, whose inventors intend to provide transepidermal drug delivery by circumventing the conventional needle and syringe. Associated chapters in this volume describe the noninvasive methodologies that are being developed. We will not discuss electroporation, since that is covered in a separate chapter. In this chapter we describe minimally invasive systems in development that will allow delivery of drugs across skin without the need for needles and without the attendant discomfort. In many cases, the technologies referred to here have been described in the patent literature, but their clinical utility has yet to be rigorously proven in scientifically controlled trials.

Virtually all the systems that we will describe rely on breaching the stratum corneum and are therefore somewhat invasive, though in all cases the systems are intended to overcome the drawbacks of conventional metal needle delivery. We have, perhaps arbitrarily, divided the technologies into those which require some form of energy input, other than that provided by the user's actions, and those that are activated by the user's actions.

Finally, we will describe new micromachining technology that is presently reinventing the notion of drug delivery. This technology, termed MEMS, (microelectromechanical systems), has allowed the construction of novel devices that, from the present vantage, appear to hold great promise. This technology has enabled the development of systems that provide transepidermal drug administration via microscopic silicon needles as effectively or more so than other more conventional delivery modes. We will also provide a glimpse into the other applications that this microtechnology will provide, in expectation that in combination with drug delivery they will enable powerful new medical devices for the future.

However, regardless of how attractive any new drug delivery concept is, its developers must answer the following crucial questions. *Will the new delivery mode be safe for the user?* This implies safety from the transmission of blood-borne pathogens, safety from infection after treatment, and safety from collateral tissue damage. *Will the delivery system be efficacious?* If the delivery system is not as effective as preexisting delivery modes, then it is highly unlikely to be accepted. That is, delivery of medication is the primary purpose of the system—to provide a cost-effective therapeutic effect, regardless of whether the delivery system uses new technology, or is painless, inexpensive, easy to use, etc. *Will the system be easy to use and convenient for the user?* Sophisticated, difficult systems, or those with complicated work flows or dose regimens, invite medication errors and reduced compliance. *What will it cost?* Drug delivery systems are intended to facilitate delivery drugs to their biological site of action. Systems that add exceptional cost without providing enhanced value in the form of better efficacy, delivery profile, targeting, or other improved therapeutic effect should be discarded, since they add cost to the consumer without providing benefit.

B. Mechanics of Skin Penetration

As described elsewhere in this volume (Chapter 1), the thickness of the stratum corneum is approximately 10 to 20 microns (2). Therefore the minimum distance breached by minimally invasive transdermal systems must be about 20 microns. Our own experiments have shown that it is difficult to produce a defined breach of this low depth, and in fact microneedle lengths 10 times higher have been found to be necessary to produce cuts of this depth. This may relate to the biomechanics of skin penetration and the density of needles in needle arrays. Drugs must penetrate to a depth of 0.12 mm to reach blood and systemic circulation (2), implying that SC disruption must allow diffusion through epidermis, or that microneedle/puncture systems must reach into the systemic circulation.

The pain of standard injections is partially, if not wholly, the result of direct mechanical stimulation of the ATP cascade. Work of Burnstock and others suggest that during tissue damage, adenosine triphosphate (ATP) is released from damaged cells. ATP is a ubiquitous cellular energy storage compound, which when released extracellularly potentiates input nociceptors (pain receptors) via direct stimulation of purinergic neurons. The effect appears to provide a local signal for tissue destruction, which in turn stimulates remedial physiological responses such as inflammation, etc. Pain may also result from tissue distension caused by drug injection, skin damage (and ATP release), or direct damage to nerves. Researchers are approaching the challenge of pain reduction by several techniques that are linked by one unifying strategy: limiting the extent of mechanical insult to skin to the first 50 microns of tissue.

The mechanical stress–strain relationships of skin complicate the practical manipulation of the stratum corneum, epidermis, and upper dermal layers. The mechanics of skin, and other soft tissue such as arteries, muscle, and ureter, do not display single-value relationships between stress and strain and are therefore classified as inelastic materials (3). When tissues in this group are held at constant strain, they display stress relaxation, and when held at constant stress they creep. Historically this relationship has required uniaxial application of mechanical

force to breach skin (as in needle penetration), and has made sensation-free clinical manipulation (containment, preparation, and breach) of skin tissue difficult
(4). In general the new skin breach technologies described in this communication
require either direct uniaxial application of mechanical energy or thermally/me-
chanically induced changes in the physical properties or structure of skin. They
differ from classical approaches in that they attempt to target the epidermis and
upper dermal layers devoid of nosiceptors while gaining access to the circulatory
and immune systems via the dermal capillary bed and epidermis, respectively.
If successful, these approaches may greatly reduce or eliminate the mechanical
stimulation of pain responses during delivery.

II. SYSTEMS FOR MINIMALLY INVASIVE
TRANSDERMAL DELIVERY

A. Energy-Induced Transdermal Delivery Systems

1. Laser-Assisted Transdermal Delivery

Studies of laser-assisted drug delivery across skin have investigated two distinctly
different processes. The first is ablative. The high energy of the laser is imparted
into the skin to form pores that permit the transit of drug through the stratum
corneum from, for example, a topically applied patch or gel. The other process
entails the formation of a laser-induced stress wave (5) which transiently increases the permeability of the stratum corneum to drugs. Zharov and Latyshev
(6) have termed these processes laser-enhanced optoacoustic impregnation and
optoacoustic impregnation, respectively. Combination of both methods into a process that ablates stratum corneum and increases its permeability through stress
wave formation has been termed laser injection (6). Though optoacoustic delivery
(also termed photomechanical delivery or laser ultrasonic transport) is technically
noninvasive, it will be discussed here in the context of laser-assisted transport.
    The objective of laser-enhanced optoacoustic impregnation is (1) to reduce
the thickness of the stratum corneum and thereby reduce the transit distance
through the skin of permeant drugs or (2) to cut holes through the stratum corneum to allow drug permeation directly into the epidermis. Effectiveness of the
ablation relates to the energy input, which correlates to wavelength, pulse length,
and pulse energy (7). High absorption of the laser energy (8) is necessary to
minimize damage to underlying structures. According to Nelson et al. (8) there
are two optimal wavelengths at which skin ablation can be achieved: a short-
wavelength ultraviolet, which is absorbed by tissue proteins, and a midinfrared
[e.g., $\lambda = 2.94$ μm, (6), $\lambda = 2.79$ μm, (8)], which is absorbed by tissue water.
During laser irradiation, the energy is absorbed by the components of the skin

as vibrational heating. Water within the irradiated area of the skin quickly reaches its boiling point, and the resulting vapor pressure elicits a microexplosion that forms the ablation, as the tissue vaporizes (8). The rapid loss of energy from the ablated site protects the surrounding skin tissues from heat-induced damage.

The ability to control the level of energy imparted to the skin allows this to be a well-controlled method of removing stratum corneum if the applied energy is sufficiently low. Nelson et al. (8) applied a pulsed erbium:YSGG laser (250 μs pulses, $\lambda = 2.79$ μm, 1 J/cm^2, 2 mm spot size) to excise porcine skin in vitro and found that electrical resistance (a surrogate for stratum corneum penetration) dropped dramatically from 200 kΩ/cm^2 to 30 kΩ/cm^2 after about 14 laser pulses were applied. This indicated that the extent of stratum corneum penetration by the laser could be controlled through the number of laser pulses applied at this energy level. As stated by these workers, "ablation with 12 small steps is easier to control than one big ablative step" (8). Under these conditions, they observed an approximately 2.9- and 2.2-fold enhancement of the permeability constants of in vitro transdermal flux of hydrocortisone or γ-interferon, respectively, through swine skin compared to untreated controls after ablating 12.6% of the skin sites. It is intuitive that larger ablation of the site should enhance drug flux. However, maximum increase in flux compared to untreated controls was no more than 4- or 5-fold for these drugs after tape stripping (equivalent to 100% ablation), implying that the theoretical maximum flux achievable in this system was approximately 5-fold over untreated skin. Thus laser ablation was very effective although only 12.6% of skin area was treated.

Laser-induced photomechanical stress waves are compressive waves that interact directly on cells (5,9,10) to increase the permeability of plasma membranes without affecting cell viability (5). In contrast to the technique previously described, the laser is applied to a target that is in contact with a drug solution (10) or the drug solution itself (6). The energy of the laser is strongly absorbed by the target or surface of the water, and this produces an ultrasonic pressure wave that propagates through the solution to the drug/skin interface. Zharov and Latyshev (6) suggest that the compressive wave drives the drug through "natural physiological skin pores." The route of delivery by this mode is uncertain, but three possible routes have been postulated: transappendageal, transcellular, and intracellular (10).

Work from Doukas's lab (e.g., 5,10,11) suggests that photomechanical waves may directly modulate the permeability of cells, including epidermis, which could thus favor the latter routes of delivery. For example, Lee et al. (11) showed that the application of a single laser-induced photomechanical wave allowed the penetration of 40,000 MW dextran molecules or 20 nm diameter latex particles to a depth of approximately 50 μm in excised rat skin. The laser pulse was produced by a Q-switched ruby laser, which ablated a black polystyrene target positioned on a drug solution contacting skin. The laser energy was absorbed by the target (i.e., the acoustic coupling medium), and the resulting photomechanical wave was propagated through the drug solution to the skin.

Similarly, 5-aminolevulinic acid has been delivered into the skin of human subjects via a single laser-induced photomechanical pulse (10). Interestingly, measurement of drug penetration was achieved by the determination of the fluorescence intensity of protoporphyrin IX, a metabolic product of 5-aminolevulinic acid that fluoresces at 634 nm, while 5-aminolevulinic acid does not fluoresce at this wavelength. After photoacoustic delivery of 5-aminolevulinic acid into the skin, the fluorescent signal was observed to increase until it reached a maximum about 6 to 12 hours later. The total fluorescence intensity observed appeared to correlate positively with the amount of applied pressure (i.e., 388, 442, or 503 bar). When 5-aminolevulinic acid was applied to skin 15 to 30 minutes after application of a photomechanical pulse, little or no fluorescence was later observed, suggesting that the increase in skin permeability caused by this process was transient and that the barrier function of the stratum corneum recovered.

Though the process by which photomechanical waves increase skin permeability is not fully known, Lee et al. (10) suggest that it is probably distinct from other energy-induced systems such as phonophoresis or sonophoresis (12,13,14), to which ultrasonic cavitation has been attributed as the means for flux enhancement. Since this entails formation of microbubbles within tissue, cellular damage is possible. In contrast, according to Lee et al. (11), photomechanical waves do not produce cavitation.

Thus the technical feasibility of laser-assisted drug delivery has been demonstrated. Many questions remain regarding the efficacy of the method, not to mention the safety of using a high-powered laser for skin treatment. We believe that this technology will have limited utility in the near future. This is because of the present cost of medical lasers and the inconvenient work flow required (1) to treat the skin, (2) to ensure that the ablation is patent and aseptic, and (3) to apply a transdermal drug delivery system such as a patch or gel. We expect that this delivery modality may fulfil niche applications within hospital settings. In this regard, several companies are developing hand-held laser systems for skin ablation, but the main interest is in developing minimally invasive sampling systems for glucose monitoring, though these might also be useful for drug delivery applications (15).

2. Direct Heat-Induced Microporation

Eppstein and colleagues have developed a novel system for producing microscopic holes in the stratum corneum to facilitate transdermal drug delivery and interstitial fluid sampling (16). It is described in a patent application that also claims mechanical puncturing via microlancets, stratum corneum ablation by focused ultrasonic energy, fluid jet injection, and electrical pulses, and it includes claims for a method for producing skin tattoos (16). The system of interest comprises a wire mesh through which current is applied and which, through the resistance, produces heat. Resultingly, small holes are burned into the stratum corneum that are stated to allow the ingress of drugs into the dermis. To our knowledge, the utility of the system has yet to be demonstrated in clinical trials.

B. Mechanical Systems

1. Tape-Stripping

The simplest method for reducing the barrier imposed by the stratum corneum is to remove it. In this and the next section we discuss methods presently used to increase skin permeability by mechanical removal of the stratum corneum and thereby to facilitate enhanced transdermal drug flux via passive delivery systems. In vivo removal of the stratum corneum by tape stripping is done by repeated application of adhesive tape to the surface of skin. The technique sees wide usage in studies of wounding and cosmetics. The method is relatively uncontrolled, and it is a cumbersome method for increasing skin permeability, but it is included here for completeness.

Pulling the tape from the skin removes the upper mostlayer of cells. Multiple strips remove substantial skin barrier as evidenced by increases in transepidermal water loss of 20 to 25 times (17). In situ occlusion of human skin to increase hydration has shown to improve the effectiveness of the removal of the skin layers; for example it reduces the number of strips required to peel off the stratum corneum from 29 to 10 (18). It was found that hydration weakened the intracellular attachments within the stratum corneum (19), thereby facilitating skin layer removal (18). The type of adhesive surface used is also critical to the process, since adhesives vary in adhesive ability and hygroscopicity (17). One study reported that tape stripping of human subjects was more effective in older subjects (59–71 years) than in younger (24–35 years), suggesting that adherence of corneocytes is less in aged patients (20). These techniques of stratum corneum removal can be somewhat painful to the subject (21). In fact tape-stripping has been claimed for *removing* noxious agents *from* skin (22).

Singer et al. (21) assessed the effect of stratum corneum removal to accelerate the anesthetic effects of EMLA cream in a clinical study of adult patients undergoing catheter insertion. EMLA is a eutectic mixture of lidocaine and prilocaine (21). It is applied topically for 60 to 90 minutes prior to catheterization in order to provide topical anesthesia at the insertion site. Successful delivery of anesthetic was measured as a reduction of pain during catheterization as scored by the patients on a visual analogue scale (VAS). It was found that VAS scores were significantly reduced when the anesthetic mixture was applied after tape stripping versus untreated skin controls, and the success rate for catheter insertion was 91% versus 74% for the controls (21). In summary, removal of the stratum corneum by tape stripping resulted in more rapid onset of EMLA anesthesia. Nonetheless, as stated above, this is a cumbersome and uncontrolled method for increasing skin permeability.

2. Suction Ablation

Suction blister formation is an alternative method for removing the stratum corneum. It is done by application of vacuum suction to a limited area of skin to produce a blister, the upper surface of which is excised to reveal a portal for entry of drugs into the dermal circulation (23) and sampling of tissue fluids (24). According to Svedman et al. (25), this technique splits the skin at the stratum lucidum (see Chapter 1), forming a bleb above the dermis and leaving the skin vasculature and nociceptors intact. Consequently, this is apparently a painless and safe process in which the subject may experience a transient tingling sensation (25) while the "erosion" is performed, prior to application of a passive transdermal patch. Such erosions have been developed by the application of negative pressure of 180 mm Hg and warming to 38°C on the surface of skin for 15 to 70 minutes (24). The resulting epidermal vesicle was excised to expose the erosion. In contrast to skin blisters produced by compression or burns, and which incur microvascular damage, ischemia, and necrosis, the method described by Svedman and colleagues (25,26) has not been found to produce tissue trauma. Consequently, the production of such erosions is painless and does not form scars.

Tests in normal human volunteers demonstrated that plasma concentration/time curves for 1-deamino-8-D-arginine vasopressin (DDAVP) administered by this method approached zero order (26), indicating that infusionlike delivery profiles might be obtained. Moreover the patency of the skin erosions was undiminished during the four days that the experiments were carried out, as indicated by sustained blood levels of DDAVP in the volunteers.

Svedman (23) holds a patent on this technology, and a commercial product, Cellpatch™, (Epiport Pain Relief, Sweden) incorporates all components of the process: suction device, epidermotome (to remove the blister), and a drug reservoir. Clinical studies have tested the feasibility of transepidermal morphine delivery by this methodology in normal healthy volunteers (27) and in postoperative patients (25). In the latter clinical trial, drug reservoirs containing 5 mL of aqueous morphine solution (20 mg/mL) were applied to the skin for 48 hours after the suction ablation treatment. In two studies, steady state levels of about 15–18 nmol/L morphine were substantially obtained within the first few hours of administration. This declined to about 10–12 nmol/L after 30 hours. C_{max} values of 17.3 $\pm$ 3.7 nmol/L and 20.9 $\pm$ 7.7 nmol/L with respective CVs of 21 and 33% were seen for the two studies. The variability seen in the latter study was attributed to weight differences between the subjects. Total drug delivered during the 48 hours was determined to be 32 ± 6 mg (i.e., dose efficiency = 32%), based on the amounts of drug recovered from the spent cells at the end of the experiments. All patients received pethidine hydrochloride via I.V. catheter, and though parenteral administration of morphine was not compared against the Cell-Patch system in this study, no discomfort was noted by the patients using the CellPatch system; all preferred the Cellpatch procedure over catheterization.

During a separate study in human subjects, these same authors evaluated the effect of molecular size of fluorescently labeled dextrans on transepidermal delivery via the CellPatch, and the effect of the size of the blister (erosion) on the transdepidermal flux of labeled dextran. A roughly linear correlation was seen between percent absorption of a 3000 MW labeled dextran versus erosion size. That is, approximately 20% absorption of a 100 μmol dose occurred through an erosion of 6 mm², and about 60% absorption of the dextran occurred through an 80 mm² erosion, the maximum size of blister that was tested.

The results of the effect of molecular weight on delivery via CellPatch were unfortunately confusing. It was found that after 24 hours, delivery through a 6 mm² erosion, 37.9% of an initial dose of 3,000 MW fluorescent dextran was delivered, whereas 20.1% of a 70,000 MW fluorescent dextran was delivered. Based on an initial drug load of 100 μMol, as stated (25), this implies that the CellPatch delivered 0.11 grams of the 3,000 MW species and 1.4 grams of the larger 70,000 MW species! These counterintuitive results warrant further investigation, since administration into the body of this amount of protein-sized agent within 24 hours would be spectacular for *any* drug delivery system.

Despite this, the concept appears to be a less painful, more patient-acceptable method of delivering drugs that would otherwise be administered by I.V. catheter. From this perspective, the system is potentially safer because there is no risk of systemic infection. (Although epidermal infection is potentially possible with the CellPatch, it would have less catastrophic effects than systemic sepsis.) If the procedure were as fast and efficient as an I.V. catheter insertion it might have utility in some hospital scenarios. However, as mentioned above, Svedman et al. (25) state that the usual suction procedure takes 30 to 90 minutes, which is probably not amenable to typical clinical work flow and cost. Open questions are how long the suction-induced blisters will remain patent and what the long-term local effects could be in the skin. These authors found erythema but no inflammation at the erosions immediately after application, and 3 to 6 months after treatment the sites showed slight, fading pigmentation, suggesting that under the conditions tested, the procedure was relatively innocuous to long-term health of the skin.

This methodology has also been shown to enable the collection of epidermal interstitial fluid at rates of 1.7 to 3.7 μL per minute. The utility of measuring interstitial glucose compared to capillary blood concentrations has been demonstrated for up to 6 days in experimental human clinical trials that used this system (24). We note that several groups have issued patents (28–30) relating to similar suction-induced sampling technologies. This application may become greater in importance than drug delivery as the analytes present in interstitial fluid become known and new miniaturized technologies for direct sensing of body constituents become available. In conclusion, suction ablation appears to be a novel, potentially safe mode for long-term systemic administration of drugs, though the work flow for application of such systems does not appear to favor general use.

3. Patch and Needle Arrays

Much of the technology to be described in this section comes from the patent literature. The language of intellectual property tends to be broadly inclusive, usually without rigorous scientific demonstration. Therefore in our descriptions of the systems we have tended to interpret their general elements to compare and contrast them.

Gerstel and Place (31) of ALZA appear to be the first to have published on the need to penetrate the skin in order for transdermal patch systems to overcome the stratum corneum barrier. Their patent described a patch system containing multiple needles for transdermal drug delivery. The device comprised a reservoir with a "plurality of projections" on its underside to provide percutaneous drug diffusion for either local or systemic delivery (Fig. 1). It is interesting that they described "percutaneous" as penetration through the skin by "puncturing, scraping or cutting the stratum corneum" without doing the same to the interior layers of the skin. The concept appears to have derived from skin penetration methods for vaccination, where local intradermal vaccine administration is sufficient for immunization.

Based on the needle technology of the time, they described the needle "projections" as 15 through 40 gauge in diameter with heights of 5 to 100 microns, this being the heights claimed to penetrate the skin to allow drug delivery without

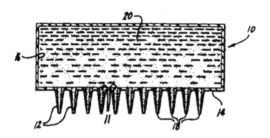

Figure 1 Drug delivery device. (From Ref. 31.)

contacting the nerves within the skin. The difficulty in manufacturing such devices using standard metal cannulae may account for the lack of published scientific studies on this concept. However, it was precient work considering that advances in silicon fabrication technology have now made it possible to test the concept (see further).

Further evolutions of this concept have surfaced in ALZA's "skin interface technology" which has been used to enhance transdermal delivery of agents, particularly protein and peptide drugs, from their E-Trans iontophoretic systems. Zuck (32) and Trautman et al. (33) described systems composed of multiple "microprotrusions" of 25 to 400 μm length on a structural support (e.g., a flat sheet, multiple sheets bolted together, or a circular coil) to enable penetration of the skin (Figs. 2 and 3). These systems were claimed for drug delivery applications with "electrotransport devices, passive devices, osmotic devices, and pressure-driven devices." Likewise these inventors, as well as Cormier and Theeuwes (34), claimed these devices to be effective for sampling of fluids from skin for diagnostic detection of body analytes, such as glucose.

Similarly, Effenhauser (35) disclosed a patch system most particularly for iontophoretic delivery that was composed of a reservoir connected to cutting edges and/or a "multitude of (hollow) needles" of about 1 to 1000 μm length (Fig. 4). Lee and Schin (36,37) invented an integrated patch system, "including

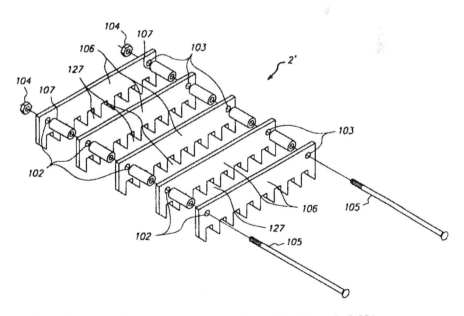

Figure 2 Device for enhancing transdermal agent flux. (From Ref. 32.)

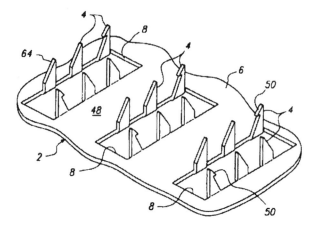

Figure 3 Device for enhancing transdermal agent flux. (From Ref. 34.)

plural skin needles" to form "the drug pathway on the skin by micropiercing by penetrating a plurality of skin needles into the skin epidermis at the treatment site and transferring the ionized drug into the skin at the treatment site by ionto-phoretic force" (Fig. 5). They claimed transdermal delivery of peptide and protein drugs by this system.

Eppstein (38) described a device that was likewise useful for reducing the barrier properties of skin or mucosa. Its claimed use was for delivery of drugs or for sampling analytes (i.e., via interstitial fluid) from the body. The device consisted of a base with a reservoir from which extended "puncturing members" of sufficient length to reduce the barrier properties of "skin or mucosa," which he claimed were 30 to 50 microns. The density of the "puncturing members" was described as being up to 50% of the surface contacting the skin (Fig. 6).

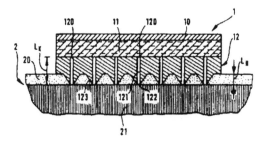

Figure 4 Transdermal system. (From Ref. 26.)

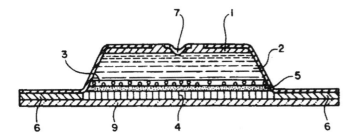

Figure 5 Transdermal administration method of protein or peptide drug and its administration thereof. (From Ref. 36.)

Interestingly, the use of vibrations (e.g., ultrasonic frequency) to facilitate "efficient and non-traumatic" penetration was described. See further regarding use of ultrasonic vibration to enhance silicon microneedle penetration through tissues.

The systems described above are intended to facilitate the transdermal delivery of drugs, including proteins, by the use of skin penetration. To our knowl-

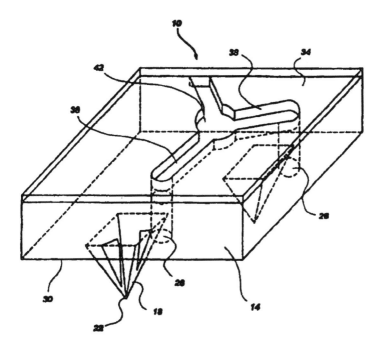

Figure 6 Multiple mechanical microporation of skin or mucosa. (From Ref. 38.)

edge, the workers from ALZA are the only ones to provide public demonstration of the utility of their systems (e.g., IBC conferences 1997–1999). However, none of the above-mentioned delivery concepts have demonstrated success in animal or human subjects that has been presented in peer reviewed publications that we are aware of.

4. Skin Puncture Devices

Some researchers have concentrated on development of devices specifically to produce mechanical disruption of the upper skin in order facilitate transdermal drug delivery. The theme of this work concentrates on the devices themselves in expectation that once the stratum corneum is breached, the way is open for any delivery mode (passive, active patches, gels, etc.). A more recent variant of this concept has been to use silicon microfabrication technology to produce microscopic holes in the stratum corneum; this will be dealt with separately below.

Jang (39,40) invented a system that comprised a roller or a series of disks from which extended needles that were intended to cut the surface of the skin (Fig. 7). Rolling the device over the skin was intended to produce multiple small cuts to provide passages for ingress for drug. Similarly, Godshall (41) and Godshall and Anderson (42) described a skin puncture system composed of a "cutter having a plurality of microprotrusions having a height chosen with respect to the layer of skin that is to be disrupted." That is, it is a bladelike apparatus with small teeth that cut only into the stratum corneum but not into the epidermis (Fig. 8). This was accomplished by a stop, which prevented the blades from cutting too deeply. Application of the protein-staining dye coomassie blue was used to

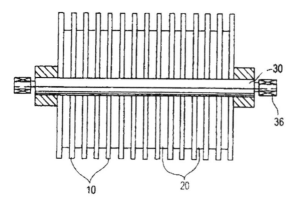

Figure 7 Skin perforating device for transdermal medication. (From Ref. 39.)

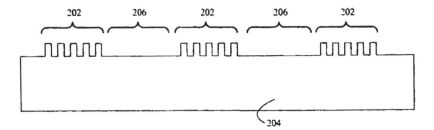

Figure 8 Apparatus for disruption of the epidermis. (From Ref. 42.)

confirm the depth of the cuts, since it was claimed that epidermis was stained specifically by the dye whereas stratum corneum was not.

Finally Kamen (43) developed a system composed of a platen or sheet (e.g., a metal sheet) into which were punched holes from which protruded jagged edges. The edges formed "micropenetrators" or cutting edges that were intended to produce an array of cuts when contacted to the skin (Fig. 9). Treatment of skin, by the device containing a "plurality of micropenetrators," was intended to improve

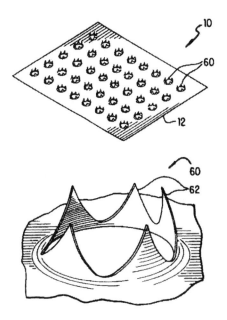

Figure 9 System for delivery of drugs by transport. (From Ref. 43.)

the delivery of a liquid drug into tissue by production of "microfissures" in skin. As mentioned earlier, none of these systems has been demonstrated to be clinically efficacious in peer-reviewed publications. Though the concepts are interesting, numerous questions must be addressed before such devices could enter general use. These were enumerated above and relate mainly to the ease of use, efficacy, and safety of these systems.

C. Microfabricated Systems

1. Overview of Fabrication Technology

Many of the previously mentioned devices appear to us to be cumbersome to use and probably also to manufacture. However, new technology for manufacturing of microscopic three-dimensional ("high aspect ratio") structures from silicon and other materials promises to revolutionize how medical devices will be manufactured. The advent of chip lithography in the microelectronics industry has resulted in the rapid expansion of instruments available for etching microstructures in silicon, platinum, germanium, and other metals. These instruments may be viewed as precision "milling" equipment, capable of etching structures with micron-scale dimensions into an otherwise monotonically flat surface. They may operate on the principles of laser ablation, ion bombardment, deep ion milling, or on the simplest acid/base reactions to produce semiconductor pathways, and in the last decade they have also been used to produce mechanical structures such as gears, valves, and rotors of microscopic size, commonly called microelectromechanical systems (MEMS) devices.

    These largely mechanical microdevices are conveniently integrated with semiconductor electronic functions to produce hybrid devices that may be operated remotely by electrical impulse, or may trigger an electronic response to a mechanical stimulus. Commercial applications of these devices have developed first in the aerospace and automotive industries, where sensitive pressure transducers and actuators are needed for precise measurement of structural flex, part fatigue, and impact energy. Those people unfortunate enough to have experienced intimate contact with an activated automotive airbag have experienced the utility of these transduction devices first-hand. Newly emerging applications in the communications industry take advantage of the microscale manipulation of light and electrical energy possible with MEMS "photonics" chips in which thousands of micron-scale mirrors may be electronically manipulated for signal transduction and amplification. These microdevices are enabling a new era of cellular communications, which in turn may indirectly introduce more consumers to the automotive airbag.

    Medical applications of MEMS were first developed around pressure transducers for catheters and heart monitors. These devices share similar architecture with signal transducers used in heavy industry applications, and do not perform any mechanical operations directly on human tissue. In essence they are integrated (or "packaged," in microelectronics parlance) directly into a larger device where they perform a purely electrical or mechanical function. The second area exploited in the medical arena is microsurgery, where small forceps, scalpels, and cutting rotors have been employed in ophthalmic surgery and angioplasty. These devices perform a strictly biomechanical function in that they contact and manipulate biological tissue directly and are used to affect mechanically the desired therapeutic outcome. The third area to exploit the medical potential of microdevices is drug delivery, in which microdevice architectures have been developed as electroactivated drug reservoirs and as the in vivo component of small drug delivery systems. The fourth area in which medical MEMS devices are being developed is cellular and molecular diagnostics, where small mixing chambers and microfluidic circuitries have been patterned to allow the collection, separation, amplification, and delivery of target species for analysis. We describe below some architectures that have been developed for the direct delivery of prophylactics and therapeutics in vivo.

2. Drug Delivery Applications

a. Overview of "Microneedles" As mentioned previously, the needle/syringe combination has become the drug delivery mainstay for drugs and vaccines deemed ineffective by other routes and has been optimized as a commodity scale product the world over. It is therefore not surprising that the needle architecture to which we are so accustomed is the focus of the first microdevice for drug delivery: "microneedles." Microneedles may be defined as needles that are 10–2000 microns in height and 10–50 microns in width. They may be solid or hollow, depending on the desired therapeutic outcome or tissue targeted, and they may be integrated with a syringe, pump, or patch for active or passive delivery to the target tissue. All architectures of these microneedles share three *hypothetical* attributes that are the subject of current research. (1) They allow the infusion or injection of drugs and vaccines into the intradermal space. (2) They are potentially less painful than standard "macroneedles." (3) They may enable the delivery of vaccines and DNA-based therapeutics directly to antigen-presenting cells or affected tissue. We describe below the architectures of some microneedle systems made to test these hypothetical attributes and the processes for their fabrication.

b. Single Hollow Microneedles Standard etch techniques result in very simple needle shapes. Single hollow microneedles may be fabricated directly from a standard silicon or metal wafer by plasma-etch (SF6 and the like) or wet-etch (potassium hydroxide or other strong base) techniques. These techniques result in microneedles of simple conical or straight-lumen shapes as shown in Figure 10 below.

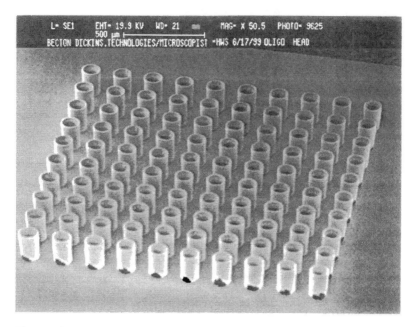

Figure 10   Silicon microneedles fabricated via wet etch of a standard 4 inch wafer with potassium hydroxide.

More complex microneedle architectures may be developed via silane deposition from plasma into silicon molds fabricated by etch and ion milling techniques (44,45,46). Figure 11 shows a microneedle "mold" fabricated from a silicon wafer via standard wet etch. This mold is the negative image of a microneedle formed from it via conformal deposition of tetramethoxysilane, tetraethoxysilane, or other silcon vapors to the mold's surface. Carefully controlled deposition leads to the hollow, single microneedle shown in Fig. 12. The body of the resulting needle may consist of polysilicon or silicon oxide, depending on the starting monomer used. Figure 13 shows other microneedle architectures fabricated by this plasma deposition technique.

The utility of these single microneedle architectures for the intradermal delivery of therapeutics and prophylactics is under investigation. Thus far there is no body of data to suggest that needles of 100–500 microns length will be efficacious for the delivery of all drugs currently delivered by standard intramuscular or venous injection, despite claims of purported painlessness. Lin and Pisano (46) showed that single silicon-processed microneedles remained intact as they penetrated Porterhouse steak (though we doubt that painlessness and mechanical stability are essential features for drug delivery products targeted to the dead bovine market).

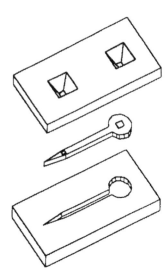

Figure 11 Silicon master mold for conformal deposition of silane vapor to form a hollow microneedle.

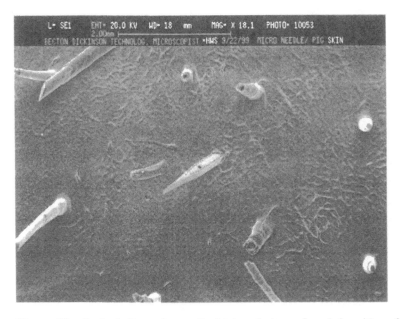

Figure 12 Single, hollow microneedle fabricated via conformal deposition of silane piercing excised porcine skin. Note size of needle in relation to hair follicles. Outlet port is at side of needle.

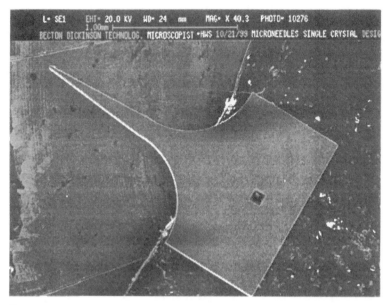

a

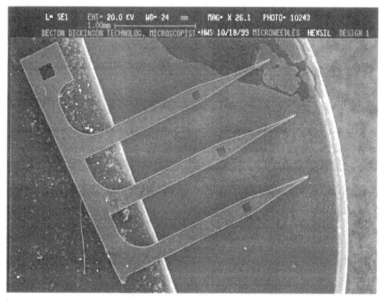

b

Figure 13   Hollow microneedle systems of varying functionality.

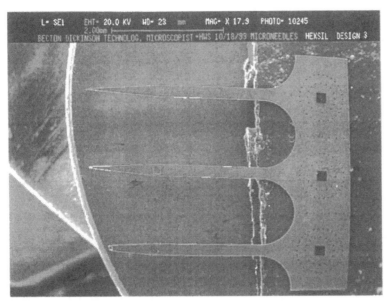

c

Figure 13   continued

c.   *Hollow Microneedle Arrays*   Arrays of closely spaced microneedles may be fabricated by wet-etch or plasma-etch or micromachining techniques. These arrayed systems have enjoyed a high profile in the past three years because they are envisioned as painless, high-flow alternatives to standard needles and single hollow microneedles alike. The microneedle array may consist of as few as two to as many as 1,000 or more needles arranged on the surface of the silicon or metal chip from which they are etched. In general, the current fabrication technology for silicon or metal microneedle arrays allows the reliable production of needles in the 100–500 micron range. Examples of such hollow microneedles arrays are shown in Figs. 13, 14, and 15.

The true utility of hollow microneedle arrays over single-needle systems has yet to be proven. McAllister et al. (47,48) have demonstrated that arrays of hollow microneedles fabricated from epoxy micromolds and electroplating were able to penetrate excised human epidermis. Similarly, they have found that the needles permitted the flow of water and showed minimal clogging when inserted into epidermis (47). These are positive first steps in demonstrating the utility of these new platforms. Frazier and colleagues (49–51) have fabricated linear arrays of metal microneedles (palladium coated on silicon). The needles had inner dimensions of 40 by 20 μm and outer dimensions of 80 by 60 μm and were arranged

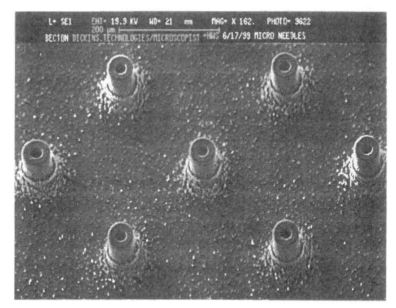

a

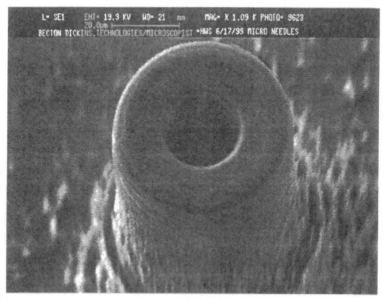

b

Figure 14 Microneedle array fabricated from silicon via wet etch with potassium hydroxide.

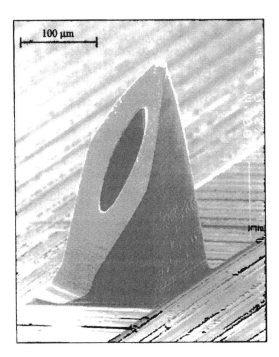

Figure 15 Single microneedle from an array micromachined from silicon wafer.

as chains of up to 25 per array. They described a prototype acrylic interface that packaged the needle arrays to the leur connector of a standard syringe. The unidirectional buckling load that each needle was calculated to withstand was 3.28 nN (or 82.1 mN or an array of 25 needles), in excess of pressure required to push the needles into skin (49). Similarly Bielen et al. (52) have described fabrication of nickle multielectrode arrays by a LIGA process (German: lithografisch, Galvanoformung, Abformung), though the application is for neural stimulation rather than transdermal drug delivery.

In general, hollow microneedle systems have several significant manufacturing and use challenges. Firstly, the quality of high-density arrays consisting of more than a dozen or so microneedles may prove to be difficult to control. Secondly, robust needles are necessary at every location in the array, for needles broken during delivery will likely result in significant leakage and misdosage. Thirdly, the spacing and number of needles within the array significantly affects the force required for penetration into the dermis. This feature becomes especially important given the inelasticity of skin and the likely nonlinearity of the force/needle relationship. Nevertheless, microneedle arrays have been shown to penetrate the upper layers of excised skin, as shown in Fig. 16.

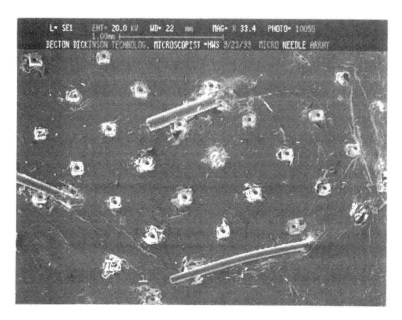

Figure 16 Machined silicon microneedle array piercing excised porcine skin. Needle holes are facing the viewer. Note size of needles and array in relation to hair follicles.

Ultrasonic methods currently under development may be applied to these devices to lower penetration forces and ease insertion by reducing the effects of skin inelasticity and the nonuniformity of the skin's stress–strain relationship. Lal and White (53) have developed micromachined silicon needles capable of producing ultrasonic velocities around 23 ms⁻¹ by bonding a silicon needle structure to a piezoelectric plate. Application of current results in a vibrating cutting edge. According to these workers, silicon is well suited to this application. It can be machined by the techniques described above, it is harder than titanium under the conditions of vibration, and it is therefore a better cutting material. Also, as a resonator material, silicon has the capacity to be driven to higher velocities than other materials before failure occurs. The resulting devices developed by these workers have been shown to cut human (lens) and plant (potato) tissue and show promise for new devices with the ability easily (and possibly with less pain) to sample tissues for many clinical applications.

The microfluidics of drug solution delivery through small aperture structures like microneedles into the confined spaces of the epidermis or upper dermal layers and the resulting effect on these tissues has not been seriously studied to date. It is not difficult to envision blister formation (and subsequent leakage) arising from the rapid delivery of drug solutions from hollow microneedles, single or arrayed. It has been posited that arrayed needles will deliver solution over a larger area per equivalent unit time, mitigating blister formation (48,49).

d. Solid Microneedle Arrays Arrays of solid ("no hole") microneedles may be fabricated by the same etch and micromachining techniques described above. Examples of solid arrays fabricated from silicon are given in Fig. 17. They may be used as "dermal enhancers" by producing openings in the stratum corneum to facilitate topical passive or iontophoretic delivery. The precise spacing and geometry of these needle systems allows the gentle abrasion of the skin to produce highly defined "furrows" (Fig. 18) through which drug solution may be driven. In situ measurements by our laboratory of transepidermal water loss in swine and humans has shown that such treatment dramatically increases the permeability of skin.

Henry et al. (54) described the fabrication and testing of a system composed of a 20 by 20 array of solid needles, 150 µm in height. They were produced by a reactive ion etch method. Those workers demonstrated insertion of the arrays into skin with relatively low penetration force (estimated at 10 newtons). The ease of penetration is presumably due to the sharpness of the needles (radii of curvature < 1 µm), a by-product of the fabrication technique. It should be mentioned that such high radii of curvature are presently not possible to obtain with today's metal grinding technologies used for conventional needles and lancets.

Simple puncture of excised human epidermis with solid microneedle arrays produced 1000-fold enhancement of calcein permeability (a fluorescent model compound) in vitro through human epidermis. When the arrays were left inserted for one hour before being removed, an increase in permeability of about 25,000 fold was observed. This suggests that the microscopic holes produced by these needle arrays may reseal if a single puncture is performed and the needles quickly removed. These workers also described the size of the holes produced by the arrays as being 1 micron. It is yet uncertain how long such holes remain patent, though elevated permeability was observed in vitro in excised human skin for up to 5 hours, the total time of the experiments. Confirmation that this treatment produced breach of the stratum corneum was demonstrated in vivo by a 50-fold drop in skin resistance (54).

Following use, less than a few percent of needles were found damaged, and the damage was localized to their tips. Likewise we have seen little or no damage following the use of solid microneedle arrays on skin. However, toxicity testing remains to be done to confirm that small quantities of residual silicon that may remain in skin after treatment are not harmful. Anecdotal findings from our laboratory and that of Henry et al. (54) indicate that solid microneedle treatment of skin is not regarded as painful by human subjects.

Eriksson et al. (55,56) described a technique they termed "microseeding," in which DNA was delivered into the skin of swine by use of oscillating solid

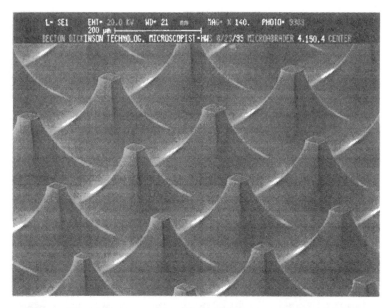

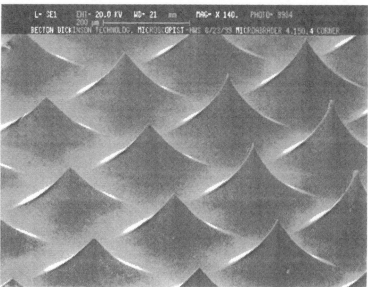

Figure 17 Solid microneedle arrays etched from silicon wafers. Needle tip geometry ranges from "mesa" to "point."

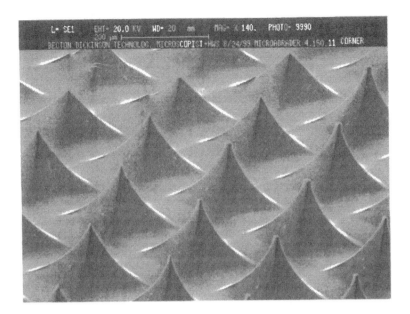

Figure 17 continued

Figure 18 Furrows trenched in porcine skin with solid microneedles.

small needles mounted on a tattooing device. The fabrication methodology for these microneedles was not given. The system they described is a potential minimally invasive delivery system for vaccines and gene therapy drugs and therefore warrants mention. The DNA solution was delivered via tubing connected to a syringe pump to the solid microneedles, which were inserted into the skin to a depth of 2 mm at a penetration rate of 7,500 times per 2.25 cm^2 per 20 seconds. Three gene transfer experiments were performed, using DNA plasmids containing the genes for human epidermal growth factor (hEGF), or β-galactosidase (β-gal) or swine influenza virus hemagglutinin. Two days after treatment, in separate experiments they found expression of hEGF or β-gal in the treated skin sites, while sites that received negative control DNA showed no detectable expression of these genes. As well, swine that were microseeded with the HA gene were protected from subsequent infection by swine influenza virus, further indicating the success of this technique for in situ delivery of DNA. It is clear that such methodology could hold promise for immunization via DNA vaccines.

In passing, we should also mention the pioneering work of Reid et al. (57) who have developed stents containing solid silicon microneedle systems they term "microprobes." They were developed to anchor the stents in arteries, and from them DNA was administered into the vessel walls as prospective systems for restenosis treatment. Similarly, they and others (58,59) have described microneedle systems for administration of genes into plant and animal tissues. In summary, this appears to be an emerging but fertile area of research for delivery of drugs and DNA into skin.

  e. Packaging of Microneedle-Based Drug Delivery Systems   Handling by patients or healthcare workers in the macroscopic world is an inherent challenge confronting microdevices that perform biomechanical functions directly on tissue. The current state of microfabrication technology allows the construction of precise architectures with the limitation of physical fragility. This is especially true of hollow high-aspect-ratio parts like microneedles. Although advances in surface-smoothing and coatings technologies are likely to improve device robustness, it is likely that the packaging of the microdevice will play a most important role in the development of procedures for their use. Integration of the microdevice in a larger system that will be handled by the practitioner and connection to a fluidic system and reservoir for solution or dry-powder-based formulations is not a trivial undertaking. It is not unreasonable to assume that device package as a whole "enables" the use of the microdevice, and that the entire drug delivery package will need to be developed for testing well in advance of clinical trials. However, the current state of development is largely focused on developing scaleable microfabrication processes and efficacious architectures for the microdevice. We may expect to see significant effort in package development for these microneedle systems in the coming decade.

III. FUTURE APPLICATIONS—MICROMECHANICAL SYSTEMS FOR DRUG DELIVERY AND DIAGNOSTICS

With the advent of micromachining technology, new opportunities exist to marry drug delivery and diagnostics, with electronic monitoring and reporting (see Ref. 60 for review). A first step toward this goal has come from the work of Santini et al. (61) who developed a micromachined chip capable of controlled drug delivery. The device comprises multiple microreservoirs, each covered by a gold foil that is electrically connected to a microelectronically controlled power source. Timed application of current results in electrolysis of the foil covering specific reservoirs. Consequently, this allows the controlled and potentially intelligent delivery of drugs from the reservoirs.

It is not difficult to imagine that this type of system could be linked to an analyte detection system, which upon activation could release drugs. This so-called closed loop delivery system has been envisioned for diabetic glucose control (e.g., Ref. 60). It appears that MEMS technology may finally enable such applications. However, since silicon machining technology is intimately associated with electronics manufacture, the opportunity exists to build easily and inexpensively coincident monitoring and communication systems onto MEMS drug delivery systems. It can be envisioned that worn or implanted "pharmacy on a chip" (61) systems can or will be developed that will "detect, deliver and report." We believe that development of such systems looms very close. Regardless of how sophisticated such systems will be, we believe that their deployment will depend on how well they satisfy the needs of drug delivery as described at the beginning of this chapter. To compete successfully, these new systems will provide therapeutic benefit through safe, efficacious delivery of drugs and/or diagnostic reporting coupled with ease of use and value-equivalent cost.

REFERENCES

1. Becton Dickinson and Co. Celebrating the First Hundred Years, 1897–1997, 1997.
2. DA Bucks. Skin structure and metabolism: relevance to the design of cutaneous therapeutics. Pharmaceutical Research 1:148–153, 1984.
3. JH McElhaney. Dynamic response of bone and muscle tissue. J Appl Physiol 21:1231, 1966.
4. P Tong, YC Fung. The stress–strain relationship for the skin. J Biomech 9:649, 1976.
5. DJ McAuliffe, S Lee, TJ Flotte, AG Doukas. Stress wave assisted transport

through the plasma membrane in vitro; Lasers in Surgery and Medicine 20: 216–222, 1997.

6. AP Zharov, AS Latyshev. Laser ultrasonic transport of drugs in living tissues. Ann NY Acad Sci 11:66–73, 1998.

7. SL Jacques, DJ McAuliffe, IH Blank, JA Parrish. Controlled removal of human stratum corneum by pusled laser to enhance percutaneous transport. U.S. patent 4,775,361, 1988.

8. JS Nelson, JL McCullough, TC Glenn, WH Wright, L-H Liaw, SL Jacques. Midinfrared laser ablation of stratum corneum enhances in vitro percutaneous transport of drugs. J Invest Dermatol 97:874–879, 1991.

9. AG Doukas, TJ Flotte. Physical characteristics and biological effects of laser-induced stress waves. Ultrasound in Med Biol 22:151–164, 1996.

10. S Lee, N Kollias, DJ McAuliffe, TJ Flotte, AG Doukas. Topical drug delivery in humans with a single photomechanical wave. Pharmaceutical Research 16:1717–1721, 1999.

11. S Lee, DJ McAuliffe, TJ Flotte, N Kollias, AG Doukas. Photomechanical transcutaneous delivery of macromolecules. J Investigative Dermatol 111: 925–929, 1998.

12. S Mitragotri, DA Edwards, D Blankschtein, R Langer. A mechanistic study of ultrasonically-enhanced drug delivery. Journal of Pharmaceutical Sciences 84:697–706, 1992.

13. S Mitragotri, D Blankschtein, R Langer. Ultrasound-mediated transdermal protein delivery. Science 269:850–853, 1995.

14. NN Byl. The use of ultrasound as an enhancer for transcutaneous drug delivery: phonophoresis. Phys Ther 75:539–553, 1995.

15. SM Reiss. Glucose- and blood-monitoring systems vie for top spot. Biophotonics Interntational, May/June pp. 43–45, 1997.

16. JA Eppstein, MR Hatch, D Yang. Microporation of human skin for drug delivery and monitoring applications. European patent application WO 97/07734, filed August 1996.

17. J-C Tsai, ND Weiner, GL Flynn, J Ferry. Properties of adhesive tapes used for stratum corneum stripping. International Journal of Pharmaceutics 72: 227–231, 1991.

18. DA Weigand, JR Gaylor. Removal of stratum corneum in vivo: an improvement on the cellophane tape stripping technique. Journal of Investigative Dermatology 60:84–87, 1973.

19. RH Wildnauer, JW Bothwell, AB Douglas. Stratum corneum biochemical properties. 1. Influence of relative humidity on normal and extracted human stratum corneum. Journal of Investigative Dermatology 56:72–78, 1971.

20. KC Moon, KP Wilhelm, HI Maibach. Sequential tape stripping of human stratum corneum: influence on pH, water content, adherence: comparison of young and aged. Clinical Research 37:759A, 1989.

21. AJ Singer, J Shallat, SM Valentine, L Doyle, V Sayage, HC Thode Jr. Cutaneous tape stripping to accelerate the anesthetic effects of EMLA cream in a randomized, controlled trial. Acad Emerg Med 5:1051–1056, 1998.

22. MR Carey. Adhesive tape application to human skin. U.S. patent 5,840,072, 1998.

23. P Svedman. Transdermal perfusion of fluids. U.S. patent 5,441,490, 1995.

24. P Svedman, C Svedman. Skin mini-erosion sampling technique: feasibility study with regard to serial glucose measurement. Pharmaceutical Research 15:883–888, 1998.

25. P Svedman, S Lundin, P Hoglund, C Hammarlund, C Malmros, N Patzar. Passive drug diffusion via standardized skin mini-erosion; methodological aspects and clinical findings with new device. Pharmaceutical Research 13: 1354–1359, 1996.

26. P Svedman, S Lundin, C Svedman. Administration of antidiuretic peptide (DDAVP) by way of suction de-epithelialised skin. Lancet 337:1506–1509, 1991.

27. D Westerling, P Hoglund, S Lundin, P Svedman. Transdermal application of morphine to healthy subjects. Br J Clin Pharmac 37:571–576, 1994.

28. H Ishibashi. Simple blood sampling device. U.S. patent 5,320,607, 1994.

29. S Saito, Y Kajiwara, A Saito. Method and apparatus of sampling suction effusion fluid. U.S. patent 5,882,317, 1999.

30. H Fujiwara, T Matsumoto. Sampling device of suction effusion fluid. U.S. patent 5,782,871, 1998.

31. MS Gerstel, VA Place. Drug delivery device. U.S. patent 3,964,482, 1976.

32. M Zuck. Device for enhancing transdermal agent flux. International patent application WO 99/29364, 1999.

33. J Trautman, P Wong, P Dadonna, H Kim, M Zuck. Device for enhancing transdermal agent flux. International patent application WO 99/29298, 1999.

34. M Cormier, F Theeuwes. Device for enhancing transdermal sampling. International patent application WO 97/48411, 1997.

35. CS Effenhauser. Transdermal system. International patent application WO 96/17648, 1996.

36. HB Lee, BC Shin. Transdermal administration method of protein or peptide drug and its administration device thereof. U.S. patent 5,250,023, 1993.

37. HB Lee, BC Shin. Device for the transdermal administration of protein or peptide drug. European patent, EP 0 509 122 B1, 1996.

38. J Eppstein. Multiple mechanical microporation of skin or mucosa. International patent application WO 98/00193, 1998.

39. K Jang. Skin perforating device for transdermal medication. U.S. patent 5,611,806, 1997.

40. K Jang. Skin perforating apparatus for transdermal medication. U.S. patent 5,843,114, 1998.

41. N Godshall. Micromechanical patch for enhancing the delivery of compounds through the skin. International patent application WO 96/37256, 1996.

42. N Godshall, R. Anderson. Method and apparatus for disruption of the epidermis. U.S. patent 5,879,326, 1999.

43. D Kamen. System for delivery of drugs by transport. International patent application WO 98/11937, 1998.

44. L Lin, AP Pisano. IC-processed microneedles. U.S. patent 5,591,139, 1997.

45. L Lin, AP Pisano. IC-processed microneedles. U.S. patent 5,855,801, 1999.

46. L Lin, AP Pisano. Silicon-processed microneedles. IEEE Journal of Microelectromechanical Systems 8:78–84, 1999.

47. DV McAllister, S Kaushik, PN Patel, JL Mayberry, MG Allen, MR Prausnitz. Solid and hollow microneedles for transdermal protein delivery. Proc Intl Symp Control Rel Bioact Mater 26:192–193, 1999.

48. DV McAllister, R Cros, SP Davis, LM Matta, MR Prausnitz, MG Allen. Three dimensional hollow microneedle and microtube arrays. Transducers 99—The 10th International Conference on Solid-State Sensors and Actuators, Vol. 2, pp. 1098–1101, Sendai, Japan, 1999.

49. JD Brazzle, S Mohanty, AB Frazier. Hollow metallic micromachined needles with multiple output ports. SPIE 3877:257–266, 1999.

50. J Brazzle, I Papautsky, AB Frazier. Micromachined needle arrays for drug delivery or fluid extraction. IEEE Engineering in Medicine and Biology, Nov/Dec: 53–58, 1999.

51. AB Frazier. Methods for preparing devices having metallic hollow microchannels on planar substrate surfaces. U.S. patent 5,876,582, 1999.

52. JA Bielen, AW Schmidt, R Weiel, WLC Rutten. Fabrication of multielectrode array structures for intra-neural stimulation: Assessment of the LIGA method. 18th Annual International Conference of the IEEE Engineering in Medicine and Biology Society, pp. 268–269, 1996.

53. A Lal, RM White. Micromachined silicon needle for ultrasonic surgery. IEEE Ultrasonics Symposium, pp. 1593–1596, 1995.

54. S Henry, D McAllister, M Allen, M Prausnitz. Microfabricated microneedles: a novel approach to transdermal drug delivery. J Pharm Sci 87: 922–925, 1998.

55. E Eriksson, F Yao, T Svensjo, T Winkler, J Slama, MD Macklin, C Andree, M McGregor, V Hinshaw, WF Swain. In vivo transfer to skin and wound by microseeding. J Surgical Research 78:85–91, 1998.

56. JS Slama, C Andree, T Svenjo, T Winkler, WF Swain, MD Macklin, E Eriksson. In vivo gene transfer with microseeding. Surg Forum 46:702, 1995.

57. ML Reid, C Wu, J Kneller, S Watkins, DA Vorp, A Nadeem, LE Weiss, K Rebello, M. Mescher, ACJ Smith, W Rosenblum, MD Feldman. Micromechanical devices for intravascular drug delivery. J Pharm Sciences 87: 1387–1394, 1998.
58. S Hashmi, P Ling, G Hashmi, ML Reid, R Gaugler, W Trimmer. Genetic transformation of nematodes using arrays of micromechanical piercing structures. Biotechniques 19:766–770, 1995.
59. W Trimmer, P Ling, C-K Chin, P Orton, R Gaugler, S Hashmi, G Hashmi, B Brunett, ML Reed. Injection of DNA into plant and animal tissues with micromechanical piercing structures. Proc. 8th International Workshop on Microelectromechanical Systems (MEMS-95), Amsterdam, pp 111–115, 1995.
60. I Fujimasa. Micromachining technology and biomedical engineering. Applied Biochemistry and Biotechnology 38:233–2242, 1993.
61. JT Santini, MJ Cima, R Langer. A controlled-release microchip. Nature 397:335–338, 1999.

11

Transdermal Drug Delivery System Regulatory Issues

Vinod P. Shah
Center for Drug Evaluation and Research, Food and Drug Administration, Rockville, Maryland, U.S.A.

I. INTRODUCTION

A transdermal drug product is intended to deliver the drug systemically to treat or prevent disorders in locations distant from the site of topical application. Adhesive patches and transdermal drug delivery systems (TDS) of defined shape and size are marketed for systemic action and are intended for the treatment or prevention of a systemic disease. Drug released from the TDS is absorbed through the stratum corneum (SC), epidermis, and dermis into blood circulation and transported to target tissues to achieve therapeutic effect. TDS are considered new drug delivery systems and often involve a demonstration of clinical safety and effectiveness of the drug (1,2). TDS products are considered controlled release dosage forms and should scientifically support in vivo and in vitro claims of controlled release features and should assure in vivo as well as in vitro reproducibility.

II. DRUG APPLICATIONS

TDS are regarded as new drugs and therefore have to be approved based on clinical safety and efficacy studies. If the toxicity and safety of the drug substance are established during the approval of a different dosage form, then the study need not be repeated for the approval of the TDS dosage form. Most of the TDS

Table 1   List of TDS Products in US Market

Drug	Number of firms	Dosage strengths
Clonidine	1	0.1, 0.2, 0.3 mg/24 h
Estradiol	6	0.025, 0.0375, 0.5, 0.075, 0.1 mg/24 h
Estradiol + Norethindrone	1	0.05 + 0.14, 0.05 + 0.25 mg/24 h
Fentenyl	1	0.6, 1.2, 1.8, 2.4 mg/24 h
Nicotine	4	7, 11, 14, 22, 28 mg/24 h
Nitroglycerine 7		0.1, 0.2, 0.3, 0.4, 0.6, 0.8 mg/h, and 15 mg/26 h
Scopolamine	1	1.0 mg/72 h
Testosterone	2	2.5, 4.0, 5.0, 6.0 mg/24 h

approved by the US Food and Drug Administration (FDA) fall into this category. In general, most of the drug substances were initially approved as oral dosage forms and subsequently as TDS. Because of this, biopharmaceutical considerations play an important role in regulatory approval of TDS drug products. Table 1 provides a list of TDS drugs and a number of manufacturers and dosage strengths as approved and marketed in the US.

A.   Regulatory Considerations for New Drug Applications (NDA) as TDS

The key considerations are

> TDS are regarded as new drugs and require scientific data to substantiate clinical safety and efficacy.
> Studies for the approval of NDA should be customized based on the critical nature of the drug, availability of marketed systemic dosage forms of the same active drug substance, medical and biopharmaceutical rationale, and literature data on a drug entity.
> If the drug substance is marketed in another form, than a comparative bio-availability employing known routes of administration and TDS should be determined.
> Clinical data to support labeling should be provided.
> TDS are regarded as controlled release dosage forms and therefore should demonstrate controlled release characteristics to support drug labeling.
> Biopharmaceutical considerations involve pharmacokinetic measurements and should define bioavailability (rate and extent) of drug absorption from TDS. The studies should also evaluate sites of drug administration for optimizing drug delivery and reproducibility of the system.

B. Regulatory Considerations for Abbreviated New Drug Applications (ANDA) as TDS

The most important study for an ANDA is a bioequivalence (BE) study. A BE study should be carried out using the test product and a reference product at the highest strength, in accordance with the Guidance: "Bioavailability and bioequivalence studies for orally administered drug products—general considerations" (3). The approaches discussed in this general guidance are also applicable to non–orally administered drug products such as TDS. Generally for modified (extended or controlled) release drug products, a replicate crossover study design with average bioequivalence criteria, a 90% confidence interval, and a bioequivalence limit of 80–125% is required for test product approval. There is less emphasis now on measurement of metabolite(s). Lower strengths of the products can be approved based on dose proportionality and in vitro dissolution profile comparison with the highest strength.

The generic product should employ a drug release mechanism similar to an approved TDS. If the release mechanism is significantly different, it may need safety and efficacy data.

The generic product should use approved adhesive and patch material. If it is different, it may need irritation study data.

The generic product should use approved inactive ingredients. If they are different, it may need safety data.

The generic product should also establish reproducibility of drug release characteristics.

In addition, the ANDA manufacturer should meet chemistry, manufacturing, and control (CMC) requirements.

Table 2 provides comparative summary information needed for NDA and ANDA approval.

Table 2 Requirements for NDA and ANDA Submissions

NDA	ANDA
Safety: Toxicity studies	
Skin irritation	Skin irritation
Cutaneous toxicity	Cutaneous toxicity
Cutaneous sensitivity	
Contact photodermatitis	
Efficacy: Clinical studies	
Bioavailability studies	Bioequivalence studies
Chemistry, manufacturing and controls	Chemistry, manufacturing, and controls
In vitro release studies	In vitro release studies

III. IN VITRO TESTING

In vitro testing of TDS is considered important in two key areas, (1) defining skin permeation kinetic studies using a diffusion cell system and cadaver skin during the drug development process, and (2) in vitro drug release kinetics, to be used for batch-to-batch release and as a compendial test.

TDS are marketed in varying strengths, sizes, and shapes and contain significantly different amounts of drug for a given rate of drug delivery. TDS contain much larger amounts of drug than the amount of drug to be delivered in the body. The amount of drug varies several fold among different brands for the same amount of drug delivery (e.g., nitroglycerin patches) (4). This makes it difficult to have the same amount or percent dissolution for a drug in a given time. Several methods are described in the US Pharmacopeia for in vitro drug release of TDS. However, some of the methods are complicated, cumbersome, and nonuniversal. The FDA has developed a simple, reproducible method of determining that in vitro release profile of a TDS (5,6). The method employs a watchglass-patch-teflon mesh sandwich assembly and the paddle method. The US Pharmacopeia has now adopted this procedure, which is referred as the paddle over disk method (Apparatus 5) (7). The method is applicable to all brands, shapes, and strengths of all marketed TDS, and is also useful for product stability indications.

When developing an in vitro release method, the following points should be considered: release procedure and its specifications should be designed to assure quality control and batch-to-batch uniformity; the in vitro release test should be capable of detecting manufacturing changes that influence product release properties; where feasible, the in vitro method could be developed to identify important changes in product performance under accelerated and room temperature stability test conditions, just as in vitro dissolution is used now for solid oral dosage forms. The method should be economical, simple, reliable, reproducible, and capable of automation; and it should avoid unnecessary method proliferation.

TDS are controlled release preparations. Because of different drug release mechanisms and different amounts of drug in the patches, it is difficult to have a single in vitro release specification for all brands of a given drug product. This is the same scenario as with oral controlled release preparations, where different brands have a different test method and specifications for a given drug.

IV. ENHANCER

For TDS products, a high flux is desired so that the drug can penetrate the stratum corneum so as to be available in sufficient amounts to the systemic circulation for therapeutic effect. Drug substances with high molecular weight and drug substances with polarity are not suitable for transdermal administration. Enhancement or penetration of selected drugs might significantly expand the list of drugs that could be delivered transdermally. From a regulatory standpoint, development of a drug for transdermal delivery poses several challenges to the drug development process, some of which are amplified with the addition of penetration enhancement methods (8).

The skin is structurally and functionally a complex, multilayered organ. The outermost layer of epidermis, the stratum corneum, is formed by continuous differentiation of adjacent viable epidermis. The SC is poorly permeable, especially to water-soluble compounds. The viable epidermis is not thought to present a significant barrier to drug penetration, except when highly lipophilic drugs are applied or when the SC is damaged. Because the SC is the dominant rate-limiting barrier to the skin, transdermal research has focused on facilitating drug transport across this barrier. The vehicle component of a TDS formulation can significantly influence drug penetration through the skin.

The primary objective in developing a TDS with chemical penetration enhancers has been to identify compounds that significantly enhance drug penetration without severe irritation or damage to the skin. Ideally, chemical permeation enhancers should be safe and nontoxic under conditions of use, pharmacologically inert, nonirritating, and nonallergic. The penetration effect should be predictable and reversible.

Unfortunately, a chemical penetration enhancer increases the permeability by reversibly damaging or by altering the nature of the SC to reduce diffusional resistance. Penetration enhancers can be used as cosolvents to increase the thermodynamic activity of the drug in the TDS, e.g., ethanol has been used as an enhancer-cosolvent in estrogen patches to increase drug penetration.

Enhancement methods usually increase the permeation not only of the drug but also of formulation excipients and of themselves as well. It is therefore important to know the fate of the enhancer in the body. In spite of being excellent penetration enhancers in TDS, many vehicles are limited because of deleterious effect on the skin, e.g., dimethyl sulfoxide (DMSO) is a powerful solvent and increases drug penetration, but at the same time it damages the structural integrity of the skin.

The enhancer effects should be evaluated to understand general toxicological implications, especially irritation potential under conditions of long-term occlusion.

A chemical enhancer should be

Safe and nontoxic
Pharmacologically inert
Nonirritating and nonallergenic

Predictable and reversible in its duration of action
Chemically and physically compatible

We need to know, about a chemical enhancer, the mechanism of action and the fate of the enhancer in the body.

TDS containing chemical enhancers are not considered pharmaceutically equivalent to TDS formulations lacking chemical enhancers. The enhancement technology imposes a sufficiently significant change in formulation, so that pharmaceutical equivalence can no longer be justified.

Physical methods for enhancing drug penetration raise different types of concerns that include destruction of the SC with high-current setting and general skin irritation.

V. RESIDUAL DRUG

Labeling should indicate the delivery rate of the drug from TDS. Knowing the amount of the drug in the TDS, the delivery rate, and the amount remaining in the TDS after it is removed, it should be possible to compute the apparent amount of the drug released and delivered to the body (2).

VI. SCALE-UP AND POSTAPPROVAL CHANGES

The manufacturer often makes changes in components and composition, manufacturing equipment and/or processing, batch size, or site of manufacturing after the product is approved. These changes often require additional testing to assure product sameness between preapproved and postapproval changes in manufacturing. Depending on the type and nature of the changes, additional in vitro testing or in vivo bioequivalence studies are required (9).

REFERENCES

1. VP Shah, JP Skelly. Regulatory considerations in transdermal drug delivery systems in the United States. In: YW Chien, ed. Transdermal Controlled Systemic Medications. New York: Marcel Dekker, 1987, pp 399–410.
2. VK Tammara, RK Baweja, A Dorantes, JP Hunt, FR Pelsor, H Sun, VP Shah, JK Wilkin. Biopharmaceutical considerations for topical and transdermal drug delivery systems. In: TP Ghosh, WR Pfister, SI Yum, eds. Transdermal and Topical Drug Delivery Systems. Buffalo Grove, IL, Interpharm Press, 1997, pp 629–658.

3. FDA Guidance for Industry: Bioavailability and bioequivalence studies for orally administered drug products—general considerations. October 2000.

4. VP Shah, SV Dighe, YC Huang, F Fang, DB Hare, LJ Lesko, RL Williams. New and generic transdermal nitroglycerin systems: regulatory considerations. Eur J Pharm Biopharm 41:189–193, 1995.

5. VP Shah, NW Tymes, L Yamamoto, JP Skelly. In vitro dissolution profile of transdermal nitroglycerin patches using paddle method. Int J Pharm 32: 243–250, 1986.

6. VP Shah, LJ Lesko, RL Williams. In vitro evaluation of transdermal drug delivery. Eur J Pharm Biopharm 41:163–167, 1995.

7 Transdermal drug delivery systems—general drug release standards. USP 23/NF 18, Apparatus 5, pp 1796–1797.

8. VP Shah, CC Peck, RL Williams. Skin penetration enhancement—clinical pharmacological and regulatory considerations. In: KA Walters, J Hadgraft, eds. Pharmaceutical Skin Penetration Enhancement. New York: Marcel Dekker, 1993, pp 417–427.

9. GA Van Buskirk et al. Scale-up of adhesive transdermal drug delivery systems. Pharm Res 14:848–852, 1997.

Index

Data in tables is indicated by *t*. Data in figures is indicated by *f*.

369